AF388951

Calixarene Complexes: Synthesis, Properties and Applications

Editors
Mario Berberan-Santos
Instituto Superior Técnico
Portugal

Paula M. Marcos
Faculdade de Farmácia da Universidade de Lisboa
Portugal

Editorial Office
MDPI
St. Alban-Anlage 66
4052 Basel, Switzerland

This is a reprint of articles from the Special Issue published online in the open access journal *Molecules* (ISSN 1420-3049) (available at: https://www.mdpi.com/journal/molecules/special_issues/calixarene).

For citation purposes, cite each article independently as indicated on the article page online and as indicated below:

LastName, A.A.; LastName, B.B.; LastName, C.C. Article Title. *Journal Name* **Year**, *Volume Number*, Page Range.

ISBN 978-3-0365-1633-2 (Hbk)
ISBN 978-3-0365-1634-9 (PDF)

Contents

About the Editors

Mario Berberan-Santos is a Full Professor of Physical Chemistry, Materials and Nanosciences at Instituto Superior Técnico, University of Lisbon (ULisboa), President of the College of Chemistry of ULisboa, and former President of the Portuguese Chemical Society. He is Fellow of the Royal Society of Chemistry, Europe Chemistry Fellow (class of 2018/2019), member of the International Academy of Mathematical Chemistry and Ferreira da Silva prize (Portuguese Chemical Society) 2020. He is a member of the Editorial Advisory Boards of several journals, including *Molecules* (MDPI). Mário Berberan Santos has carried out theoretical and experimental work on the structure of nano- and supramolecular systems, on the photophysics of molecular probes, fluorescence polarization, radiative and nonradiative (FRET) electronic energy transfer and transport, on the kinetics of diffusion-influenced processes, and on the photophysics of fullerenes and derivatives. Current interests include the luminescence of fullerenes and nanostructured systems, as well as luminescence relaxation and other kinetic processes, and applications such as optical sensing and OLEDs. He is the author/editor of three books and over 220 articles and book chapters.

Paula M. Marcos graduated in Chemistry at Faculdade de Ciências da Universidade de Lisboa (FCUL) and got her PhD in Organic Chemistry in 1998 from the same university. She is Professor at Faculdade de Farmácia da Universidade de Lisboa (FFUL) and Researcher at Centro de Química Estrutural (CQE). At present, she is a member of the Topical Advisory Panel of *Molecules* for the Organic Chemistry section, and a Review Editor on the Editorial Board of Supramolecular Chemistry section of Frontiers in Chemistry. Her research is focused on macrocyclic and supramolecular chemistry, calixarenes, host–guest chemistry, ion binding (including metal and organic cations, and biological and environmental relevant anions), organic ion-pair recognition (such as biogenic amines and amino acids), NMR and UV-vis absorption spectroscopy.

Preface to "Calixarene Complexes: Synthesis, Properties and Applications"

There is a growing interest in host systems for ionic and neutral species based on calixarenes. These cyclic oligomers show important host–guest properties, which has led to numerous applications in a broad range of fields, including organocatalysis, sensing, extraction and separation, and, more recently, biomedical applications. The relatively easy functionalization of their upper and lower rims and the presence of a pre-organized cavity, available in different sizes and conformations, make calixarenes attractive building blocks for the construction of supramolecular assemblies. This book presents the most recent developments in the calixarene field, including host–guest properties, as well as new synthetic methods and applications.

Mario Berberan-Santos, Paula M. Marcos
Editors

Article

Conventional vs. Microwave- or Mechanically-Assisted Synthesis of Dihomooxacalix[4]arene Phthalimides: NMR, X-ray and Photophysical Analysis [§]

Alexandre S. Miranda [1,2], Paula M. Marcos [1,3,*], José R. Ascenso [4], M. Paula Robalo [4,5], Vasco D. B. Bonifácio [2], Mário N. Berberan-Santos [2], Neal Hickey [6] and Silvano Geremia [6]

[1] Centro de Química Estrutural, Faculdade de Ciências da Universidade de Lisboa, Edifício C8, 1749-016 Lisboa, Portugal; miranda.m.alexandre@gmail.com
[2] IBB-Institute for Bioengineering and Biosciences, Instituto Superior Técnico, Universidade de Lisboa, 1049-001 Lisboa, Portugal; vasco.bonifacio@tecnico.ulisboa.pt (V.D.B.B.); berberan@tecnico.ulisboa.pt (M.N.B.-S.)
[3] Faculdade de Farmácia, Universidade de Lisboa, Av. Prof. Gama Pinto, 1649-003 Lisboa, Portugal
[4] Centro de Química Estrutural, Instituto Superior Técnico, Complexo I, Av. Rovisco Pais, 1049-001 Lisboa, Portugal; jose.ascenso@ist.utl.pt (J.R.A.); mprobalo@deq.isel.ipl.pt (M.P.R.)
[5] Área Departamental de Engenharia Química, Instituto Superior de Engenharia de Lisboa, Instituto Politécnico de Lisboa, Rua Conselheiro Emídio Navarro, 1, 1959-007 Lisboa, Portugal
[6] Centre of Excellence in Biocrystallography, Department of Chemical and Pharmaceutical Sciences, University of Trieste, via L. Giorgieri 1, 34127 Trieste, Italy; nhickey@units.it (N.H.); sgeremia@units.it (S.G.)
* Correspondence: pmmarcos@fc.ul.pt
§ Dedicated to Prof. Placido Neri on the occasion of his 60th birthday.

Citation: Miranda, A.S.; Marcos, P.M.; Ascenso, J.R.; Robalo, M.P.; Bonifácio, V.D.B.; Berberan-Santos, M.N.; Hickey, N.; Geremia, S. Conventional vs. Microwave- or Mechanically-Assisted Synthesis of Dihomooxacalix[4]arene Phthalimides: NMR, X-ray and Photophysical Analysis §. *Molecules* 2021, 26, 1503. https://doi.org/10.3390/molecules26061503

Academic Editor: Alejandro Baeza Carratalá

Received: 18 February 2021
Accepted: 4 March 2021
Published: 10 March 2021

Abstract: Direct *O*-alkylation of *p-tert*-butyldihomooxacalix[4]arene (**1**) with *N*-(bromopropyl)- or *N*-(bromoethyl)phthalimides and K_2CO_3 in acetonitrile was conducted under conventional heating (reflux) and using microwave irradiation and ball milling methodologies. The reactions afforded mono- and mainly distal di-substituted derivatives in the cone conformation, in a total of eight compounds. They were isolated by column chromatography, and their conformations and the substitution patterns were established by NMR spectroscopy (^{1}H, ^{13}C, COSY and NOESY experiments). The X-ray structures of four dihomooxacalix[4]arene phthalimide derivatives (**2a**, **3a**, **3b** and **5a**) are reported, as well as their photophysical properties. The microwave (MW)-assisted alkylations drastically reduced the reaction times (from days to less than 45 min) and produced higher yields of both 1,3-di-substituted phthalimides (**3a** and **6a**) with higher selectivity. Ball milling did not reveal to be a good method for this kind of reaction.

Keywords: dihomooxacalix[4]arenes; phthalimide derivatives; conventional synthesis; microwave irradiation; ball milling; NMR spectroscopy; X-ray diffraction; electronic absorption and fluorescence studies

1. Introduction

Calixarene-based molecules are amongst the most investigated frameworks in host-guest and supramolecular chemistry [1,2]. Their relatively easy functionalization at either the upper or lower rim, a pre-organized cavity available in different sizes and conformations, and ion-binding sites are the key for their great diversity of applications. The recognized importance of anions in biological, chemical and environmental processes continues to attract chemists for the design and use of new synthetic anion receptors [3,4]. Calixarenes possessing (thio)urea moieties have been developed as receptors, with anion interactions established by hydrogen bonding only.

Since the first articles on the use of domestic microwave ovens for organic synthesis in 1986 [5,6], this technology has become a very important tool in a wide range of chemical reactions [7–10]. The microwave (MW) technique provides uniform heating of the reagents throughout the reaction vessel, allowing a rapid and homogeneous heat transfer. MW heating drastically reduces reaction times compared to conventional heating (from hours or even days to some minutes), saving a huge amount of energy and time. Higher product yields, higher reaction selectivity and low waste are other advantages of this technique, as well as the use of less or even no solvent and catalyst. Ball milling (BM), another environmentally friendly methodology applied to organic synthesis under solvent-free conditions, has also been developed mainly in the last decade [11–13]. Both techniques have been employed in the synthesis of calixarenes and related macrocyclic hosts [14], although ball milling is much less extensive.

The syntheses of *p*-alkylcalix[*n*]arenes (R = *t*-Bu, *n* = 4 and 8; R = Me, *n* = 6) were the first calixarene reactions carried out under MW irradiation [15,16]. Other examples reported are the preparation of water-soluble azocalix[4]arenes [17], the copper(I)-catalysed syntheses of calix[4]arene glycoconjugates [18] and calix[4]arene tetraazide derivatives [19], as well as calix[4]arenes containing chiral unsymmetrical urea moieties at the lower rim [20]. The alkylation of *p*-*tert*-butylcalix[4 and 6]arenes in the presence of K_2CO_3 and MeCN or DMF are among the most investigated MW-assisted reactions [21–23]. Calix[6]arenes functionalized at the lower and upper rims were also successfully obtained [24]. All these reactions were shown to be faster and more efficient compared to conventional methods. The formation of *p*-benzylcalix[5 and 7]arenes from *p*-benzylcalix[6 or 8]arenes in the presence of KOH, formaldehyde and molecular sieves [25], and the synthesis of a *p*-nitrocalix[6]arene [26] are the only examples of ball milling methods employed in calixarene synthesis found in the literature.

In the course of our research on anion binding by ureido-dihomooxacalix[4]arenes [27–31], we were interested in studying the introduction of urea binding sites into the 1,3-positions of the calixarene lower rim, through shorter spacers (propyl or ethyl) compared to the ones obtained in our previous dihomooxa receptors. The two remaining phenolic OH groups will participate in the formation of hydrogen bonds, contributing to keep the macrocycle in the cone conformation. It is known that *O*-alkylation reactions of calixarenes by conventional methods can take a long time and produce low yield products. Thus, MW irradiation and ball milling were employed in this work to obtain the target 1,3-disubstituted phthalimides, which have been prepared as precursors to urea-terminated anion-binding macrocycles. Other differently substituted derivatives were also obtained. Their solution and solid state conformational analysis and the determination of some photophysical properties are also reported.

2. Results and Discussion

2.1. Conventional vs. MW- and Mechanically-Assisted Synthesis

With the aim of introducing urea groups at the 1,3-positions of the *p*-*tert*-butyldihomooxacalix[4]arene (**1**) via a propyl spacer, the parent compound **1** was treated with 2 equiv of *N*-(3-bromopropyl)phthalimide and 2 equiv of K_2CO_3 in refluxing acetonitrile for 4 days (Scheme 1). This reaction afforded, according to the proton NMR spectrum, a mixture composed of distal 1,3-diphthalimide **3a** as the major product (65%), along with minor amounts of mono **2a** (13%) and tetra-phthalimide **4** (12%), and a further 3% of proximal 3,4-diphthalimide **3b** and 6% of unreacted **1** (Table 1, entry 1). A trace amount of mono **2b** was also observed. When the reaction was conducted for 5 days, the percentage of formation of **3a** increased (76%), as well as that of **3b** (15%), while minor amounts of tetra **4**, mono **2a** and no starting material were obtained (Table 1, entry 2).

	Y_1	Y_2	Y_3	Y_4
2a	H	H	H	(CH$_2$)$_3$Pht
2b	(CH$_2$)$_3$Pht	H	H	H
3a	(CH$_2$)$_3$Pht	H	(CH$_2$)$_3$Pht	H
3b	H	H	(CH$_2$)$_3$Pht	(CH$_2$)$_3$Pht

Scheme 1. Synthesis of propylphthalimide derivatives **2** and **3**.

Table 1. Conditions and product percentages of reaction of **1** with Br(CH$_2$)$_3$Pht.

Entry	Method	Reaction Time	1:RX:Base Ratio	Reaction Mixture (%) [1]				
				1	**2a**	**3a**	**3b**	**4**
1	Reflux	4 days	1:2:2	6	13	65	3	12
2	Reflux	5 days	1:2:2	—	4	76	15	5
3	MW	20 min	1:2:2	—	10	86	2	2
4	MW	30 min	1:2:2	—	3	93	2	2
5	BM	2 h	1:2:2	70	30	—	—	—
6	BM	12 h	1:4:4	27	46	19	4	4

[1] Percentage of product formation in the reaction mixture determined by ^{1}H NMR.

In view of the long reaction time, the alkylation of parent **1** was investigated under MW irradiation and ball milling methodologies. The reaction of **1** with 2 equiv of the previous alkylating agent and base in acetonitrile after 20 min of irradiation (160 °C) afforded 86% of 1,3-diphthalimide **3a**, accompanied by 10% of mono **2a** and minor amounts (<2%) of proximal di-phthalimide **3b** and tetra **4** (Table 1, entry 3). The increase in the irradiation time to 30 min (Table 1, entry 4) produced an increase of 1,3-diphthalimide **3a** (93%) and a decrease of mono **2a** (3%). The use of solvent-free ball milling during 2 h with a rotation speed of 500 rpm, 200 stainless steel balls and keeping the same reagent ratio as before gave a mixture composed of 70% of unreacted **1** and 30% of mono **2a** (Table 1, entry 5). Replacing stainless steel balls with zirconium oxide balls and increasing the rotation speed to 650 rpm led to a similar result. After several attempts to improve the reaction yield, we found that using 4 equiv of alkylating agent and base, and prolonging the reaction time to 12 h, raised the yield of mono **2a** to 46%, although 27% of the starting material was still present (Table 1, entry 6). The formation of the previous distal and proximal diphthalimides **3a** and **3b**, as well as of the tetra derivative **4**, was also observed, with distal **3a** being obtained as the main compound (19%).

The separation of the reaction mixture into the pure compounds was achieved by column chromatography and by recrystallization. The tetra-phthalimide derivative **4** was not purified, as it had already been obtained by us before [29].

Concerning the synthesis of the 1,3-diethylphthalimide **6a** (Scheme 2), a urea precursor with an even shorter spacer (only two carbon atoms), several difficulties were found in its preparation. The alkylation of parent **1** with *N*-(2-bromoethyl)phthalimide and K$_2$CO$_3$ was tried under different conditions. Firstly, the reaction was carried out using 2 equiv of the reactant and base for 3 days in acetonitrile under reflux. According to the proton NMR integration, a mixture composed mainly of unreacted **1** (61%) and mono-phthalimide **5a** (21%), along with minor amounts of mono **5b** (8%) and 1,3-diphthalimide **6a** (11%), was obtained (Table 2, entry 1). When the reaction was conducted for the same time, but in the presence of KI (to promote the halogen exchange of the electrophile in situ) and a 25% excess of alkylating agent, the percentage of unreacted **1** decreased (30%) and those of mono **5a** (45%) and 1,3-di **6a** increased (16%), although the latter only slightly (Table 2, entry 2). To improve the reactivity of the nucleophile, **1** was also previously heated with the base in MeCN for about 30 min. Four equiv of reactant and base were necessary to raise the percentage of **6a** to 30% and to decrease that of **1** to 19% (Table 2, entry 3). A final attempt under extreme conditions (6 equiv of Br(CH$_2$)$_2$Pht and base for 7 days in refluxing MeCN) gave identical results (Table 2, entry 4). Substitution and elimination reactions are almost always in competition with each other. The peak assignments in the proton NMR spectrum of the reaction mixture compatible with bromo- and iodo-ethylphthalimides, as well as *N*-ethenylphthalimide in an approximately 40:60 ratio, indicate that elimination should predominate over substitution, explaining the low yields obtained even under extreme conditions [32].

	Y_1	Y_2	Y_3	Y_4
5a	H	H	H	$(CH_2)_2Pht$
5b	$(CH_2)_2Pht$	H	H	H
6a	$(CH_2)_2Pht$	H	$(CH_2)_2Pht$	H
6b	H	H	$(CH_2)_2Pht$	$(CH_2)_2Pht$

Scheme 2. Synthesis of ethylphthalimide derivatives **5** and **6**.

Table 2. Conditions and product percentages of reaction of **1** with $Br(CH_2)_2Pht$.

Entry	Method	Reaction Time	1:RX:Base:KI Ratio	Reaction Mixture (%) [1]				
				1	5a	5b	6a	6b
1	Reflux	3 days	1:2:2:0	61	21	6	11	1
2	Reflux	3 days	1:2.5:2:2	30	45	7	16	2
3	Reflux	3 days	1:4:4:2	19	44	4	30	3
4	Reflux	7 days	1:6:6:4	18	37	12	30	3
5	MW	6 + 12 min	1:4:2:2	16	58	—	26	—
6	MW	6 + 36 min	1:4:2:2	—	59	—	41	—
7	MW	12 + 36 min	1:4:2:2	—	38	—	62	—
8	MW	12 + 36 min	1:4:2:0	—	43	—	57	—
9	MW	18 + 36 min	1:4:4:0	—	—	—	100	—
10	MW	6 + 36 min	1:4:4:0	—	—	—	100	—
11	BM	12 h	1:4:4:0	82	18	—	—	—

[1] Percentage of product formation in the reaction mixture determined by ^{1}H NMR.

The alkylation reaction with *N*-(2-bromoethyl)phthalimide was also studied by MW irradiation, under a range of conditions. The irradiation of parent **1** with 2 equiv of base in acetonitrile for 6 min at 110 °C, followed by the addition of 4 equiv of reagent and 2 equiv of KI and a further 12 min irradiation at 150 °C, afforded a mixture of mono **5a** as the main product, along with 1,3-di **6a** and unreacted **1** (Table 2, entry 5). An increase in the irradiation time at 150 °C (36 min) led to the disappearance of **1** and to an increase of **6a** (Table 2, entry 6). The effect of KI on the reaction yield was investigated in two MW-assisted alkylations carried out with and without KI (Table 2, entries 7 and 8, respectively). Given that just a slight improvement (5%) of **6a** yield was obtained in the presence of KI, the next assay was performed without KI, but with 4 equiv of base, an irradiation time of 18 min (110 °C), followed by a further 36 min (150 °C), resulting **6a** as the unique product (Table 2, entry 9). The reduction of the time at 110 °C to 6 min provided 1,3-diphthalimide **6a** as the unique product as well (Table 2, entry 10). As expected, the MW-assisted reactions showed several advantages compared to the conventional synthesis, mainly in the case of the alkylation with *N*-(bromoethyl)phthalimide. Beyond the removal of KI reactant and the reduction in solvent quantity, the MW irradiation provided drastically shorter reaction times. The alkylation of parent **1** was complete after 42 min under MW conditions, while under reflux for 7 days the conversion was still incomplete. Higher reaction selectivity was also observed, with distal 1,3-disubstituted phthalimide **6a** being the only derivative obtained. Ball milling was likewise performed, but gave the poorest results (Table 2, entry 11).

It was not possible to isolate the mono-phthalimide **5b** as a pure compound, despite the several columns and preparative chromatographies done. Recrystallization attempts in different solvents have also failed. Similarly, distal 1,3-diphthalimide **6a** and proximal 3,4-diphthalimide **6b** could not be separated from each other, as they have the same R_f value in all the solvents tested. A similar situation was reported before for distal and proximal di-substituted calix[4]arene derivatives [19].

Besides the NMR assignment of all the previous derivatives, the proton spectra of both reaction mixtures still present a set of small signals, which could not be identified. Since a total of nine possible *O*-alkylating compounds can be obtained [33], the formation of other differently substituted phthalimide derivatives cannot be excluded.

2.2. NMR Conformational Analysis

The conformation and the substitution pattern of the phthalimide derivatives were established by proton, carbon-13, COSY and NOESY NMR spectroscopy in chloroform at room temperature.

2.2.1. Mono-Propylphthalimide **2a**

The absence of symmetry in compound **2a** is reflected by its proton and carbon-13 NMR spectra. The ^{1}H NMR spectrum (Figure 1a) displays four singlets for the *tert*-butyl groups, five AB quadruplets for the CH_2 bridge protons, four pairs of doublets for the aromatic protons and three singlets for the OH groups of the calixarene skeleton, besides several multiplets for the $-OCH_2CH_2CH_2N-$ methylene protons and phthalimide group (two signals). The proton assignments were confirmed by cross-peak correlations in a COSY spectrum. The ^{13}C NMR spectrum shows 28 downfield resonances arising from the aromatic carbon atoms and phthalimide group, three downfield resonances arising from the methylene carbon atoms of $-OCH_2CH_2CH_2N-$ and CH_2OCH_2 groups, and 12 of the expected 13 upfield resonances arising from the quaternary (3 lines in a 1:1:2 ratio) and methyl carbon atoms (4 lines) of the *tert*-butyl groups, and the methylene carbon atoms of the $ArCH_2Ar$ (3 lines) and $-OCH_2CH_2CH_2N-$ (2 lines) groups. All resonances were assigned by DEPT experiments. The three $ArCH_2Ar$ resonances appear at 30.1, 31.3, and 32.8 ppm, indicating a cone conformation for **2a** [34].

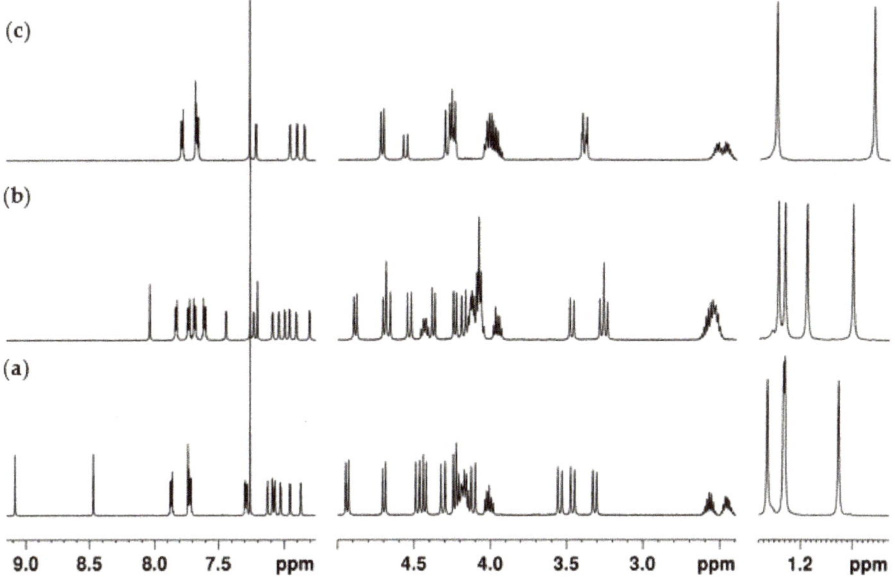

Figure 1. ^{1}H NMR spectra (500 MHz, CDCl$_3$, 22 °C) of compounds: (a) **2a**, (b) **3a** and (c) **3b**.

The position of the phthalimide group was confirmed by proton-proton correlations observed in the NOESY spectrum. The preferential formation of **2a** over **2b** is expected, as described by us before [33]. The simultaneous observation of two NOE effects between the OH$_1$ proton at 9.08 ppm (position 29) and the two axial methylene protons at positions 10 and 16, as well as those observed between the OH$_2$ group at 8.47 ppm (position 30) and the axial CH_2 protons at positions 4 and 10, and between the OH_3 group at 7.73 ppm (position 27) and the axial CH_2 protons at positions 2, 22 and 4, are conclusive for the position of the substituent group (Figure 2). In addition to those effects, two other relevant NOEs were observed between OH_1 and -$OCH_2CH_2CH_2N$- groups and OH_1 and OH_2 groups.

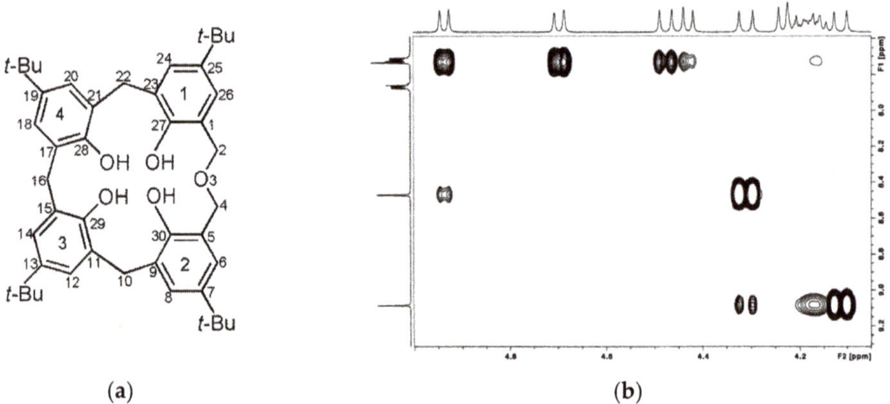

(a)   (b)

Figure 2. (a) Chemical structure of **1** with carbon atoms numbering; (b) Section of the NOESY spectrum of **2a** (500 MHz, $CDCl_3$, 22 °C).

Concerning mono-propylphthalimide **2b**, it could be identified as a trace compound in a set of fractions obtained by column chromatography (Figure S1, Table S1).

2.2.2. 1,3-Dipropylphthalimide **3a**

The ^1H NMR spectrum of **3a** (Figure 1b) is also asymmetric and shows identical number and type of signals as **2a** for the majority of the protons of the calixarene skeleton. Two singlets for the OH groups and four multiplets for the phthalimide aromatic protons can, in addition, be observed. The ^{13}C NMR spectrum displays 32 downfield, four midfield and 13 of the expected 15 upfield resonances. The presence of the three $ArCH_2Ar$ resonances in the range 29.8–32.9 ppm indicates a cone conformation for **3a**.

The 1,3-substitution pattern was also confirmed by a NOESY experiment. The preferential formation of distal disubstituted compounds is expected, as previously reported for the formation of other dihomooxacalix[4]arenes [28,33]. Two simultaneous NOE effects between OH_1 proton at 8.04 ppm (position 29) and the two axial methylene protons at positions 10 and 16 were observed. Moreover, both OH groups (at 8.04 and 7.20 ppm) give NOE effects on both -$OCH_2CH_2CH_2N$- groups (positions 28 and 30), and OH_2 (position 27) also gives effects on the axial CH_2 protons at positions 2 and 22.

2.2.3. 3,4-Dipropylphthalimide **3b**

Derivative **3b** exhibits symmetric proton and carbon-13 NMR spectra, matching with a proximal disubstituted compound in the cone conformation. The ^1H NMR spectrum (Figure 1c) shows two singlets for the *tert*-butyl groups, three AB quadruplets in a 2:2:1 ratio for the CH_2 bridge protons, two pairs of doublets for the aromatic protons, one singlet for the OH groups, four multiplets for the -$OCH_2CH_2CH_2N$- and two multiplets for the phthalimide groups. The ^{13}C NMR spectrum displays 16 downfield, two midfield and 8 upfield resonances. The $ArCH_2Ar$ resonances appear at 30.8 (one carbon atom) and 31.6 (two carbon atoms), indicating a cone conformation for **3b**.

The positions of the phthalimide groups were confirmed by two simultaneous NOE effects observed between the -$OCH_2CH_2CH_2N$- protons (position 28) and the two axial CH_2 protons at positions 16 and 22, as well as between the OH proton (position 27) and the axial CH_2 proton at position 2.

2.2.4. Mono-Ethylphthalimide **5a**

Similarly to mono-propylphthalimide **2a**, derivative **5a** also presents asymmetric proton (Figure 3a) and carbon-13 spectra, being the lower number of multiplets for the CH_2 protons of the pendant group the only observed difference. The same 12 upfield carbon resonances are observed (as in mono **2a**), as there is no overlapping of the quaternary carbon atoms (4 lines) of the *tert*-butyl groups.

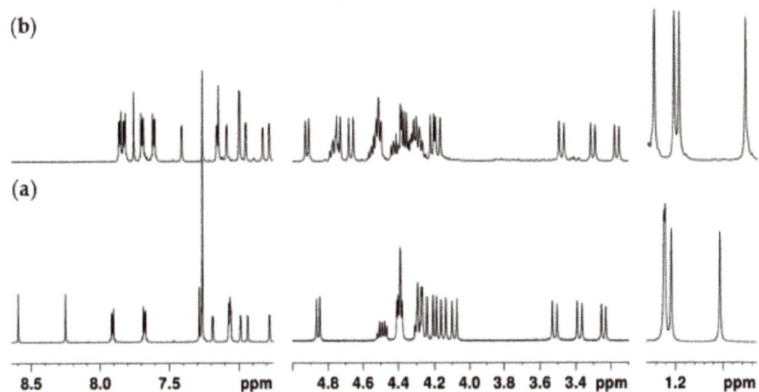

Figure 3. ^{1}H NMR spectra (500 MHz, CDCl$_3$, 22 °C) of compounds: (**a**) **5a** and (**b**) **6a**.

Concerning the phthalimide group position, the NOESY spectrum of **5a** shows identical effects to those exhibited for **2a**. NOEs between each of the three OH groups (position 29, 30 and 27) and the two adjacent axial CH_2 protons (positions 10/16, 4/10 and 2/22, respectively) are observed. Furthermore, the -OCH_2CH_2N- group gives NOE effects on both OH_1 and OH_3 protons (positions 29 and 27), and each OH group gives NOE on its OH neighbor ($OH_1 \leftrightarrow OH_2 \leftrightarrow OH_3$).

A pure mixture of mono-phthalimides **5b** and **5a** was obtained by preparative chromatography (Figure S1, Table S1). However, further separation of mono **5b** from the mixture was not possible to achieve.

2.2.5. 1,3- and 3,4-Diethylphthalimides **6a** and **6b**

As mentioned before, 1,3-diphthalimide **6a** and 3,4-diphthalimide **6b** could not be separated from each other, as they have the same R_f value. In the case of the MW-assisted reaction, a 100% conversion of parent **1** into **6a** was obtained. However, after several purification attempts it was not possible to obtain pure **6a**, as revealed by the ^{1}H spectrum (Figure 3b).

Both ^{1}H and ^{13}C NMR spectra of **6a** show spectral patterns compatible with an asymmetric molecule and very similar to those obtained for **3a**. The presence of the three $ArCH_2Ar$ carbon-13 resonances at 29.8, 30.0 and 32.7 ppm indicates a cone conformation for **6a**. The 1,3-substitution was also corroborated by proton-proton correlations observed in the NOESY spectrum. Concerning derivative **6b**, it was possible to clearly identify in the ^{1}H spectrum two singlets for the *tert*-butyl groups, two pairs of doublets for the aromatic protons and one singlet for the OH groups (Figure S1, Table S1). The overlapping of signals (with those of **6a**) prevents the assignment of the methylenic region.

2.3. X-ray Structural Analysis

Small single crystals of the mono-substituted **2a** and **5a** and the di-substituted **3a** and **3b** were analyzed by X-ray diffraction at 100 K using synchrotron radiation. With the exclusion of the symmetric **3b**, all the other molecules are inherently chiral due to the presence of the phthalimide substituents on the lower rim, which are not located symmetrically with respect to the oxa bridge inserted in the calixarene cone (Figure 4). As all their space groups are centrosymmetric (C2/c for **2a** and **3a** and P-1 for **5a**), racemic mixtures of the two inherently chiral enantiomers are present in these crystals.

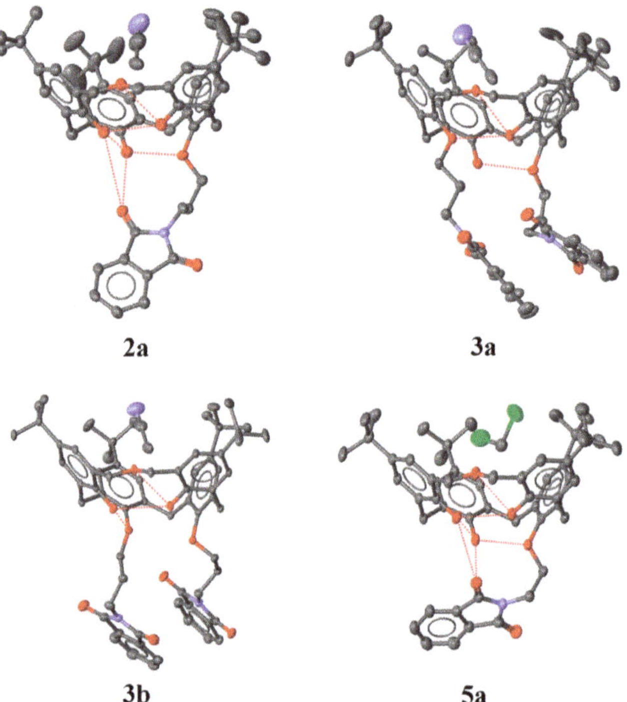

Figure 4. Thermal ellipsoid representations of **2a**, **3a**, **3b** and **5a**. Only one of the crystallographically independent molecules is shown for **2a** (4 molecules in the asymmetric unit) and for **3a** (3 molecules in the asymmetric unit). The atomic species are represented in CPK colors. Ellipsoids are drawn at 50% probability, except for **2a**, in which the ellipsoids are represented at 30% probability. Hydrogen atoms, disordered fragments and solvent molecules located outside the calixarene cavities are omitted for clarity.

A different number of crystallographically independent molecules was observed for the analyzed structures: four for **2a**, three for **3a**, and one for **3b** and **5a**. In order to compare the different structures in a coherent manner, the asymmetric units with all the molecules having the same inherent chirality were selected. The same labeling scheme was applied to these molecules, namely, the phenyl ring labels A–D are ordered anti-clockwise, as viewed from above the calix cone, with A and B across the oxa bridge. In this scheme, ring C always has a phthalimide substituent and ring B always has a hydroxyl group, while rings A and D have a hydroxyl group (**2a** and **5a**), except for the di-substituted derivatives, in which the second phthalimide substituent is present on ring A (**3a**) or ring D (**3b**) (Figure 5).

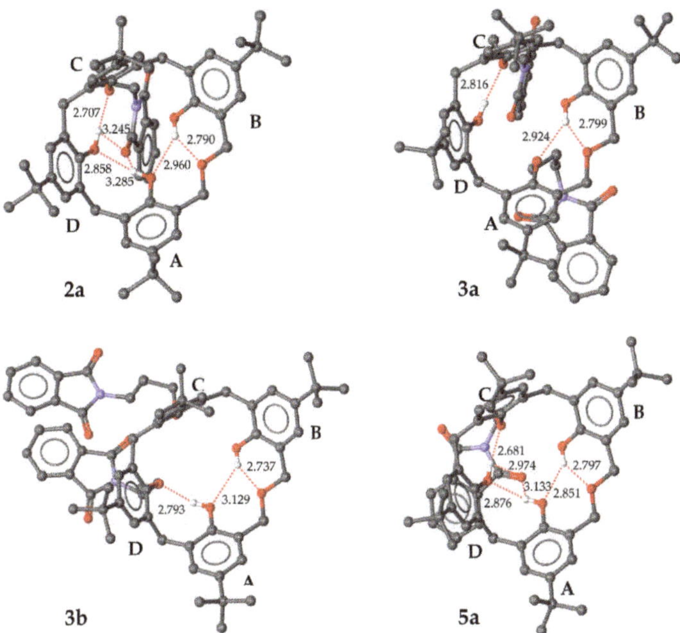

Figure 5. Comparison of the intramolecular hydrogen bonds in **2a**, **3a**, **3b** and **5a**. Only one of the crystallographically independent molecules is shown for **2a** (4 molecules in the asymmetric unit) and for **3a** (3 molecules in the asymmetric unit). The atomic species are represented in CPK colors. Ellipsoids are drawn at 50% probability, except for **5a**, in which the ellipsoids are represented at 30% probability. Hydrogen atoms, atom disorder and solvent molecules are omitted for clarity.

The X-ray structures show that all the dihomooxacalix[4]arenes adopt comparable cone conformations (Table 3 and Figure 6). All the canting angles, defined as the dihedral angles between the mean planes of the phenyl rings and the mean plane of the five methylene bridging carbon atoms, lean outwards from the center of the cone with similar values for corresponding phenyl rings, irrespective of the substituent pattern on the lower rim and the consequent differences in the intramolecular H-bond network (Figure 5). In particular, the two phenyl rings adjacent to the oxa bridge show large canting angles (mean values with standard deviation in parenthesis: A = 138(5)° and B = 148(3)°), while the other two are smaller (C = 102(4)° and D = 123(6)°) (see Table 3). This observation indicates that the conformation of the cone in these compounds is not determined by the different substitution pattern and different intramolecular H-bond networks. However, it should be noted that with respect to the cone conformation, all these structures have in common a bifurcated H-bond between the OH-donor group on ring B and acceptor oxygens atoms of the oxa bridge and the hydroxyl/alkoxy group on ring A (Figure 5, Table S2). This means that in all crystallographically independent molecules, the oxa bridge conformation has the oxygen atom oriented towards the center of the cone. The di-substituted derivatives show an additional intramolecular H-bond involving the second OH donor hydroxyl groups on ring A or D with O-acceptor atoms on ring D or C for **3a** or **3b**, respectively (Figure 5, Table S2). On the other hand, both of these H-bonds (A-D and D-C) are observed in the mono-substituted **2a** and **5a**, which have hydroxyl functions on the A and D rings. However, these are both bifurcated H-bonds, involving an additional O acceptor atom of a phthalimide carbonyl group. These bifurcated H-bonds are quite asymmetric, with the second, phthalimide group interaction characterized by long O•••O distances (Figure 5). Although these interactions are weak, they are important to determine the orientation of the phthalimide arms on the lower rims. For example, it is of interest to note that in **2a**, the four crystallographically independent molecules all have the same orientation of the phthalimide arms (Figure 6a). The intramolecular H-bonds involving the phthalimide groups are weaker in **2a**, in which the linker is longer, with respect to **5a** (propyl vs. ethyl). Interestingly, when the linker is even longer, as in the case of a butyl derivative, this potential intramolecular H-bond interaction with the phthalimide group is completely lost (Figure 6b) [35].

Table 3. Comparison of cone conformations: Dihedral angles (canting angles) between corresponding aryl planes of the calixarene cones (A, B, C and D) and the mean planes of the bridging methylene carbon atoms for various dihomooxacalix[4]arenes.

Molecule		A	B	C	D
2a *	(I)	138.4	154.7	105.0	126.1
	(II)	141.0	146.1	107.9	132.5
	(III)	138.6	149.5	102.3	122.4
	(IV)	128.0	148.8	98.5	129.9
3a **	(I)	137.02	145.72	100.30	118.38
	(II)	134.54	146.88	100.23	118.73
	(III)	135.17	147.47	101.34	116.39
3b		140.96	146.90	94.82	120.85
5a		144.25	149.98	106.82	118.39

* Four crystallographically independent molecules in the asymmetric unit; ** Three crystallographically independent molecules in the asymmetric unit. See Figure 5 for description of aryl rings A, B, C and D.

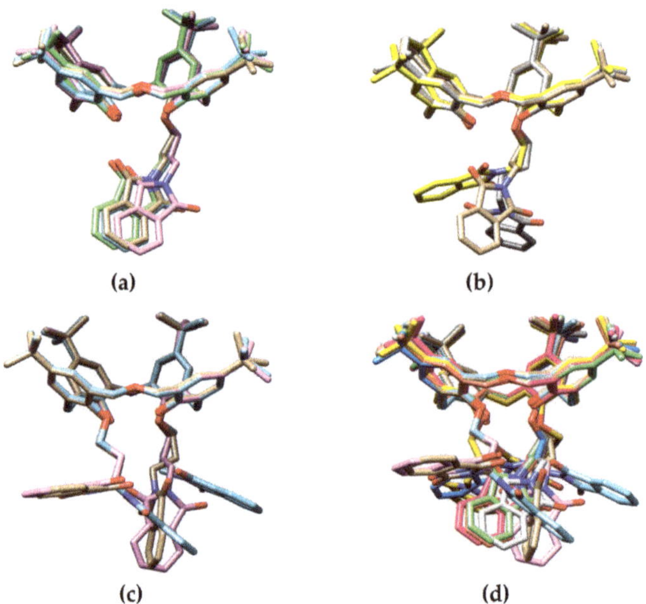

Figure 6. Comparison of the conformations of the phthalimide arms. All molecules exhibit very similar cone conformations: (**a**) Overlay of the 4 independent molecules of **2a**; (**b**) Overlay of **5a** (ethyl linker) one of the 4 independent molecules of **2a** (propyl linker) and the analogous derivative with butyl linker [35]; (**c**) Overlay of the 3 independent molecules of **3a**; (**d**) Overlay of the 9 independent molecules of **2a**, **3a**, **3b** and **5a**. The atomic species are represented in CPK colors. Hydrogen atoms, atom disorder and solvent molecules are omitted for clarity.

The effect on the phthalimide arm orientation of the different linkers is also evidenced by the different α angles which the planes of the phthalimide groups make with the mean planes of the five calixarene methylene groups (Figure 6b and Table S3). For **2a** it is almost orthogonal (mean value 77°), while for **5a** it is more parallel (22°). With regard to the di-substituted derivatives **3a** and **3b**, the presence of the second phthalimide arm in place of a hydroxyl group reduces the number of H-bond donors, which reduces the overall strength of potential H-bond interactions with the phthalimide arms. The loss of these interactions increases the conformational freedom of the substituents on the lower rim (Figure 6c,d). In fact, for **3a**, the three crystallographically independent molecules exhibit three different conformations (Figure 6c, Table S3). The presence of a second phthalimide group allows the possibility of intramolecular π-stacking interactions between these aromatic groups, as observed for **3b** and for one crystallographically independent molecule of **3a** (Figures 4 and 6c). On the other hand, intermolecular π-stacking between two phthalimide groups is also observed for the other two molecules of **3a** and for the mono-substituted **2a** derivative (Figure 7). In **2a**, for example, the four molecules can be described as two dimers formed by the intermolecular π-stacking of phthalimide groups.

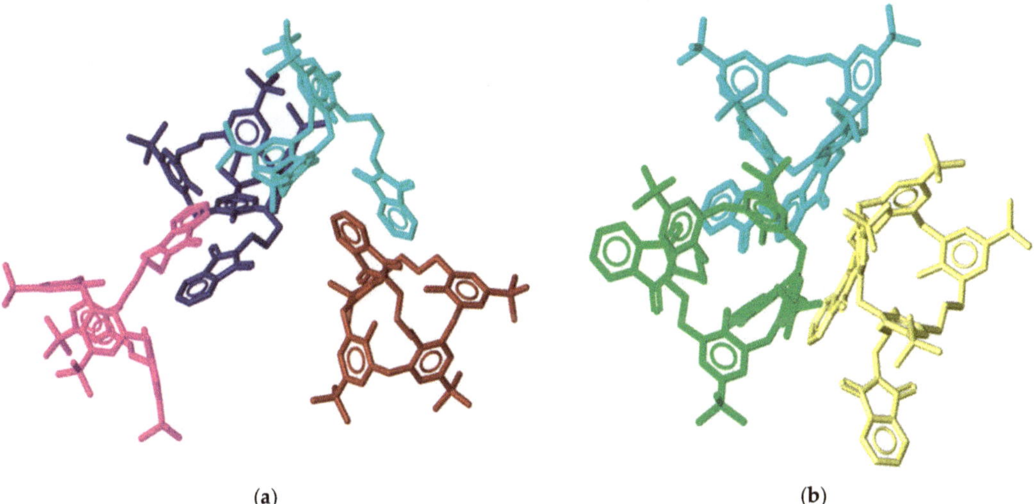

(a)   (b)

Figure 7. Illustration of intermolecular π-stacking interactions observed in the crystal packing of **2a** (a) and **3a** (b). In **2a**, the molecules are present as dimeric species through π-stacking of the phthalimide groups of the 4 independent molecules (violet-blue and cyan-brown). In **3a**, in addition to an intramolecular π-stacking interaction observed in one of the independent molecules (cyan), there is an intermolecular π-stacking interaction between the other two (green-yellow). For clarity, independent molecules are represented in arbitrarily chosen different colors, while hydrogen atoms, atom disorder and solvent molecules are omitted.

The importance of the intramolecular H-bonds network on the open conformation of the calixarene cavity is evidenced by the comparison with the dihomooxa calixarene structures reported with four alkoxy groups on the low rim [28,30,31]. In all structures with four substituents on the lower rim, one phenyl group leans towards the center of the cavity, closing the aperture with the *tert*-butyl groups. In fact, no solvent guest molecules were observed in these crystal structures. On the contrary, in the present structure the presence of hydroxyl groups, responsible for the above described H-bond network, pre-organizes the cavity to host guest molecules. In fact, all the reported molecules contain a solvent guest molecule in the calixarene cavity (Figure 4).

2.4. Photophysical Properties

Parent calixarenes bear a mesitol-like intrinsic fluorophore. This opens the possibility of using its fluorescence for structural and dynamic information, and for sensing applications [30,31,36]. In the present case, compounds **2a**, **3a**, **3b** and **5a** contain not only mesitol-like and alkylated mesitol-like fluorophores, but also a second chromophore (phthalimide) that can engage in intramolecular electronic energy transfer, acting as acceptor towards the ring fluorophores, as will be shown. To evaluate the changes caused by the introduction of the phthalimide substituent, parent **1** and model compounds 1,3,5-tri-*tert*-butyl-2-methoxybenzene (**A**) and *N*-butylphthalimide (**B**) were also studied. The absorption and emission spectra of **1**, of model compounds and of dihomooxa phthalimide derivatives in dichloromethane are shown in Figure 8. All of them present a well-defined absorption in the ultraviolet region, with phthalimide derivatives slightly red shifted compared to parent **1** and displaying a shoulder at the long wavelength side (Figure 8b). The absorption of **1** is in turn red-shifted with respect to that of **A**. These results indicate that the macrocycle dominates the absorption in the near UVC region (270–290 nm), while the phthalimide groups contribute to the absorption in the UVB region (290–320 nm). Dihomooxa compounds also display a well-defined emission in the UV region (Figure 8c) that extends into the visible owing to the (minor) phthalimide moiety contribution. Indeed, **B** exhibits a broad spectrum peaking at about 400 nm and extending from the UVA to the yellow region, in agreement with data reported for other *N*-phthalimides [37]. The weak 400 nm band is more evident for proximal di-phthalimide **3b**, probably owing to the proximity of the substituent groups at the lower rim. The relevant photophysical properties of all compounds studied are collected in Table 4.

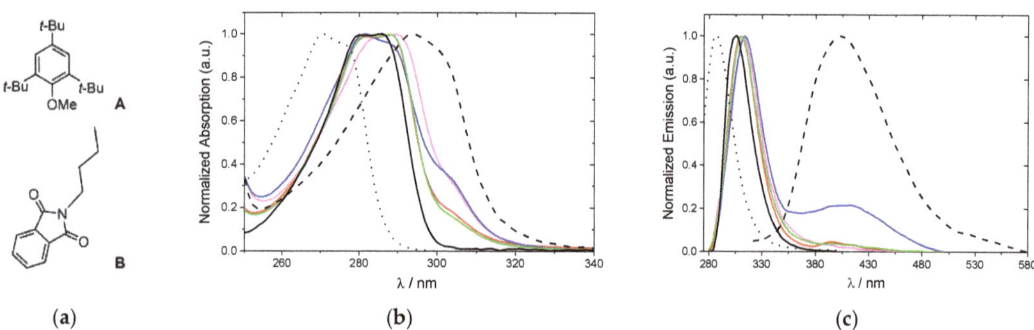

Figure 8. (a) Chemical structures of the model compounds **A** and **B**. Normalized absorption (b) and emission (c) spectra of **1** (black line), **A** (dotted line), **B** (dashed line), **2a** (red line), **3a** (pink line), **3b** (blue line) and **5a** (green line) in CH_2Cl_2. [dihomooxa] = 5×10^{-5} M; [**A**] = 8×10^{-4} M; [**B**] = 2×10^{-4} M. λ_{ex} = 270 nm.

Table 4. Photophysical properties of the studied compounds in CH_2Cl_2 at 293 K.

	$\lambda_{max,abs}$ (nm)	ε (M^{-1} cm^{-1})	$\lambda_{max,em}$ (nm)	Stokes Shift [a] (nm)	τ_f (ns)	Φ_F [b]
A	271	5.0×10^2	288	17	4.50 [c]	7.5×10^{-3}
B	294	8.2×10^2	403	109	0.26 [d]	1.8×10^{-3}
1	286	3.6×10^4	306	20	1.50 [c]	1.3×10^{-1}
2a	288	2.0×10^4	310	22	—	1.1×10^{-2}
3a	290	1.9×10^4	314	22	1.32 [d]	1.7×10^{-3}
3b	282	1.7×10^4	314	30	3.57 [d]	2.5×10^{-3} [e]
5a	288	2.9×10^4	310	22	—	1.4×10^{-3}

[a] Computed as $\lambda_{max,em} - \lambda_{max,abs}$; [b] Against quinine sulfate Φ_F = 0.546 in H_2SO_4 0.5 M; [c] λ_{ex} = 280 nm and λ_{em} = 306 nm; [d] λ_{ex} = 295 nm and λ_{em} = 400 nm; [e] After subtraction of the fluorescence of the phthalimide moiety (peaking at 400 nm, see Figure 8c).

The significant spectral overlap between the absorption and the fluorescence of the calixarene chromophores, on the one hand, and between the absorption of the phthalimides and the fluorescence of the calixarene chromophores, on the other hand (Figure S2), along with the close distance between chromophores, makes possible the existence of resonance energy transfer, both of the homotransfer (reversible) type and of the heterotransfer (irreversible) type (Figure 9).

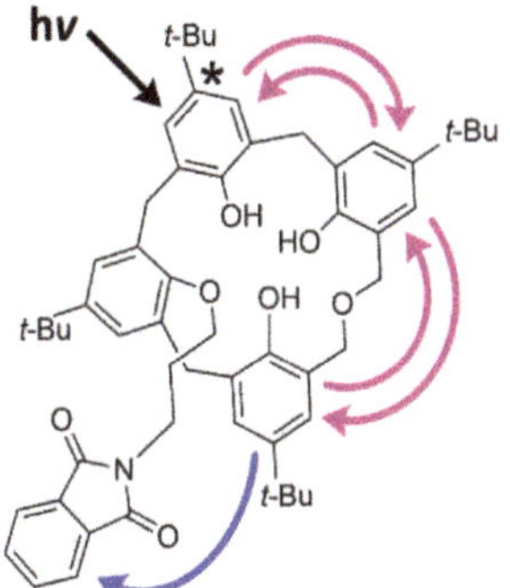

Figure 9. Schematic representation for compound **2a** of homotransfer (reversible FRET) within the calixarene followed by heterotransfer (irreversible FRET) from the calixarene to the phthalimide group (only one of many possible excitation outcomes is shown).

The dominant energy transfer mechanism is likely Förster resonance energy transfer (FRET) type. The computed Förster radii are 1.4 nm for homotransfer and 1.3 nm for heterotransfer. Given that average distances between neighboring chromophores are of the order of 0.5 to 0.7 nm for homotransfer, but probably somewhat higher for heterotransfer (conformation and structure dependent, between 0.6 and 1.0 nm), it is expected that efficient and fast (a few picoseconds, on average) energy hopping takes place in the calixarene skeleton, before the decay of the substituted phenoxy chromophores occurs either by the intrinsic radiative and nonradiative channels, or by irreversible FRET to the phthalimide group(s) (Figure 9). Evidence for homotransfer is obtained from fluorescence anisotropy measurements of both **1** (model compound for the calixarene chromophore) and **A** in a rigid medium (Zeonex film). The respective excitation fluorescence anisotropy spectra are shown in Figure 10, as well as the ratio of the two anisotropies.

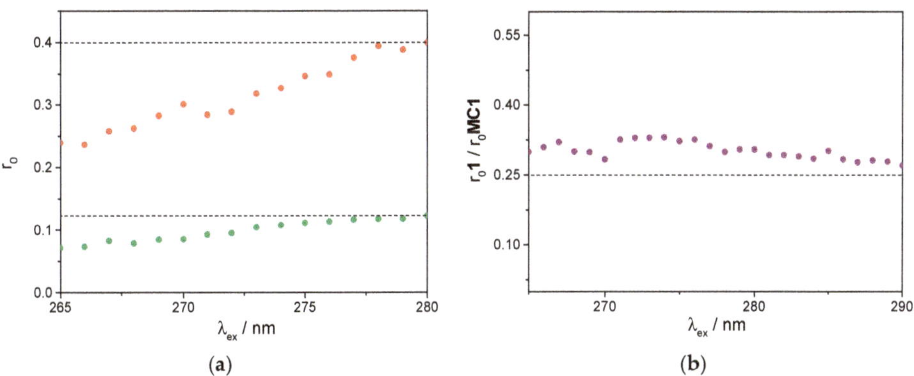

Figure 10. Fluorescence excitation anisotropy spectra of **1** (green) and **A** (red) in a Zeonex film at room temperature (**a**), and the respective anisotropy ratio (**b**), as a function of excitation wavelength. Slits were 8 nm for excitation and 14 nm for emission. λ_{ex} = 270 nm.

The anisotropy $r_0(\lambda_{exc})$ of the model compound **A** (Figure 10) attains the value of 0.40 at the $S_1 \leftarrow S_0$ absorption onset (280 nm), corresponding to collinear absorption and emission transition dipole moments [38] (^1L_b band), as expected. The excitation wavelength-dependence of the anisotropy, with a progressive decrease along the first band down to 0.24 at 260 nm, is compatible with a mixed polarisation, where the effect of the orthogonally-polarised ^1L_a band increases when decreasing the excitation wavelength, as observed by Weber [39] for both *p*-cresol and tyrosine (but unlike the results reported in [40]). The fluorescence anisotropy excitation spectrum of parent **1** (Figure 10) follows the same trend, albeit with lower values, starting from 0.12 at 280 nm. The drop in anisotropy observed for this calixarene can be entirely ascribed to fast energy migration (FRET homotransfer). Given that there are four equivalent chromophores, with nearly isotropic and uncorrelated orientations, the anisotropy is expected to reduce to $r_0(\lambda_{exc})/4$ owing to FRET [41]. As seen in Figure 10, this is indeed borne out by the experiment, with a ratio of 0.30 ± 0.03 being obtained between 265 and 290 nm. The slight excess over 0.25 can be ascribed to a residual degree of orientational anisotropy and orientational correlation. The fluorescence quantum yield decrease (by one or two orders of magnitude) in the case of compounds **2a**, **3a**, **3b** and **5a** results mainly from long-range quenching by the phthalimides owing to irreversible FRET. The two compounds bearing two acceptor phthalimides, **3a** and **3b**, display similar but lower donor fluorescence quantum yields when compared to the monophthalimide compound **2a**, as expected. The enhanced phthalimide emission in the **3b** case may result from the existence of two acceptors both in close proximity of the calixarene fluorophores and in a non self-quenching geometry. The monophthalimide **5a** differs from all the others by the number of carbon atoms of the spacer attaching the phthalimide (C_2 vs. C_3). The lower fluorescence quantum yield of **5a** (in fact, the lowest of all) is in accordance with the shortest (C_2) chain, hence the shortest average donor-acceptor distance (down to 0.6 nm).

3. Material and Methods

3.1. Synthesis

All chemicals were reagent grade and were used without further purification. Microwave-assisted syntheses were performed with a CEM Discover SP microwave synthesizer system (2.45 GHz, 300 W) equipped with a non-contact infrared temperature sensor. The reaction temperature was controlled by variable microwave irradiation (0–150 W) and cooling by nitrogen current. The mechanosyntheses were performed on a PM100 Planetary Ball Mill (Retsch, Haan, Germany) using a 50 mL stainless steel or zirconium oxide reactor, and 200 balls with 5 mm of diameter each, of the corresponding materials. Chromatographic separations were performed on Merck silica gel 60 (particle size 40–63 μm, 230–400 mesh). Melting points were measured on a Stuart Scientific apparatus and are uncorrected. FTIR spectra were recorded on a Shimadzu Model IRaffinity-1 spectrophotometer (Kyoto, Japan). ^1H and ^{13}C NMR spectra were recorded on a Bruker Avance III 500 MHz spectrometer (Bruker, Billerica, MA, USA), with TMS as internal reference. The conventional COSY 45 and the phase-sensitive NOESY experiments were collected as 256×2 K complex points. Elemental analysis was determined on a Fisons EA 1108 microanalyser (Milano, Italy).

3.1.1. Conventional Procedure for the Synthesis of **2a**, **3a** and **3b**

A mixture of *p-tert*-butyldihomooxacalix[4]arene (1 g, 1.47 mmol), *N*-(3-bromopropyl)-phthalimide (0.81 g, 2.95 mmol) and K_2CO_3 (0.41 g, 2.95 mmol) in CH_3CN (25 mL) was refluxed and stirred under N_2 for 5 days. After cooling, the solvent was evaporated under reduced pressure and the residue was dissolved in CH_2Cl_2 (50 mL) and washed with 1 M HCl (2×30 mL), NH_4Cl saturated solution (2×30 mL) and brine (30 mL). The organic layer was dried over Na_2SO_4, filtered and the solvent evaporated to dryness. The crude product was subjected to flash chromatography on silica gel (eluent CH_2Cl_2/MeOH, 99:1). According to TLC analysis, the fractions were combined into three sets, corresponding to derivatives **2a**, **3a** and **3b**, respectively.

7,13,19,25-Tetra-*tert*-butyl-28-[(3-phthalimidopropyl)oxy]-27,29,30-tri-hydroxy-2,3-dihomo-3-oxacalix[4]arene (**2a**): This compound was obtained in $\approx$ 2% yield (25 mg). An analytically pure sample was obtained by recrystallization from CH_2Cl_2/*n*-hexane.

mp 171–173 °C; IR (KBr) 1717 cm^{-1} (CO), 3377 cm^{-1} (OH); ^{1}H NMR (CDCl$_3$, 500 MHz) δ 1.13, 1.230, 1.233, 1.26 [4s, 36H, C(CH$_3$)], 2.45, 2.57 (2m, 2H, OCH$_2$CH_2CH$_2$N), 3.31, 4.48 (ABq, 2H, J = 12.9 Hz, ArCH$_2$Ar), 3.46, 4.31 (ABq, 2H, J = 13.8 Hz, ArCH$_2$Ar), 3.54, 4.11 (ABq, 2H, J = 13.8 Hz, ArCH$_2$Ar), 4.01, 4.17 (2m, 2H, OCH$_2$CH$_2$CH_2N), 4.14–4.21 (m, 2H, OCH_2CH$_2$CH$_2$N), 4.23, 4.94 (ABq, 2H, J = 9.3 Hz, CH$_2$OCH$_2$), 4.43, 4.70 (ABq, 2H, J = 10.0 Hz, CH$_2$OCH$_2$), 6.87, 6.95, 7.03, 7.07, 7.09, 7.13, 7.28, 7.30 (8d, 8H, ArH), 7.72, 7.87 (2m, 4H, ArH-Pht), 7.73, 8.47, 9.08 (3s, 3H, OH); ^{13}C NMR (CDCl$_3$, 125.8 MHz) δ 29.4 (OCH$_2$CH$_2$CH$_2$N), 30.1, 31.3, 32.8 (ArCH$_2$Ar), 31.1, 31.4, 31.5, 31.6 [C(CH$_3$)$_3$], 33.8, 33.9, 34.2 [C(CH$_3$)$_3$], 35.2 (OCH$_2$CH$_2$CH$_2$N), 71.7, 72.2 (CH$_2$OCH$_2$), 74.2 (OCH$_2$CH$_2$CH$_2$N), 122.6, 122.7, 126.5, 126.7, 127.4, 128.4, 131.3, 132.2, 132.8, 141.5, 142.4, 143.5, 147.8, 147.9, 149.3, 151.3, 152.7 (Ar), 123.4, 133.9 (ArH-Pht), 123.7, 125.2, 125.5, 125.6, 125.9, 126.7, 127.4, 128.1 (ArH), 168.3 (CO); Anal. Calcd for C$_{56}$H$_{67}$O$_7$N: C, 77.66; H, 7.80; N, 1.62. Found: C, 77.18; H, 7.60; N, 1.38.

7,13,19,25-Tetra-*tert*-butyl-27,29-bis[(3-phthalimidopropyl)oxy]-28,30-di-hydroxy-2,3-dihomo-3-oxacalix[4]arene (**3a**): This compound was obtained in 54% yield (0.84 g). An analytically pure sample was obtained by recrystallization from CH$_2$Cl$_2$/MeOH. mp 172–174 °C; IR (KBr) 1709 cm^{-1} (CO), 3400 cm^{-1} (OH); ^{1}H NMR (CDCl$_3$, 500 MHz) δ 1.10, 1.19, 1.23, 1.24 [4s, 36H, C(CH$_3$)], 2.55 (m, 4H, OCH$_2$CH_2CH$_2$N), 3.25, 4.53 (ABq, 2H, J = 12.8 Hz, ArCH$_2$Ar), 3.27, 4.67 (ABq, 2H, J = 12.8 Hz, ArCH$_2$Ar), 3.47, 4.18 (ABq, 2H, J = 13.1 Hz, ArCH$_2$Ar), 3.96, 4.13 (2m, 2H, OCH$_2$CH$_2$CH_2N), 4.08, 4.43 (2m, 2H, OCH$_2$CH$_2$CH_2N), 4.04–4.15 (2m, 4H, OCH_2CH$_2$CH_2N), 4.23, 4.89 (ABq, 2H, J = 9.5 Hz, CH$_2$OCH$_2$), 4.37, 4.70 (ABq, 2H, J = 10.1 Hz, CH$_2$OCH$_2$), 6.81, 6.91, 6.96, 7.00, 7.04, 7.09, 7.23, 7.44 (8d, 8H, ArH), 7.20, 8.04 (2s, 2H, OH), 7.61, 7.68, 7.73, 7.83 (4m, 8H, ArH-Pht); ^{13}C NMR (CDCl$_3$, 125.8 MHz) δ 29.3, 30.0 (OCH$_2$CH$_2$CH$_2$N), 29.8, 29.9, 32.9 (ArCH$_2$Ar), 31.2, 31.4, 31.5, 31.6 [C(CH$_3$)$_3$], 33.8, 34.1 [C(CH$_3$)$_3$], 35.4, 35.7 (OCH$_2$CH$_2$CH$_2$N), 71.9, 72.6 (CH$_2$OCH$_2$), 72.7, 74.2 (OCH$_2$CH$_2$CH$_2$N), 122.2, 127.6, 128.0, 128.8, 129.2, 132.2, 132.3, 132.5, 132.6, 135.3, 141.2, 141.9, 146.4, 147.0, 149.4, 150.0, 152.8, 154.3 (Ar), 123.1, 123.2, 133.6, 133.8 (ArH-Pht), 123.7, 124.7, 125.1, 125.2, 125.6, 126.0, 127.7, 129.5 (ArH), 168.3, 168.4 (CO). Anal. Calcd for C$_{67}$H$_{76}$O$_9$N$_2$: C, 76.40; H, 7.27; N, 2.66. Found: C, 76.01; H, 7.63; N, 2.67.

7,13,19,25-Tetra-*tert*-butyl-28,29-bis[(3-phthalimidopropyl)oxy]-27,30-di-hydroxy-2,3-dihomo-3-oxacalix[4]arene (**3b**): This compound was obtained in 7% yield (0.11 g). An analytically pure sample was obtained by recrystallization from CH$_2$Cl$_2$/MeOH. mp 145–147 °C; IR (KBr) 1715 cm^{-1} (CO), 3397 cm^{-1} (OH); ^{1}H NMR (CDCl$_3$, 500 MHz) δ 1.06, 1.25 [2s, 36H, C(CH$_3$)], 2.46, 2.52 (2m, 4H, OCH$_2$CH_2CH$_2$N), 3.38, 4.29 (ABq, 4H, J = 13.7 Hz, ArCH$_2$Ar), 3.39, 4.56 (ABq, 2H, J = 13.0 Hz, ArCH$_2$Ar), 3.93–4.05 (m, 4H, OCH$_2$CH$_2$CH_2N), 4.03, 4.25 (2m, 4H, OCH_2CH$_2$CH$_2$N), 4.25, 4.71 (ABq, 4H, J = 9.7 Hz, CH$_2$OCH$_2$), 6.84, 6.90, 6.96, 7.21 (4d, 8H, ArH), 7.66, 7.78 (2m, 8H, ArH-Pht), 7.68 (s, 2H, OH); ^{13}C NMR (CDCl$_3$, 125.8 MHz) δ 29.4 (OCH$_2$CH$_2$CH$_2$N), 30.8, 31.6 (ArCH$_2$Ar), 31.2, 31.6 [C(CH$_3$)$_3$], 33.8, 34.0 [C(CH$_3$)$_3$], 35.4 (OCH$_2$CH$_2$CH$_2$N), 71.7 (CH$_2$OCH$_2$), 73.1 (OCH$_2$CH$_2$CH$_2$N), 123.17, 133.8 (ArH-Pht), 123.22, 128.2, 132.2, 132.6, 133.4, 141.7, 146.4, 151.5, 152.1 (Ar), 124.7, 125.2, 125.8, 128.0 (ArH), 168.3 (CO); Anal. Calcd for C$_{67}$H$_{76}$O$_9$N$_2$: C, 76.40; H, 7.27; N, 2.66. Found: C, 76.34; H, 7.62; N, 2.63.

3.1.2. Conventional Procedure for the Synthesis of **5a** and **6a**

A mixture of **1** (1 g, 1.47 mmol) and K$_2$CO$_3$ (0.81 g, 5.9 mmol) in MeCN (25 mL) was stirred and gently warmed under N$_2$ for 30 min. After cooling, *N*-(2-bromoethyl)-phthalimide (1.58 g, 5.9 mmol) and KI (0.49 g, 2.95 mmol) were added, and the reaction mixture was refluxed and stirred under N$_2$ for 3 days. By the same workup described above, the crude product obtained was subjected to flash chromatography on silica gel (eluent gradient from *n*-hexane/ethyl acetate, 90:10 to 70:30). According to TLC analysis, the fractions were combined into two sets, corresponding mainly to derivatives **5a** and **6a**, respectively.

7,13,19,25-Tetra-*tert*-butyl-28-[(2-phthalimidoethyl)oxy]-27,29,30-tri-hydroxy-2,3-dihomo-3-oxacalix[4]arene (**5a**): This compound was obtained in 24% yield (0.30 g). An analytically pure sample was obtained by recrystallization from CH_2Cl_2/MeOH. mp 176–178 °C; IR (KBr) 1717 cm^{-1} (CO), 3375 cm^{-1} (OH); ^{1}H NMR (CDCl$_3$, 500 MHz) δ 1.11, 1.21, 1.222, 1.224 [4s, 36H, C(CH$_3$)], 3.24, 4.25 (ABq, 2H, J = 12.8 Hz, ArCH$_2$Ar), 3.38, 4.15 (ABq, 2H, J = 13.7 Hz, ArCH$_2$Ar), 3.52, 4.09 (ABq, 2H, J = 13.6 Hz, ArCH$_2$Ar), 4.20, 4.86 (ABq, 2H, J = 9.1 Hz, CH$_2$OCH$_2$), 4.28, 4.40 (ABq, 2H, J = 10.0 Hz, CH$_2$OCH$_2$), 4.29, 4.49 (2m, 2H, OCH$_2$CH$_2$N), 4.39 (t, 2H, OCH$_2$CH$_2$N), 6.78, 6.94, 6.99, 7.06, 7.07, 7.08, 7.19, 7.26 (8d, 8H, ArH), 7.28, 8.25, 8.60 (3s, 3H, OH), 7.68, 7.91 (2m, 4H, ArH-Pht); ^{13}C NMR (CDCl$_3$, 125.8 MHz) δ 29.8, 31.2, 32.2 (ArCH$_2$Ar), 31.1, 31.4, 31.5, 31.6 [C(CH$_3$)$_3$], 33.77, 33.81, 33.9, 34.2 [C(CH$_3$)$_3$], 37.5 (OCH$_2$CH$_2$N), 71.6, 71.9 (CH$_2$OCH$_2$), 72.0 (OCH$_2$CH$_2$N), 122.2, 122.7, 126.4, 126.7, 127.3, 127.9, 131.5, 132.5, 132.5, 141.2, 142.2, 143.3, 147.8, 147.9, 148.6, 151.3, 152.5 (Ar), 123.4, 133.7 (ArH-Pht), 123.7, 125.2, 125.3, 125.5, 125.7, 126.7, 127.3, 128.2 (ArH), 168.5 (CO); Anal. Calcd for C$_{55}$H$_{65}$O$_7$N: C, 77.52; H, 7.69; N, 1.64. Found: C, 77.12; H, 8.01; N, 1.65. Although this compound had been synthesised before by a slightly different procedure [34], neither the mp nor the ^{1}H and ^{13}C NMR chemical shifts are totally correct.

7,13,19,25-Tetra-*tert*-butyl-27,29-bis[(3-phthalimidoethyl)oxy]-28,30-di-hydroxy-2,3-dihomo-3-oxacalix[4]arene (**6a**): This compound was obtained in 18% yield (0.27 g). ^{1}H NMR (CDCl$_3$, 500 MHz) δ 1.06, 1.20, 1.21, 1.25 [4s, 36H, C(CH$_3$)], 3.17, 4.37 (ABq, 2H, J = 13.0 Hz, ArCH$_2$Ar), 3.30, 4.67 (ABq, 2H, J = 13.0 Hz, ArCH$_2$Ar), 3.48, 4.18 (ABq, 2H, J = 13.3 Hz, ArCH$_2$Ar), 4.21, 4.92 (ABq, 2H, J = 9.5 Hz, CH$_2$OCH$_2$), 4.25–4.34 (m, 2H, OCH$_2$CH$_2$N), 4.32, 4.77 (2m, 2H, OCH$_2$CH$_2$N), 4.38, 4.74 (ABq, 2H, J = 10.3 Hz, CH$_2$OCH$_2$), 4.42, 4.55 (2m, 2H, OCH$_2$CH$_2$N), 4.51 (m, 2H, OCH$_2$CH$_2$N), 6.78, 6.82, 6.95, 7.00, 7.09, 7.16, 7.41 (7d, 8H, ArH), 7.14, 7.75 (2s, 2H, OH), 7.61, 7.69, 7.82, 7.85 (4m, 8H, ArH-Pht); ^{13}C NMR (CDCl$_3$, 125.8 MHz) δ 29.8, 30.0, 32.7 (ArCH$_2$Ar), 31.1, 31.4, 31.5, 31.6 [C(CH$_3$)$_3$], 33.77, 33.79, 34.1 [C(CH$_3$)$_3$], 37.4, 38.4 (OCH$_2$CH$_2$N), 70.1, 71.7, 72.8 (CH$_2$OCH$_2$ and OCH$_2$CH$_2$N), 122.0, 127.4, 127.8, 128.6, 129.0, 132.1, 132.3, 132.4, 134.7, 141.0, 141.9, 146.4, 147.1, 149.5, 149.7, 152.5, 154.0 (Ar), 123.1, 123.3, 133.6, 133.9 (ArH-Pht), 123.3, 124.8, 125.2, 125.3, 125.6, 126.0, 127.5, 129.6 (ArH), 168.1, 168.5 (CO).

3.1.3. Microwave- and Mechanically-Assisted Syntheses

MW procedure for the synthesis of **2a**, **3a** and **3b**: A mixture of **1** (0.30 g, 0.44 mmol), *N*-(3-bromopropyl)phthalimide (0.24 g, 0.89 mmol) and K$_2$CO$_3$ (0.12 g, 0.89 mmol) was suspended in CH$_3$CN (3 mL) and irradiated under stirring for 30 min at 160 °C. The products were isolated and purified by the same workup and purification steps described in 3.1.1.

MW procedure for the synthesis of **6a**: A mixture of **1** (0.30 g, 0.44 mmol) and K$_2$CO$_3$ (0.24 g, 1.77 mmol) was suspended in CH$_3$CN (3 mL) and irradiated under stirring for 6 min at 110 °C. After cooling, *N*-(2-bromoethyl)phthalimide (0.45 g, 1.77 mmol) was added and the reaction mixture irradiated for a further 36 min at 150 °C. The product was obtained by the same steps described in 3.1.2.

Mechanochemical reaction protocol: 50 mg of parent **1**, *N*-(bromoalkyl)phthalimide and K$_2$CO$_3$ were loaded into the selected reactor containing the corresponding 200 balls. The reactor was then adjusted into the planetary ball mill and the mixture was grinded at 500 rpm, with rotation inversion cycles of 30 min (2.5 min. pause between inversion cycles). Different reagent ratios and reaction times were tested, as mentioned before (see 2.1.). The products were obtained by the same workup described previously.

3.2. Determination of the Crystallographic Structures of 2a, 3a, 3b and 5a

Single crystals suitable for X-ray investigation were obtained by slow evaporation of acetonitrile solutions containing compound **2a**, **3a** or **3b**, while dichloromethane/ethanol solvent mixtures were used for **5a**. Data collection was carried out at the XRD1 beamline of the Elettra synchrotron (Trieste, Italy), employing the rotating-crystal method with a Dectris Pilatus 2M area detector. Single crystals were dipped in paratone cryoprotectant, mounted on a loop and flash-frozen under a nitrogen stream at 100 K. Diffraction data were indexed and integrated using the XDS package [42], while scaling was carried out with

XSCALE [43]. Structures were solved using the SHELXT program [44] and the refinement was performed with SHELXL-14 [45], operating through the WinGX GUI [46], by full-matrix least-squares (FMLS) methods on F^2. Non-hydrogen atoms were anisotropically refined with the exception of non-hydrogen atoms with a low occupancy factor, which were refined isotropically. Hydrogen atoms, located on the difference Fourier maps, were added at the calculated positions and refined using the riding model. Crystallographic data and refinement details are reported in Table S4.

3.3. Fluorescence Studies

Absorption and fluorescence studies were performed using an UV-3101PC UV-Vis-NIR spectrophotometer (Shimadzu, Kyoto, Japan) and a Fluorolog F112A fluorimeter (Spex Industries, Edison, NJ, USA) in right-angle configuration, respectively. The absorption spectra were recorded between 240 and 340 nm and the emission ones between 285 and 600 nm, and using quartz cells with an optical path length of 1 cm. Details concerning the photophysical properties determination are given in the Supplementary Materials.

4. Conclusions

The alkylation of *p-tert*-butyldihomooxacalix[4]arene (**1**) with *N*-(bromopropyl)- or *N*-(bromoethyl)phthalimides and K_2CO_3 in MeCN was carried out by conventional heating and by microwave irradiation and ball milling methodologies. Eight compounds were identified and five isolated (**2a**, **3a**, **3b**, **5a** and **6a**), including mono- and mainly 1,3-disubstituted derivatives in the cone conformation. A microwave-assisted procedure was developed, affording both 1,3-disubstituted phthalimides (**3a** and **6a**) as major/unique products. The reaction times were drastically reduced (30–42 min) and the selectivity increased when compared to refluxing conditions (several days for completion). Minor amounts of solvent and less reactants used were other advantages of this environmentally friendly technique. The results obtained open new perspectives for using this MW-assisted protocol in the synthesis of new calixarenes. Mechanically-assisted synthesis was not shown to be a good method for these alkylation reactions. The conformation and the substitution patterns were confirmed by NMR spectroscopy (^{1}H, ^{13}C, COSY and NOESY) and X-ray diffraction. The X-ray structural analysis of **2a**, **3a**, **3b** and **5a** reveals a common open conformation of the calixarene cone, determined by an intramolecular H-bond which involves the oxygen acceptor of the oxa bridge oriented towards the center of the macrocycle. The lower rim substitution patterns, which determine the intramolecular H-bond network, strongly influence the orientation and π-stacking interactions of the phthalimide arms. The photophysical properties of **2a**, **3a**, **3b** and **5a** were also determined. In particular, it is found that upon excitation of the arene fluorophores, fast energy migration (FRET homotransfer) within the ring ensues, followed by irreversible energy transfer (FRET heterotransfer) to the phthalimide chromophores. These phthalimide derivatives will be used as precursors to urea-terminated anion-binding macrocycles. The ureido-dihomooxacalix[4]arenes bearing propyl or ethyl spacers are expected to be better receptors than those with butyl spacers, mainly for ion pairs, as they will be in a closer proximity. This study is currently under investigation.

Supplementary Materials: The following are available online, ^{1}H NMR assignments of compounds **2b**, **5b** and **6b**; Crystallographic data and refinement details; Photophysical properties determination; ^{1}H, ^{13}C, COSY and NOESY spectra of compounds **2a**, **3a**, **3b**, **5a** and **6a**.

Author Contributions: A.S.M.: Investigation, acquisition and analysis of data, editing. P.M.M.: Conceptualization, supervision, acquisition, analysis and interpretation of data, writing, review and editing. J.R.A.: Analysis and interpretation of NMR data, writing and review. M.P.R.: Methodology, analysis of data, writing and review. V.D.B.B.: Methodology, analysis of data, writing and review. M.N.B.-S.: Analysis, interpretation and writing of photophysics data. N.H. and S.G.: Analysis of structural data, writing and review. All authors have read and agree to the published version of the manuscript.

Funding: The authors thank Fundação para a Ciência e a Tecnologia, Projects ref. UIDB/00100/2020, UIDB/04565/2020 and PTDC/MEC-ONC/29327/2017. A. S. Miranda thanks a PhD Grant ref. SFRH/BD/129323/2017.

Institutional Review Board Statement: Not applicable.

Informed Consent Statement: Not applicable.

Data Availability Statement: Not applicable.

Conflicts of Interest: The authors declare no conflict of interest.

Sample Availability: Samples of the compounds are not available from the authors.

References

1. Gutsche, C.D. *Calixarenes, an Introduction*, 2nd ed.; Monographs in Supramolecular Chemistry; The Royal Society of Chemistry: Cambridge, UK, 2008.
2. Neri, P.; Sessler, J.L.; Wang, M.-X. (Eds.) *Calixarenes and Beyond*; Springer International Publishing: Cham, Switzerland, 2016.
3. Busschaert, N.; Caltagirone, C.; Van Rossom, W.; Gale, P.A. Applications of supramolecular anion recognition. *Chem. Rev.* **2015**, *115*, 8038–8155. [CrossRef]
4. Gale, P.A.; Howe, E.N.W.; Wu, X. Anion receptor chemistry. *Chem* **2016**, *1*, 351–422. [CrossRef]
5. Gedye, R.; Smith, F.; Westaway, K.; Ali, H.; Baldisera, L.; Laberge, L.; Rousell, J. The use of microwave ovens for rapid organic synthesis. *Tetrahedron Lett.* **1986**, *27*, 279–282. [CrossRef]
6. Giguere, R.J.; Bray, T.L.; Duncan, S.M. Application of commercial microwave ovens to organic synthesis. *Tetrahedron Lett.* **1986**, *27*, 4945–4948. [CrossRef]
7. Strauss, C.R. A strategic, "green" approach to organic chemistry with microwave assistance and predictive yield optimization as core, enabling technologies. *Aust. J. Chem.* **2009**, *62*, 3–15. [CrossRef]
8. Caddick, S.; Fitzmaurice, R. Microwave enhanced synthesis. *Tetrahedron* **2009**, *65*, 3325–3355. [CrossRef]
9. Gawande, M.B.; Shelke, S.N.; Zboril, R.; Varma, R.S. Microwave-assisted chemistry: Synthetic applications for rapid assembly of nanomaterials and organics. *Acc. Chem. Res.* **2014**, *47*, 1338–1348. [CrossRef]
10. Kumar, A.; Kuang, Y.; Liang, Z.; Sun, X. Microwave chemistry, recent advancements, and eco-friendly microwave-assisted synthesis of nanoarchitectures and their applications: A review. *Mater. Today Nano* **2020**, *11*, 100076. [CrossRef]
11. Stolle, A.; Szuppa, T.; Leonhardt, S.E.S.; Ondruschka, B. Ball milling in organic synthesis: Solutions and challenges. *Chem. Soc. Rev.* **2011**, *40*, 2317–2329. [CrossRef]
12. James, S.L.; Adams, C.J.; Bolm, C.; Braga, D.; Collier, P.; Friscie, T.; Grepioni, F.; Harris, K.D.M.; Hyett, G.; Jones, W.; et al. Mechanochemistry: Opportunities for new and cleaner synthesis. *Chem. Soc. Rev.* **2012**, *41*, 413–447. [CrossRef] [PubMed]
13. Wang, G.W. Mechanochemical organic synthesis. *Chem. Soc. Rev.* **2013**, *42*, 7668–7700. [CrossRef]
14. Rahman, M.; Santra, S.; Kovalev, I.S.; Kopchuk, D.S.; Zyryanov, G.V.; Majee, A.; Chupakhin, O.N. Gree synthetic approaches for practically relevant (hetero)macrocycles: An overview. *AIP Conf. Proc.* **2020**, *2280*, 030014.
15. Baozhi, L.; Gengliang, Y.; Jinsong, Z.; Kefang, D. Microwave-assisted synthesis of *p*-alkylcalix[n]arene catalysed by KOH. *Eur. J. Chem.* **2005**, *2*, 70–74.
16. Takagaki, M.; Hosoda, A.; Mori, H.; Miyake, Y.; Kimura, K.; Taniguchi, H.; Nomura, E. Rapid and convenient laboratory method for the preparation of *p-tert*-butylcalix[4]arene using microwave irradiation. *Green Chem.* **2008**, *10*, 978–981. [CrossRef]
17. Agrawal, Y.K.; Desai, N.C.; Mehta, N.D. Microwave-assisted synthesis of azocalixarenes. *Synth. Commun.* **2007**, *37*, 2243–2252. [CrossRef]
18. Cecioni, S.; Lalor, R.; Blanchard, B.; Praly, J.P.; Imberty, A.; Matthews, S.E.; Vidal, S. Achieving high affinity towards a bacterial lectin through multivalent topological isomers of calix[4]arene glycoconjugates. *Chem. Eur. J.* **2009**, *15*, 13232–13240. [CrossRef] [PubMed]
19. Fatykhova, G.A.; Burilov, V.A.; Dokuchaeva, M.N.; Solov'eva, S.E.; Antipin, I.S. Synthesis of tetraazide derivatives of *p-tert*-butylcalix[4]arene using copper-catalyzed nucleophilic aromatic substitution. *Dokl. Chem.* **2018**, *479*, 64–67. [CrossRef]
20. Wang, X.; Zhao, Z.; Chen, B.; Li, X.; Liu, M. Microwave-assisted synthesis of bidentate chiral unsymmetrical urea derivatives of *p-tert*-butylcalix[4]arene and their anion recognition properties. *J. Chem. Res.* **2015**, *39*, 303–306. [CrossRef]
21. Nayak, S.K.; Choudhary, M.K. Microwave-assisted synthesis of 1,3-dialkyl ethers of calix[4]arenes: Application to the synthesis of cesium selective calix[4]crown-6 ionophores. *Tetrahedron Lett.* **2012**, *53*, 141–144. [CrossRef]
22. Burilov, V.A.; Nugmanov, R.I.; Ibragimova, R.R.; Solovieva, S.E.; Antipin, I.S.; Konovalov, A.I. Microwave-assisted alkylation of *p-tert*-butylcalix[4]arene lower rim: The effect of alkyl halides. *Mendeleev Commun.* **2013**, *23*, 113–115. [CrossRef]
23. Bakic, M.T.; Klaric, D.; Espinosa, M.S.; Kazazic, S.; Frkanec, L.; Babay, P.A.; Galic, N. Syntheses of ester and amide derivatives of calix[6]arene and their complexation affinities towards La^{3+}, Eu^{3+}, and Yb^{3+}. *Supramol. Chem.* **2019**, *31*, 723–731. [CrossRef]
24. Galán, H.; de Mendoza, J.; Prados, P. Microwave-assisted synthesis of a nitro-*m*-xylylenedioxycalix[6]arene building block functionalized at the upper rim. *Eur. J. Org. Chem.* **2010**, 7005–7011. [CrossRef]

25. Atwood, J.L.; Hardie, M.J.; Raston, C.L.; Sandoval, C.A. Convergent synthesis of *p*-benzylcalix[7]arene: Condensation and UHIG of *p*-benzylcalix[6 or 8]arenes. *Org. Lett.* **1999**, *1*, 1523–1526. [CrossRef]
26. Chennakesavulu, K.; Raju, G.B. Mechanochemical synthesis of *p*-nitro calix[6]arene. *Asian J. Chem.* **2010**, *22*, 4947–4949.
27. Marcos, P.M.; Teixeira, F.A.; Segurado, M.A.P.; Ascenso, J.R.; Bernardino, R.J.; Michel, S.; Hubscher-Bruder, V. Bidentate urea derivatives of *p-tert*-butyldihomooxacalix[4]arene: Neutral receptors for anion complexation. *J. Org. Chem.* **2014**, *79*, 742–751. [CrossRef]
28. Marcos, P.M.; Teixeira, F.A.; Segurado, M.A.P.; Ascenso, J.R.; Bernardino, R.J.; Brancatelli, G.; Geremia, S. Synthesis and anion binding properties of new dihomooxacalix[4]arene diurea and dithiourea receptors. *Tetrahedron* **2014**, *70*, 6497–6505. [CrossRef]
29. Augusto, A.S.; Miranda, A.S.; Ascenso, J.R.; Miranda, M.Q.; Félix, V.; Brancatelli, G.; Hickey, N.; Geremia, S.; Marcos, P.M. Anion recognition by partial cone dihomooxacalix[4]arene-based receptors bearing urea groups: Remarkable affinity for benzoate ion. *Eur. J. Org. Chem.* **2018**, 5657–5667. [CrossRef]
30. Miranda, A.S.; Serbetci, D.; Marcos, P.M.; Ascenso, J.R.; Berberan-Santos, M.N.; Hickey, N.; Geremia, S. Ditopic receptors based on dihomooxacalix[4]arenes bearing phenylurea moieties with electron-withdrawing groups for anions and organic ion pairs. *Front. Chem.* **2019**, *7*, 758. [CrossRef]
31. Miranda, A.S.; Marcos, P.M.; Ascenso, J.R.; Berberan-Santos, M.N.; Schurhammer, R.; Hickey, N.; Geremia, S. Dihomooxacalix[4]-arene-based fluorescent receptors for anion and organic ion pair recognition. *Molecules* **2020**, *25*, 4708. [CrossRef] [PubMed]
32. Barboso, S.; Carrera, A.G.; Matthews, S.E.; Arnaud-Neu, F.; Bohmer, V.; Dozol, J.F.; Rouquette, H.; Schiwing-Weill, M.J. Calix[4]arenes with CMPO functions at the lower rim. Synthesis and extraction properties. *J. Chem. Soc. Perkin Trans. 2* **1999**, 719–723. [CrossRef]
33. Marcos, P.M.; Ascenso, J.R.; Pereira, J.L.C. Synthesis and NMR conformational studies of p-tert-butyldihomooxacalix[4]arene derivatives bearing pyridyl pendant groups at the lower rim. *Eur. J. Org. Chem.* **2002**, 3034–3041. [CrossRef]
34. Jaime, C.; de Mendoza, J.; Prados, P.; Nieto, P.; Sanchez, C. Carbon-13 NMR chemical shifts. A single rule to determine the conformation of calix[4]arenes. *J. Org. Chem.* **1991**, *56*, 3372–3376. [CrossRef]
35. Liu, Y.; Sun, J.; Yan, C.G. Synthesis and crystal structures of *p-tert*-butyldihomooxa-calix[4]arene mono-Schiff bases. *J. Incl. Phenom. Macrocycl. Chem.* **2017**, *87*, 157–166. [CrossRef]
36. Miranda, A.S.; Martelo, L.M.; Fedorov, A.A.; Berberan-Santos, M.N.; Marcos, P.M. Fluorescence properties of *p-tert*-butyldihomooxacalix[4]arene derivatives and the effect of anion complexation. *New J. Chem.* **2017**, *41*, 5967–5973. [CrossRef]
37. Griesbeck, A.G.; Görner, H. Laser flash photolysis study of *N*-alkylated phathalimides. *J. Photochem. Photobiol.* **1999**, *129*, 111–119. [CrossRef]
38. Valeur, B.; Berberan-Santos, M.N. *Molecular Fluorescence. Principles and Applications*, 2nd ed.; Wiley-VCH: Weinheim, Germany, 2012; p. 195.
39. Weber, G. Fluorescence-polarization spectrum and electronic-energy transfer in tyrosine, tryptophan and related compounds. *Biochem. J.* **1960**, *75*, 335–345. [CrossRef] [PubMed]
40. Fornander, L.H.; Feng, B.; Beke-Somfai, T.; Nordén, B. UV transition moments of tyrosine. *J. Phys. Chem. B* **2014**, *118*, 9247–9257. [CrossRef]
41. Berberan-Santos, M.N.; Canceill, J.; Brochon, J.C.; Jullien, L.; Lehn, J.-M.; Pouget, J.; Tauc, P.; Valeur, B. Multichromophoric cyclodextrins. 1. Synthesis of *O*-naphthoyl beta cyclodextrins and investigation of excimer formation and energy hopping. *J. Am. Chem. Soc.* **1992**, *114*, 6427–6436. [CrossRef]
42. Kabsch, W. XDS. *Acta Crystallogr. Sect. D Biol. Crystallogr.* **2010**, *66*, 125–132. [CrossRef] [PubMed]
43. Kabsch, W. Integration, scaling, space-group assignment and post-refinement. *Acta Crystallogr. Sect. D Biol. Crystallogr.* **2010**, *66*, 133–144. [CrossRef]
44. Sheldrick, G.M. SHELXT–integrated space-group and crystal-structure determination. *Acta Crystallog. Sect. A Found. Crystallogr.* **2015**, *71*, 3–8. [CrossRef]
45. Sheldrick, G.M. A short history of SHELX. *Acta Crystallog. Sect. A Found. Crystallogr.* **2008**, *64*, 112–122. [CrossRef]
46. Farrugia, L.J. WinGX and ORTEP for windows: An update. *J. Appl. Crystallog.* **2012**, *45*, 849–854. [CrossRef]

Article

Self-Assembly of Discrete Porphyrin/Calix[4]tube Complexes Promoted by Potassium Ion Encapsulation

Massimiliano Gaeta [1], Elisabetta Rodolico [1], Maria E. Fragalà [1], Andrea Pappalardo [1], Ilenia Pisagatti [2], Giuseppe Gattuso [2], Anna Notti [2,*], Melchiorre F. Parisi [2,*], Roberto Purrello [1,*] and Alessandro D'Urso [1,*]

[1] Dipartimento di Scienze Chimiche, Università degli Studi di Catania, Viale A. Doria 6, 95125 Catania, Italy; gaetamassimiliano@libero.it (M.G.); elisabetta.rod27@gmail.com (E.R.); me.fragala@unict.it (M.E.F.); andrea.pappalardo@unict.it (A.P.)

[2] Dipartimento di Scienze Chimiche, Biologiche, Farmaceutiche ed Ambientali, Università degli Studi di Messina, Viale F. Stagno d'Alcontres, 31, 98166 Messina, Italy; ipisagatti@unime.it (I.P.); ggattuso@unime.it (G.G.)

* Correspondence: anotti@unime.it (A.N.); mparisi@unime.it (M.F.P.); rpurrello@unict.it (R.P.); adurso@unict.it (A.D.)

Citation: Gaeta, M.; Rodolico, E.; Fragalà, M.E.; Pappalardo, A.; Pisagatti, I.; Gattuso, G.; Notti, A.; Parisi, M.F.; Purrello, R.; D'Urso, A. Self-Assembly of Discrete Porphyrin/Calix[4]tube Complexes Promoted by Potassium Ion Encapsulation. *Molecules* **2021**, *26*, 704. https://doi.org/10.3390/molecules26030704

Academic Editors: Mario Berberan-Santos, Paula M. Marcos and M. Amparo F. Faustino

Received: 17 October 2020
Accepted: 23 January 2021
Published: 29 January 2021

Publisher's Note: MDPI stays neutral with regard to jurisdictional claims in published maps and institutional affiliations.

Abstract: The pivotal role played by potassium ions in the noncovalent synthesis of discrete porphyrin-calixarene nanostructures has been examined. The *flattened-cone* conformation adopted by the two cavities of octa-cationic calix[4]tube **C4T** was found to prevent the formation of complexes with well-defined stoichiometry between this novel water-soluble calixarene and the tetra-anionic phenylsulfonate porphyrin **CuTPPS**. Conversely, preorganization of **C4T** into a C_{4v}-symmetrical scaffold, triggered by potassium ion encapsulation (**C4T@K$^+$**), allowed us to carry out an efficient hierarchical self-assembly process leading to 2D and 3D nanostructures. The stepwise formation of discrete **CuTPPS/C4T@K$^+$** noncovalent assemblies, containing up to 33 molecular elements, was conveniently monitored by UV/vis spectroscopy by following the absorbance of the porphyrin Soret band.

Keywords: noncovalent synthesis; hierarchical control; calixarenes; calix[4]tubes; metallo-porphyrins

1. Introduction

Porphyrins, owing to their redox [1,2] and opto-electronic properties [3–5], relative ease of derivatization [6] and propensity to self-organize in architectures of different size and topology [7–9], are very attractive building blocks for the synthesis of functional nanomaterials useful for light harvesting [10,11], sensing [12], catalysis [13], imaging [14] and photodynamic therapy [15] applications. Rods [16–18], wires [19–21], tubes [22], sheets [23], spheres [24] and rings [25] are examples of porphyrin-based nanostructure motifs reported to date.

In aqueous solution, however, one of the main obstacles to the development of discrete nanostructures is created by the pronounced tendency of porphyrins to spontaneously self-aggregate (via π–π stacking interactions), precluding the formation of arrays of well-defined shape and size. To overcome this, early endeavors have mainly focused on targeted covalent derivatizations or the formation of coordination bonds [26]. More recently, the rational design of porphyrin-based supramolecular assemblies has been successfully carried out in the presence of templating agents such as: polyelectrolytes [27,28], peptides [29], inorganic molecules bearing metal-coordination centers [30–33] and macrocyclic compounds [34–36], by taking advantage of single or multiple metal coordination, hydrogen bonding, π–π stacking, electrostatic and hydrophobic interactions.

Multi-charged water-soluble calix[n]arenes [37–40], because of their remarkable affinity towards charge- and shape-complementary substrates [41–44], have been successfully employed as templating agents for the assembly of hybrid porphyrin-calixarene nanostructures in aqueous media [45,46] and in the solid state [47–50]. We have shown that

21

both anionic [51] and cationic [52] calix[4]arenes quantitatively interact with oppositely charged porphyrins, under rigid hierarchical rules, providing assemblies with predictable sequence and stoichiometry. By replacing the calixarene framework with ditopic or tritopic bis- [53] or tris-calixarene [54] scaffolds we were able to control the dimensionality of the assembly, hierarchically forming 2D and 3D noncovalent architectures of considerable size. We were also able to induce chirality in some of these multi-component assemblies by using appropriate enantiopure agents [52,54,55]. The self-assembly in aqueous solution is mainly driven by electrostatic interactions between differently charged components as well as solvophobic effects and other noncovalent weak forces, all of which ultimately contribute to the thermodynamic stability of the species formed. The stability and kinetic inertness of these multicomponent assemblies have been assessed by light scattering, diffusion NMR studies [56] and a number of single-crystal XRD analyses [51,56].

As a follow up to these studies, to test the limits of the noncovalent approach to the synthesis of porphyrin-calixarene nanostructures we have synthesized a water-soluble congener of the known *p-tert*-butylcalix[4]tube **1** (Scheme 1), first reported by Beer and coworkers [57], and now wish to report the profound effect played by a single potassium ion on the overall self-assembly process carried out in water in the presence of copper(II) *meso-tetrakis*-(4-sulfonatophenyl)porphyrin tetrasodium salt (**CuTPPS**). Unlike **CT4** (Scheme 1), the corresponding potassium complex (**C4T@K⁺**), because of the preorganization of its cavities and the rigidity of its ditopic tubular structure, is able to promote the noncovalent assembly of discrete porphyrin/calixtube nanostructures with a stoichiometry as high as 17:16.

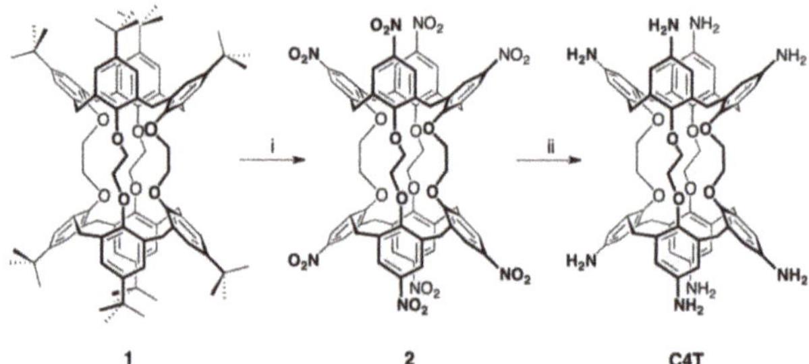

Scheme 1. The synthesis of the octa-amino calix[4]tube C4T. Reaction conditions: (i) HNO_3, $CHCl_3$/AcOH, r.t., 72 h; (ii) H_2, Raney/Ni, DMF, r.t., 24 h.

2. Results and Discussion

Octa-amino calix[4]tube **C4T** was synthesized in overall 70% yield from *p-tert*-butylcalix [4]tube **1** [57] by exhaustive nitration followed by reduction of the resultant nitro derivative **2** (Scheme 1).

Analogously to the parent *p-tert*-butylated derivative **1** [57] and a number of other calix[4]tubes [58–62], the two cavities of octa-amino calix[4]tube **C4T** adopt a preferential *flattened-cone* conformation (C_{2v} symmetry) in solution (DMF-d_7, 298 K). Consistent with this, the ^{1}H NMR spectrum displays pairs of equally intense broad singlets for the ArH, OCH_2 and NH_2 groups and a single AX-system for the $ArCH_2Ar$ bridging moieties (Figure 1a). Signal doubling is also seen in the ^{13}C NMR spectrum of **C4T**.

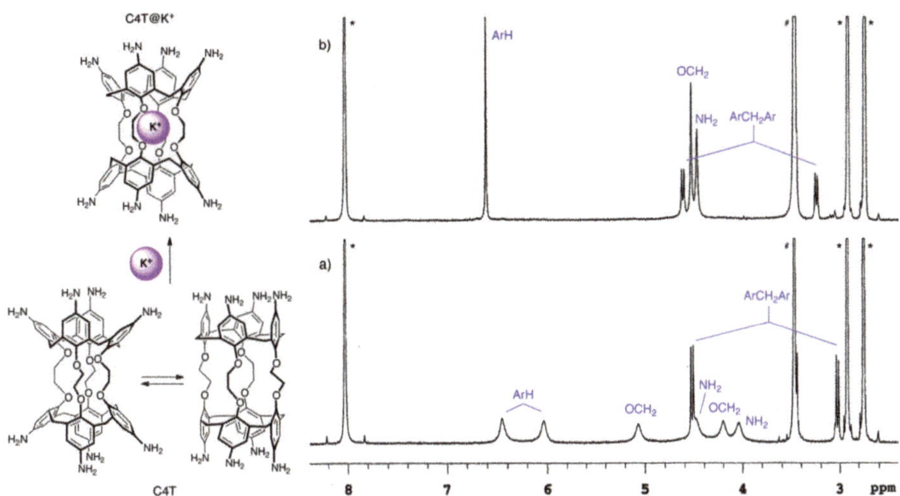

Figure 1. ^{1}H NMR (DMF-d_7, 298 K) of: (**a**) octa-amino calix[4]tube **C4T**; (**b**) the potassium complex of the octa-amino calix[4]tube **C4T@K$^+$**. The hashtag and the asterisks indicate the H$_2$O and the residual solvent peaks, respectively.

To assess the ability of calixtube **C4T** to act as a templating agent for the noncovalent synthesis of porphyrin-based supramolecular architectures, a 2.0 μM aqueous solution (pH = 3.0 [63]) of **C4T** was titrated with increasing aliquots of an aqueous solution of copper(II) *meso-tetrakis*-(4-sulfonatophenyl)porphyrin tetrasodium salt (**CuTPPS**). Earlier studies carried out on several water-soluble calix[4]arenes [51,52], bis-calix[4]arenes [52,55] and tris-calix[4]arenes [54], and a number of metallo-porphyrins (Chart 1), had shown that the formation of discrete calixarene/porphyrin assemblies proceeds in a step-wise hierarchical fashion [64] with the display of clear-cut spectral changes (absorption or emission). More specifically, the formation of any complex/assembly of well-defined stoichiometry is always pinpointed by a specific *break-point* on a diagram where the absorbance values of the porphyrin Soret band are plotted vs. the [porphyrin] × 4/[calixarene] ratio (*vide infra*). That is, an experimental data point of the titration curve where the slope variation is larger than 10%. Different slopes indicate the presence in solution of discrete assemblies, each of these characterized by a different molar extinction coefficient. In other words, the presence of *break-points* confirms that the species formed are not in equilibrium with each other, otherwise a straight line would only be observed over the entire course of a titration experiment.

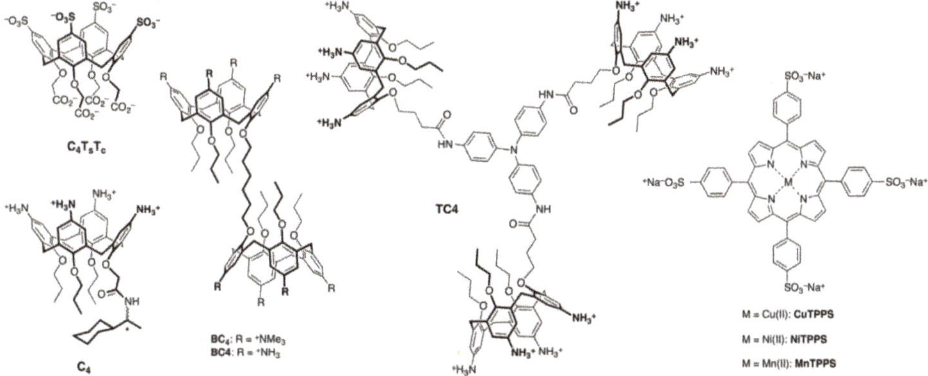

Chart 1. Calix[4]arene derivatives and metallo-porphyrins used in previous studies [51–55].

In the case of the ditopic calix[4]tube **C4T** and the tetratopic porphyrin **CuTPPS**, however, no such trend was observed (Figure 2), despite the fact that the absorption data collected over the course of the titration displayed—with respect to the blank experiment carried out in the absence of **C4T**—hypochromicity and a broadening of the **CuTPPS** Soret band and, therefore, a clear indication of porphyrin–calixtube interactions (Figures S2 and S3, see the Supplementary Materials). The porphyrin–calix[4]tube assembly process was best analyzed by plotting the **CuTPPS** absorbance values at 412 nm vs. the [**CuTPPS**] × 4/[**C4T**] ratio. The trace joining the experimental data-points (Figure 2, trace b) is straight (i.e., no detectable *break-points*) up to the equivalence ([**CuTPPS**] = [**C4T**] = 2 μM) and then its slope becomes steeper upon further addition of **CuTPPS** (up to a 3 μM total concentration). This absorbance trend is consistent with an initial formation of **CuTPPS/C4T** complexes with a 1:1 stoichiometry, followed by a marked hyperchromic effect due to the absorbance of the porphyrin molecules present in excess in solution; the slope of the second segment of trace b approaches the one observed in **CuTPPS** solutions of increasing concentration (Figure 2, trace a). One possible explanation for this unexpected lack of stoichiometric complementarity between this oppositely charged pair is that the **C4T** cavities are forced to adopt a *flattened-cone* conformation by the four ethylene linkages (see above), resulting in poor preorganization. These structural features, because of steric hindrance and weaker electrostatic interactions, prevent the simultaneous binding of four calix[4]tube molecules to a single porphyrin and, as a result, the formation of a stable **CuTPPS/C4T** complex with a 1:4 stoichiometry and a cruciform structure (*vide infra*).

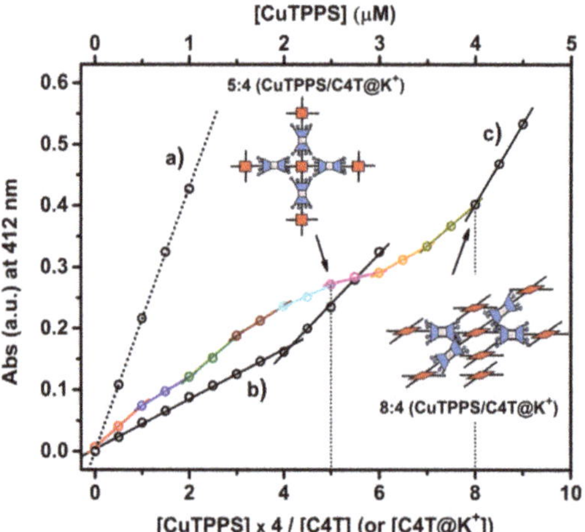

Figure 2. Variation in the absorbance of the CuTPPS Soret band (λ_{max} = 412 nm) observed upon: (i) increase of the porphyrin concentration in water (dotted black trace labelled as (**a**)); (ii) portion-wise addition of CuTPPS to a 2 μM aqueous solution of calix[4]tube C4T at pH = 3.0 (solid black trace labelled as (**b**)); (iii) portion-wise addition of CuTPPS to a 2 μM aqueous solution of C4T@K⁺ at pH = 3.0 (multicolored trace labelled as (**c**)).

Given the known proclivity of *p-tert*-butylcalix[4]tube **1** to encapsulate potassium ions within the cryptand-like binding site, provided by the four dioxyethylene bridges, Beer's findings [57,58] were exploited to "freeze" the cavities of the octa-amino calix[4]tube into a more favorable *cone* conformation. Accordingly, stirring of **C4T** in DMF in the presence of a large excess of KI, afforded the corresponding **C4T@K⁺** potassium complex (Figure 1b).

To test whether the potassium complex of calix[4]tube **C4T** was able to promote supramolecular assembly as a result of its cavities being preorganized in a C_{4v}-symmetry, the templating-agent potentials of **C4T@K⁺** were reassessed under the conditions described earlier. The UV/vis titration was similarly carried out by adding increasing amounts of **CuTPPS** to a solution of **C4T@K$^+$** (2 μM) in water at pH = 3.0 [63]. As seen above, the spectra acquired in the course of the titration, showed that the absorbance of the porphyrin Soret band broadens and undergoes a hypochromic effect (Figure S4). However, compared with the **CuTPPS/C4T** system (Figure 2, trace b), in this case, the assembly process proceeds under stoichiometric control up to a [**CuTPPS**] = 4 μM, as unambiguously proven by the presence of several *break-points* coinciding with **CuTPPS/C4T@K$^+$** complexes of precise stoichiometry (i.e., 1:4-, 2:4-, 3:4-, 4:4-, 5:4-, 6:4-, 7:4- and 8:4-(**CuTPPS/C4T@K$^+$**); see Figure 2, trace c). As the titration proceeds and further aliquots of porphyrin are added to the solution (i.e., for [**CuTPPS**] > 4 μM), the absorbance sharply increases and the related slope is seen to closely match the one detected for **CuTPPS** on its own (Figure 2, compare traces a with the last segment of trace c). These findings indicate that the excess porphyrin molecules now present in solution are no longer interacting with the supramolecular complex. As for the largest species obtained under these conditions, the absorption data of Figure 2 (trace c) suggests the formation of an 8:4-(**CuTPPS/C4T@K$^+$**) supramolecular structure, similar to those observed elsewhere by single-crystal X-ray analysis [48–51], likely obtained as a result of the stacking of three additional porphyrin molecules above and/or below the plane containing the parent 5:4-(**CuTPPS/C4T@K$^+$**) assembly [65], the latter having radially grown around the central **CuTPPS** unit in a step-wise and hierarchical fashion. Compared to the case analyzed at the beginning, where the two C_{2v}-*symmetrical* cavities of **C4T** were seen to prevent efficient porphyrin binding, the result observed in the presence of the ditopic C_{4v}-arranged **C4T@K$^+$** calixtube complex is quite dramatic in terms of stoichiometric control of the assembly process.

Similar to the case of bis-calix[4]arene **BC$_4$** (Chart 1, [53]), the 5:4-(**CuTPPS/C4T@K$^+$**) assembly is a "fork-point" precursor, key to the subsequent 2D or 3D syntheses of larger porphyrin/calixtube supramolecular architectures (Figure 3).

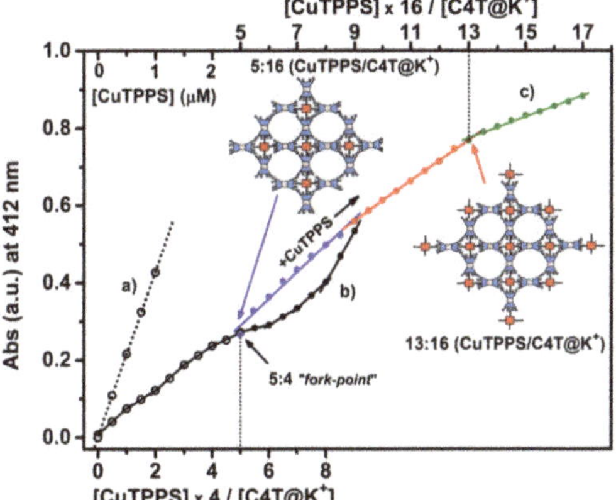

Figure 3. Variation in the absorbance of the **CuTPPS** Soret band (λ_{max} = 412 nm) observed upon: (i) increase of the porphyrin concentration in water (dotted black trace labelled as (**a**)); (ii) portion-wise addition of **CuTPPS** to a 2 μM aqueous solution of **C4T@K$^+$** at pH 3.0 (solid black trace labelled as (**b**)); (iii) initial portion-wise addition of **CuTPPS** to a 2 μM aqueous solution of **C4T@K$^+$** at pH 3.0 up to the formation of the 5:4-(**CuTPPS/C4T@K$^+$**) assembly (see the black arrow of trace (**b**)), followed by portion-wise addition of **C4T@K$^+$** (up to a final [**C4T@K$^+$**] = 8 μM) to form the 5:16-(**CuTPPS/C4T@K$^+$**) assembly (see the blue arrow of trace (**c**)) and then final increment of the **CuTPPS** concentration from 2.5 to 8.5 μM to yield the 17:16-(**CuTPPS/C4T@K$^+$**) assembly (multicolored portion of trace (**c**)).

Addition of porphyrins leads to the above-mentioned 8:4-(**CuTPPS/C4T@K$^+$**) species, where the extra **CuTPPS** molecules are stacked above and/or below the planar 5:4-(**CuTPPS/C4T@K$^+$**) assembly (second section of trace b), whereas an increase in the calixtube concentration (in a parallel titration experiment) from 2 to 8 µM prompts a planar growth of the assembly, yielding the 5:16-(**CuTPPS/C4T@K⁺**) species (Figure 3). As expected, given the absence of 'free' porphyrin molecules in solution, the formation of the latter proceeds with no variations in the absorbance at 412 nm, (blue arrow close to trace c in Figure 3, see also Figure S5). On the other hand, once the 5:16-(**CuTPPS/C4T@K⁺**) species has formed, a total of twelve cavities of calix[4]tube become available. As a result, upon subsequent increase of the **CuTPPS** concentration from 2.5 to 8.5 µM the solution undergoes—with respect to the blank, see trace a—a noticeable hypochromic effect (Figure 3, blue and red segments of trace c) which is consistent with the formation of an assembly with a 13:16 **CuTPPS/C4T@K⁺** stoichiometry (Figure 3, red arrow close to trace c). In agreement with the even larger hypochromicity observed in the green segment of trace *c* (Figure 3), the formation of a new discrete species (i.e., 17:16-(**CuTPPS/C4T@K⁺**) is finally observed. This latter is probably formed by further piling of four **CuTPPS** molecules above and/or below the planar 13:16-(**CuTPPS/C4T@K⁺**) assembly. After reaching the 17:16 molar ratio the titration experiment was halted because of an incipient precipitate formation in solution.

3. Materials and Methods

3.1. General

Commercial reagent grade chemicals were used as received without any further purification. Solvents were dried by standard methods. Melting points were determined on a Kofler hot stage apparatus and are uncorrected. ^{1}H and ^{13}C NMR spectra were acquired at 25 °C in DMF-d_7, at 500 and 125 MHz respectively. Chemical shifts are reported in ppm and are referenced to the solvent residual peak (δ_H = 2.75 ppm and δ_C = 29.8 ppm). The ESI-MS spectrum of **C4T@K⁺** (dissolved in H$_2$O at pH = 3.0) was recorded on an ES-MS Thermo-Finnigan LCQ-DECA instrument (positive ion mode) using a low declustering potential. UV/vis measurements were carried out at room temperature on a JASCO V-530 spectrophotometer. Quartz cuvettes with 1 cm path-length were used for all measurements. Routinely, 3 to 5 different solutions were used for each determination.

p-tert-Butylcalix[4]tube **1** was prepared according to a known literature procedure [57]. Copper(II) *meso-tetrakis*-(4-sulfonatophenyl)porphyrin tetrasodium salt (**CuTPPS**) was synthesized from the corresponding metal oxides (CuO) by heterogeneous metal-insertion in water, according to a previously reported procedure [65].

CuTPPS, **C4T** and **C4T@K⁺** stock solutions (about 4×10^{-4}, 3.2×10^{-4} and 5.4×10^{-4} M, respectively) were prepared in ultrapure water obtained from Elga Veolia Purelab flex and their concentrations were calculated spectrophotometrically (UV/vis in H$_2$O) by looking at the maximum intensity of the porphyrin Soret band $\lambda_{max}(\varepsilon)$ = 412 nm (416,000 M^{-1} cm^{-1}). **CuTPPS/C4T@K⁺** assemblies were obtained at room temperature by adding increasing aliquots of **CuTPPS** (so that the concentration of the porphyrin in the titrating solution was 0.25 µM higher after each addition) to a 2 µM aqueous solution (pH = 3.0) of the **C4T@K⁺** complex, up to desired molar ratio [**CuTPPS**]/[**C4T@K⁺**]. In water, under acidic conditions (pH = 3.0), the octa-amino calix[4]tube **C4T** is converted to its fully protonated octa-ammonium form(see Figure S1).

3.2. Syntheses of Calix[4]tube *C4T* and the Potassium Inclusion Complex *C4T@K⁺*

3.2.1. Octa-Nitro Calix[4]tube 2

Fuming HNO$_3$ (4.5 mL) was slowly added to a cooled (T = -15 °C) solution of *p-tert*-butylcalix[4]tube **1** (630 mg, 0.45 mmol) in CHCl$_3$ (180 mL) and AcOH (13.5 mL) and the mixture was then left to stir vigorously at room temperature. Addition of the same amounts of AcOH and HNO$_3$ was repeated after 24 and 48 h. After 72 h, the solvent was evaporated under vacuum and the resulting residue was triturated with CH$_3$OH (30 mL). Several recrystallizations from DMF of the solid thus obtained yielded derivative **2** as an off-white powder (554 mg, 94%). M.p. > 230 °C. ^{1}H NMR (DMF-d_7) δ 8.62 (s, ArH, 8H), 7.64 (s, ArH, 8H), 5.51 (s, OCH$_2$, 8H), 4.85 and 4.03 (AX system, *J* = 13.6 Hz, ArCH$_2$Ar,

16H), 4.84 (s, OCH$_2$, 8H) ppm. ^{13}C NMR (DMF-d_7) δ 164.1, 162.1, 143.7, 143.1, 125.8, 124.5, 73.8, 73.4, 31.6 ppm. Anal. Calcd for C$_{64}$H$_{48}$N$_8$O$_{24}$: C, 58.54; H, 3.68; N, 8.53. Found: C, 58.23; H, 3.92; N, 8.37.

3.2.2. Octa-Amino Calix[4]tube **C4T**

A suspension of **2** (100 mg, 0.076 mmol) and Raney/Nickel in DMF (50 mL) was stirred under H$_2$ (1 atm) at room temperature for 24 h. The catalyst was filtered off and the eluate was evaporated to dryness under vacuum to provide a residue that upon treatment with MeOH gave **C4T** as a brown powder that was collected by suction filtration (61 mg, 75%). M.p. > 230 °C. ^{1}H NMR (DMF-d_7): δ 6.45 (br s, ArH, 8H), 6.03 (br s, ArH, 8H), 5.06 (br s, OCH$_2$, 8H), 4.51 and 3.02 (AX system, J = 12.3 Hz, ArCH$_2$Ar, 16 H), 4.47 (br s, NH$_2$, 8H), 4.23 (br s, OCH$_2$, 8H), 4.04 (br s, NH$_2$, 8H) ppm. ^{13}C NMR (DMF-d_7): δ 150.6, 148.5, 143.3, 142.7, 135.9, 133.0, 115.07, 114.96, 73.7, 32.6 ppm. Anal. Calcd for C$_{64}$H$_{64}$N$_8$O$_8$: C, 71.62; H, 6.01, N, 10.44. Found: C, 71.31; H, 6.23; N, 10.29.

3.2.3. Formation of the **C4T@K$^+$** Complex

A suspension of **C4T** (10 mg, 9 µmol) and KI (100 mg, 600 µmol) in dry DMF (30, mL) was kept under vigorous stirring at room temperature for 12 h. Excess of the inorganic salt was removed by suction filtration and the organic eluate was concentrated to dryness under vacuum to afford **C4T@K$^+$** as a solid residue. ^{1}H NMR (DMF-d_7): δ 6.60 (s, ArH, 16H), 4.63 and 3.26 (AX system, J = 13.8 Hz, ArCH$_2$Ar, 16H), 4.53 (br s, OCH$_2$, 16H), 4.48 (br s, NH$_2$, 16H) ppm; ESI(+)-MS: m/z 287 ([M + K + 4H + Cl]$^{4+}$, 17%).

4. Conclusions

Our findings show how minute structural modifications, of no apparent significance, may dramatically change the overall outcome of a self-assembly process, by either preventing or promoting the association of complex supramolecular nanostructures. Here, a single potassium ion is seen to produce an allosteric effect which triggers the subsequent non-covalent assembly of octa-cationic calix[4]tube **C4T** and the complementary tetra-anionic metallo-porphyrin **CuTPPS**. K$^+$ binds to the cryptand-like binding site, formed by the four dioxoethylene bridging moieties of the calix[4]tube and, by acting as an "effector", promotes the conformational change of its two cavities from C_{2v} to C_{4v}. As a result, **C4T@K$^+$** is then able to act as an efficient templating agent, providing a variety of 2D and 3D **CuTPPS/C4T@K$^+$** assemblies of predictable sequence, topology and stoichiometry. The noncovalent synthesis of these species follows a strict hierarchical pattern which can, in principle, be exploited to introduce a given property (e.g., chirality [52]) at a specific location in the assembly so as to modulate the reactivity of the entire nanostructure.

We are currently investigating the ability of other ionic and neutral species to act as effectors of discrete noncovalent porphyrin-calixarene functional materials.

Supplementary Materials: The following are available online, UV/Vis titration spectra, NMR and MS spectra.

Author Contributions: Conceptualization, A.D., R.P., A.N. and M.F.P.; methodology, A.D., R.P. and A.N.; validation, M.G., E.R., A.P. and A.N.; formal analysis, A.D., M.G. and G.G.; investigation, E.R., M.G. and I.P.; resources, A.D., M.F.P., G.G. and M.E.F.; data curation, E.R., M.G. and AN.; writing—original draft preparation, M.G., A.D. and A.N.; writing—review and editing, M.F.P. and R.P.; visualization, A.D. and M.F.P.; supervision, A.D., R.P., A.N. and M.F.P.; project administration, A.D.; funding acquisition, A.D., M.F.P., G.G. and A.P. All authors have read and agreed to the published version of the manuscript.

Funding: Università degli Studi di Catania, Dipartimento di Scienze Chimiche (programma ricerca di ateneo UNICT 2016–18 linea 1 and linea 2 and programma ricerca di ateneo UNICT 2020–22 linea 2); Università degli Studi di Messina, Dipartimento di Scienze Chimiche, Biologiche, Farmaceutiche ed Ambientali (FFABR_UniME_2019 project) and Ministero dell'Università e della Ricerca (MUR), PRIN projects 2017YJMPZN-005 and 20179BJNA2.

Institutional Review Board Statement: Not applicable.

Informed Consent Statement: Not applicable.

Data Availability Statement: Not available.

Conflicts of Interest: The authors declare no conflict of interest.

Sample Availability: Samples of the compounds **2** and **C4T** are available from the authors.

References and Notes

1. Bhyrappa, P.; Sankar, M.; Varghese, B. Mixed substituted porphyrins: Structural and electrochemical redox properties. *Inorg. Chem.* **2006**, *45*, 4136–4149. [CrossRef] [PubMed]
2. Fonda, H.N.; Gilbert, J.V.; Cormier, R.A.; Sprague, J.R.; Kamioka, K.; Connolly, J.S. Spectroscopic, photophysical, and redox properties of some meso-substituted free-base porphyrins. *J. Phys. Chem.* **1993**, *97*, 7024–7033. [CrossRef]
3. Yella, A.; Lee, H.W.; Tsao, H.N.; Yi, C.; Chandiran, A.K.; Nazeeruddin, M.K.; Diau, E.W.G.; Yeh, C.Y.; Zakeeruddin, S.M.; Grätzel, M. Porphyrin-sensitized solar cells with cobalt (II/III)-based redox electrolyte exceed 12 percent efficiency. *Science* **2011**, *334*, 629–634. [CrossRef]
4. Jurow, M.; Schuckman, A.E.; Batteas, J.D.; Drain, C.M. Porphyrins as molecular electronic components of functional devices. *Coord. Chem. Rev.* **2010**, *254*, 2297–2310. [CrossRef]
5. Scandola, F.; Chiorboli, C.; Prodi, A.; Iengo, E.; Alessio, E. Photophysical properties of metal-mediated assemblies of porphyrins. *Coord. Chem. Rev.* **2006**, *250*, 1471–1496. [CrossRef]
6. Hiroto, S.; Miyake, Y.; Shinokubo, H. Synthesis and functionalization of porphyrins through organometallic methodologies. *Chem. Rev.* **2017**, *117*, 2910–3043. [CrossRef]
7. Beletskaya, I.; Tyurin, V.S.; Tsivadze, A.Y.; Guilard, R.; Stern, C. Supramolecular chemistry of metalloporphyrins. *Chem. Rev.* **2009**, *109*, 1659–1713.
8. Drain, C.M.; Varotto, A.; Radivojevic, I. Self-organized porphyrinic materials. *Chem. Rev.* **2009**, *109*, 1630–1658.
9. Medforth, C.J.; Wang, Z.; Martin, E.M.; Song, Y.; Jacobsen, J.L.; Shelnutt, J.A. Self-assembled porphyrin nanostructures. *Chem. Commun.* **2009**, *47*, 7261–7277. [CrossRef]
10. Otsuki, J. Supramolecular approach towards light-harvesting materials based on porphyrins and chlorophylls. *J. Mater. Chem. A* **2018**, *6*, 6710–6753. [CrossRef]
11. Balaban, T.S. Tailoring porphyrins and chlorins for self-assembly in biomimetic artificial antenna systems. *Acc. Chem. Res.* **2005**, *38*, 612–623. [CrossRef] [PubMed]
12. Paolesse, R.; Nardis, S.; Monti, D.; Stefanelli, M.; Di Natale, C. Porphyrinoids for chemical sensor applications. *Chem. Rev.* **2017**, *117*, 2517–2583. [CrossRef] [PubMed]
13. Wang, H.; Song, Y.; Medforth, C.J.; Shelnutt, J.A. Interfacial synthesis of dendritic platinum nanoshells templated on benzene nanodroplets stabilized in water by a photocatalytic lipoporphyrin. *J. Am. Chem. Soc.* **2006**, *128*, 9284–9285. [CrossRef] [PubMed]
14. Rabiee, N.; Yaraki, M.T.; Garakani, S.M.; Garakani, S.M.; Ahmadi, S.; Lajevardi, A.; Bagherzadeh, M.; Rabiee, M.; Tayebi, L.; Tahriri, M.; et al. Recent advances in porphyrin-based nanocomposites for effective targeted imaging and therapy. *Biomaterials* **2020**, *232*, 119707.
15. Tian, J.; Huang, B.; Nawaz, M.H.; Zhang, W. Recent advances of multi-dimensional porphyrin-based functional materials in photodynamic therapy. *Coord. Chem. Rev.* **2020**, *420*, 213410. [CrossRef]
16. Doan, S.C.; Shanmugham, S.; Aston, D.E.; McHale, J.L. Counterion Dependent Dye Aggregates: Nanorods and nanorings of tetra(*p*-carboxyphenyl)porphyrin. *J. Am. Chem. Soc.* **2005**, *127*, 5885–5892. [CrossRef]
17. Schwab, A.D.; Smith, D.E.; Rich, C.S.; Young, E.R.; Smith, W.F.; de Paula, J.C. Porphyrin nanorods. *J. Phys. Chem. B* **2003**, *107*, 11339–11345. [CrossRef]
18. Guo, P.; Chen, P.; Liu, M. One-dimensional porphyrin nanoassemblies assisted via graphene oxide: Sheetlike functional surfactant and enhanced photocatalytic behaviors. *ACS Appl. Mater. Interfaces* **2013**, *5*, 5336–5345. [CrossRef]
19. Koepf, M.; Conradt, J.; Szmytkowski, J.; Wytko, J.A.; Allouche, L.; Kalt, H.; Balaban, T.S.; Weiss, J. Highly linear self-assembled porphyrin wires. *Inorg. Chem.* **2011**, *50*, 6073–6082. [CrossRef]
20. Fathalla, M.; Neuberger, A.; Li, S.-C.; Schmehl, R.; Diebold, U.; Jayawickramarajah, J. Straightforward self-assembly of porphyrin nanowires in water: Harnessing adamantane/cyclodextrin interactions. *J. Am. Chem. Soc.* **2010**, *132*, 9966–9967. [CrossRef]
21. Lee, S.J.; Hupp, J.T.; Nguyen, S.T. Growth of narrowly dispersed porphyrin nanowires and their hierarchical assembly into macroscopic columns. *J. Am. Chem. Soc.* **2008**, *130*, 9632–9633. [CrossRef]
22. Wang, Z.; Medforth, C.J.; Shelnutt, J.A. Porphyrin nanotubes by ionic self-assembly. *J. Am. Chem. Soc.* **2004**, *126*, 15954–15955. [CrossRef] [PubMed]
23. Wang, Z.; Li, Z.; Medforth, C.J.; Shelnutt, J.A. Self-assembly and self-metallization of porphyrin nanosheets. *J. Am. Chem. Soc.* **2007**, *129*, 2440–2441. [CrossRef]
24. Zhang, H.; Zhang, B.; Zhu, M.; Grayson, S.M.; Schmehl, R.; Jayawickramarajah, J. Water-soluble porphyrin nanospheres: Enhanced photo-physical properties achieved via cyclodextrin driven double self-inclusion. *Chem. Commun.* **2014**, *50*, 4853–4855. [CrossRef] [PubMed]

25. Aratani, N.; Kim, D.; Osuka, A. Discrete cyclic porphyrin arrays as artificial light-harvesting antenna. *Acc. Chem. Res.* **2009**, *42*, 1922–1934. [CrossRef] [PubMed]

26. Imamura, T.; Fukushima, K. Self-assembly of metallopyridylporphyrin oligomers. *Coord. Chem. Rev.* **2000**, *198*, 133–156. [CrossRef]

27. Frühbeißer, S.; Gröhn, F. Catalytic activity of macroion–porphyrin nanoassemblies. *J. Am. Chem. Soc.* **2012**, *134*, 14267–14270. [CrossRef]

28. Onouchi, H.; Miyagawa, T.; Morino, K.; Yashima, E. Assisted formation of chiral porphyrin homoaggregates by an induced helical poly(phenylacetylene) template and their chiral memory. *Angew. Chem.* **2006**, *45*, 2381–2384. [CrossRef]

29. Liu, K.; Xing, R.; Chen, C.; Shen, G.; Yan, L.; Zou, Q.; Ma, G.; Mçhwald, H.; Yan, X. Peptide-induced hierarchical long-range order and photocatalytic activity of porphyrin assemblies. *Angew. Chem.* **2015**, *54*, 500–505. [CrossRef]

30. Oliveras-González, C.; Di Meo, F.; González-Campo, A.; Beljonne, D.; Norman, P.; Simón-Sorbed, M.; Linares, M.; Amabilino, D.B. Bottom-up hierarchical self-assembly of chiral porphyrins through coordination and hydrogen bonds. *J. Am. Chem. Soc.* **2015**, *137*, 15795–15808. [CrossRef]

31. Boccalon, M.; Iengo, E.; Tecilla, P. Metal-organic transmembrane anopores. *J. Am. Chem. Soc.* **2012**, *134*, 20310–20313. [CrossRef] [PubMed]

32. Ono, K.; Yoshizawa, M.; Kato, T.; Watanabe, K.; Fujita, M. Porphine dimeric assemblies in organic-pillared coordination cages. *Angew. Chem.* **2007**, *46*, 1803–1806. [CrossRef] [PubMed]

33. Fujita, N.; Biradha, K.; Fujita, M.; Sakamoto, S.; Yamaguchi, K. A Porphyrin prism: Structural switching triggered by guest inclusion. *Angew. Chem.* **2001**, *40*, 1718–1721. [CrossRef]

34. Sasaki, K.; Nakagawa, H.; Zhang, X.; Sakurai, S.; Kano, K.; Kuroda, Y. Construction of porphyrin–cyclodextrin self-assembly with molecular wedge. *Chem. Commun.* **2004**, *4*, 408–409. [CrossRef] [PubMed]

35. Kano, K.; Nishiyabu, R.; Asada, T.; Kuroda, Y. Static and dynamic behavior of 2:1 inclusion complexes of cyclodextrins and charged porphyrins in aqueous organic media. *J. Am. Chem. Soc.* **2002**, *124*, 9937–9944. [CrossRef]

36. Feiters, M.C.; Fyfe, M.C.T.; Martínez-Díaz, M.-V.; Menzer, S.; Nolte, R.J.M.; Stoddart, J.F.; van Ken, P.J.M.; Williams, D.J. A supramolecular analog of the photosynthetic special pair. *J. Am. Chem. Soc.* **1997**, *119*, 8119–8120. [CrossRef]

37. Tian, H.-W.; Liu, Y.-C.; Guo, D.-S. Assembling features of calixarene-based amphiphiles and supra-amphiphiles. *Mater. Chem. Front.* **2020**, *4*, 46–98. [CrossRef]

38. Pisagatti, I.; Barbera, L.; Gattuso, G.; Parisi, M.F.; Geremia, S.; Hickey, N.; Notti, A. Guest-length driven high fidelity self-sorting in supramolecular capsule formation of calix[5]arenes in water. *Org. Chem. Front.* **2019**, *6*, 3804–3809. [CrossRef]

39. Giuliani, M.; Morbioli, I.; Sansone, F.; Casnati, A. Moulding calixarenes for biomacromolecule targeting. *Chem. Commun.* **2015**, *51*, 14140–14159. [CrossRef]

40. Arena, G.; Pappalardo, A.; Pappalardo, S.; Gattuso, G.; Notti, A.; Parisi, M.F.; Pisagatti, I.; Sgarlata, C. Complexation of biologically active amines by a water-soluble calix[5]arene. *J. Therm. Anal. Calorim.* **2015**, *121*, 1073–1079.

41. Pisagatti, I.; Barbera, L.; Gattuso, G.; Villari, V.; Micali, N.; Fazio, E.; Neri, F.; Parisi, M.F.; Notti, A. Tuning the aggregation of an amphiphilic anionic calix[5]arene by selective host-guest interactions with bola-type dications. *New J. Chem.* **2019**, *43*, 7936–7940. [CrossRef]

42. Manganaro, N.; Lando, G.; Gargiulli, C.; Pisagatti, I.; Notti, A.; Pappalardo, S.; Parisi, M.F.; Gattuso, G. Unique binding behavior of water-soluble polycationic oxacalix[4]arene tweezers towards the paraquat dication. *Chem. Commun.* **2015**, *51*, 12657–12660. [CrossRef] [PubMed]

43. Guo, D.-S.; Liu, Y. Supramolecular chemistry of *p*-sulfonatocalix[*n*]arenes and its biological applications. *Acc. Chem. Res.* **2014**, *47*, 1925–1934. [CrossRef]

44. Gattuso, G.; Notti, A.; Pappalardo, S.; Parisi, M.F.; Pisagatti, I.; Patanè, S. Encapsulation of monoamine neurotransmitters and trace amines by amphiphilic anionic calix[5]arene micelles. *New J. Chem.* **2014**, *38*, 5983–5990. [CrossRef]

45. D'Urso, A.; Fragalà, M.E.; Purrello, R. From self-assembly to noncovalent synthesis of programmable porphyrins' arrays in aqueous solution. *Chem. Commun.* **2012**, *48*, 8165–8176. [CrossRef]

46. Guo, D.-S.; Chen, K.; Zhang, H.-Q.; Liu, Y. Nano-supramolecular assemblies constructed from water-soluble bis(calix[5]arenes) with porphyrins and their photoinduced electron transfer properties. *Chem. Asian J.* **2009**, *4*, 436–445. [CrossRef]

47. Alex, J.M.; McArdle, P.; Crowley, P.B. Supramolecular stacking in a high Z′ calix[8]arene-porphyrin assembly. *CrystEngComm* **2020**, *22*, 14–17. [CrossRef]

48. Brancatelli, G.; De Zorzi, R.; Hickey, N.; Siega, P.; Zingone, G.; Geremia, S. New multicomponent porous architecture of self-assembled porphyrins/calixarenes driven by nickel ions. *Cryst. Growth Des.* **2012**, *12*, 5111–5117. [CrossRef]

49. De Zorzi, R.; Guidolin, N.; Randaccio, L.; Geremia, S. A bifunctionalized porous material containing discrete assemblies of copper-porphyrins and calixarenes metallated by ion diffusion. *CrystEngComm* **2010**, *12*, 4056–4058. [CrossRef]

50. De Zorzi, R.; Guidolin, N.; Randaccio, L.; Purrello, R.; Geremia, S. Nanoporous crystals of calixarene/porphyrin supramolecular complex functionalized by diffusion and coordination of metal ions. *J. Am. Chem. Soc.* **2009**, *131*, 2487–2489. [CrossRef]

51. Di Costanzo, L.; Geremia, S.; Randaccio, L.; Purrello, R.; Lauceri, R.; Sciotto, D.; Gulino, F.G.; Pavone, V. Calixarene-porphyrin supramolecular complexes: pH-tuning of the complex stoichiometry. *Angew. Chem.* **2001**, *40*, 4245–4247. [CrossRef]

52. D'Urso, A.; Nicotra, P.F.; Centonze, G.; Fragalà, M.E.; Gattuso, G.; Notti, A.; Pappalardo, A.; Pappalardo, S.; Parisi, M.F.; Purrello, R. Induction of chirality in porphyrin-(bis)calixarene assemblies: A mixed covalent-non-covalent vs a fully non-covalent approach. *Chem. Commun.* **2012**, *48*, 4046–4048. [CrossRef] [PubMed]

53. D'Urso, A.; Cristaldi, D.A.; Fragalà, M.E.; Gattuso, G.; Pappalardo, A.; Villari, V.; Micali, N.; Pappalardo, S.; Parisi, M.F.; Purrello, R. Sequence, stoichiometry, and dimensionality control in porphyrin/biscalix[4]arene self-assemblies in aqueous. *Chem. Eur. J.* **2010**, *16*, 10439–10446. [PubMed]

54. D'Urso, A.; Marino, N.; Gaeta, M.; Rizzo, M.S.; Cristaldi, D.A.; Fragalà, M.E.; Pappalardo, S.; Gattuso, G.; Notti, A.; Parisi, M.F.; et al. Porphyrin stacks as an efficient molecular glue to induce chirality in hetero-component calixarene-porphyrin assemblies. *New J. Chem.* **2017**, *41*, 8078–8083. [CrossRef]

55. Gaeta, M.; Sortino, G.; Randazzo, R.; Pisagatti, I.; Notti, A.; Fragalà, M.E.; Parisi, M.F.; D'Urso, A.; Purrello, R. Long-range chiral induction by a fully noncovalent approach in supramolecular porphyrin–calixarene assemblies. *Chem. Eur. J.* **2020**, *26*, 3515–3518. [CrossRef]

56. Gulino, F.G.; Lauceri, R.; Frish, L.; Evan-Salem, T.; Cohen, Y.; De Zorzi, R.; Geremia, S.; Di Costanzo, L.; Randaccio, L.; Sciotto, D.; et al. Noncovalent synthesis in aqueous solution and spectroscopic characterization of multi-porphyrin complexes. *Chem. Eur. J.* **2006**, *12*, 2722–2729. [CrossRef]

57. Schmitt, P.; Beer, P.D.; Drew, M.G.B.; Sheen, P.D. Calix[4]tube: A tubular receptor with remarkable potassium ion selectivity. *Angew. Chem.* **1997**, *36*, 1840–1842. [CrossRef]

58. Matthews, S.E.; Schmitt, P.; Felix, V.; Drew, M.G.B.; Beer, P.D. Calix[4]tubes: A new class of potassium-selective ionophore. *J. Am. Chem. Soc.* **2002**, *124*, 1341–1353. [CrossRef]

59. Buhdka, J.; Lhotak, P.; Stibor, I.; Michlova, V.J.; Sykova, J.; Cisarova, I. A biscalix[4]arene-based ditopic hard/soft receptor for K$^+$/Ag$^+$complexation. *Tetrahedron Lett.* **2002**, *43*, 2857–2861. [CrossRef]

60. Matthews, S.E.; Felix, V.; Drew, M.G.B.; Beer, P.D. Halo-derivatised calix[4]tubes. *Org. Biomol. Chem.* **2003**, *1*, 1232–1239. [CrossRef]

61. Puchnin, K.; Zaikin, P.; Cheshkov, D.; Vatsouro, I.; Kovalev, V. Calix[4]tubes: An approach to functionalization. *Chem. Eur. J.* **2012**, *18*, 10954–10968. [CrossRef] [PubMed]

62. Puchnin, K.; Cheshkov, D.; Zaikin, P.; Vatsouro, I.; Kovalev, V. Tuning conformations of calix[4]tubes by weak intramolecular interactions. *New J. Chem.* **2013**, *37*, 416–424. [CrossRef]

63. The pH of the solution was adjusted to 3.0, by dropwise addition of an aqueous 0.5 M HCl solution, to ensure protonation of the eight amino groups of **C4T** or **C4T@K**$^+$ and ultimately solubility in water.

64. Moschetto, G.; Lauceri, R.; Gulino, F.G.; Sciotto, D.; Purrello, R. Non-Covalent synthesis in aqueous solution of discrete multi-porphyrin aggregates with programmable stoichiometry and sequence. *J. Am. Chem. Soc.* **2002**, *124*, 14536–14537. [PubMed]

65. Herrmann, O.; Mehdi, S.H.; Corsini, A. Heterogeneous metal-insertion: A novel reaction with porphyrins. *Can. J. Chem.* **2006**, *56*, 1084–1087. [CrossRef]

Article

Terpyridine-Functionalized Calixarenes: Synthesis, Characterization and Anion Sensing Applications

Nicola Y. Edwards [1,*], David M. Schnable [1], Ioana R. Gearba-Dolocan [2] and Jenna L. Strubhar [1]

[1] Department of Chemistry and Biochemistry, Misericordia University, Dallas, PA 18612, USA; dschnable@gmail.com (D.M.S.); jennastrubhar@gmail.com (J.L.S.)

[2] Department of Chemistry, The University of Texas at Austin, Austin, TX 78712, USA; gearba@austin.utexas.edu

* Correspondence: nedwards@misericordia.edu; Tel.: +1-(570)-674-6340

Abstract: Lanthanide complexes have been developed and are reported herein. These complexes were derived from a terpyridine-functionalized calix[4]arene ligand, chelated with Tb^{3+} and Eu^{3+}. Synthesis of these complexes was achieved in two steps from a calix[4]arene derivative: (1) amide coupling of a calix[4]arene bearing carboxylic acid functionalities and (2) metallation with a lanthanide triflate salt. The ligand and its complexes were characterized by NMR (^1H and ^{13}C), fluorescence and UV-vis spectroscopy as well as MS. The photophysical properties of these complexes were studied; high molar absorptivity values, modest quantum yields and luminescence lifetimes on the ms timescale were obtained. Anion binding results in a change in the photophysical properties of the complexes. The anion sensing ability of the Tb(III) complex was evaluated via visual detection, UV-vis and fluorescence studies. The sensor was found to be responsive towards a variety of anions, and large binding constants were obtained for the coordination of anions to the sensor.

Keywords: calix[4]arene; anion; sensors; lanthanides; fluorescence studies; UV–vis studies; quantum yields; lifetime measurements

Citation: Edwards, N.Y.; Schnable, D.M.; Gearba-Dolocan, I.R.; Strubhar, J.L. Terpyridine-Functionalized Calixarenes: Synthesis, Characterization and Anion Sensing Applications. *Molecules* **2021**, *26*, 87. https://dx.doi.org/10.3390/molecules26010087

Academic Editors: Paula M. Marcos and Mario Berberan-Santos
Received: 9 November 2020
Accepted: 23 December 2020
Published: 27 December 2020

Publisher's Note: MDPI stays neutral with regard to jurisdictional claims in published maps and institutional affiliations.

1. Introduction

1.1. Lanthanide Complexes as Anion Sensors

Developing sensors for the detection and quantification of anions is a major area of interest in supramolecular chemistry. This is due to the pivotal roles that these sensors play in many arenas including environmental, biological and medicinal arenas [1–6]. Lanthanide-based complexes have been developed and used extensively as sensors for a variety of analytes; many of these sensors are now used in commercial settings [7–10]. These complexes are standouts in this arena; anion sensing is no exception to this [11–13]. For a lanthanide complex—or any compound—to act as an anion sensor, two criteria must be met: (1) There must be binding of the sensor to anion, and (2) The sensor must display a readout (color, luminescence, etc.), and this readout must be modulated upon anion binding. Anions readily coordinate to lanthanide centers due to their potent Lewis acidity [14–16]. Lanthanide complexes boast excellent photophysical properties such as relatively long-lived luminescence (on the ms timescale). Furthermore, there are changes in the photophysical properties of these complexes that occur upon anion coordination which can be harnessed in a sensing scheme.

The terpyridine moiety is an effective sensitizer of lanthanide luminescence, and hence this moiety has been incorporated into several lanthanide-based anion sensors [17–19]; none of these sensors utilize the calixarene scaffold [20–23]. The calixarene scaffold has been frequently used in the construction of anion (and other) sensors as it imparts a high degree of pre-organization of the recognition groups leading to enhanced binding affinities of target molecules or ions [24,25]. The current work details the development of calixarene-based sensors for anions that have been synthesized in two steps from a calixarene derivative. A calix[4]arene scaffold with carboxylic acid functionalities at the lower rim was coupled to an amine derivative bearing a terpyridine moiety. The resulting compound was metallated by stirring it overnight with the appropriate lanthanide(III) triflate salt to form either the Eu(III) or Tb(III) complex. These lanthanide complexes have large molar absorptivity values, long luminescent lifetimes and modest quantum yields. Furthermore, these complexes were shown by visual detection, fluorescence and UV-vis studies to sense a variety of anions.

1.2. Sensor Design

Parker, Gunnlaugsson and others have pioneered the development of anion sensors based on antenna-sensitized lanthanide luminescence [26–30]. Two sensing paradigms that were developed and used for the basis of our study are described as follows: (1) Coordinatively-unsaturated metal complexes are used, in which anion binding leads to a change in the photophysical properties of the metal complex, due to the displacement of solvent molecules from the metal center. (2) The sensitizer is designed so that it is equipped with anion binding moieties; anion coordination leads to a change in the readout of the luminescence of the sensor due to changes that occur to the sensitizer upon anion binding [14].

Anion receptors that were constructed using the calix[4]arene scaffold, containing linkers with amide functionalities and pyridine moieties, were previously reported by our group [31]. We showed by ^{1}H NMR studies that these receptors were responsive to carboxylates and $H_2PO_4^-$ with binding constants that ranged from 14 to 275 M^{-1} (in DMSO). Building on this earlier model system, we set our sights on developing sensors in which the linkers appended from the calix[4]arene scaffold terminated in terpyridine moieties. These terpyridine serve the dual purpose of binding lanthanide salts and acting as the sensitizing antenna (Scheme 1).

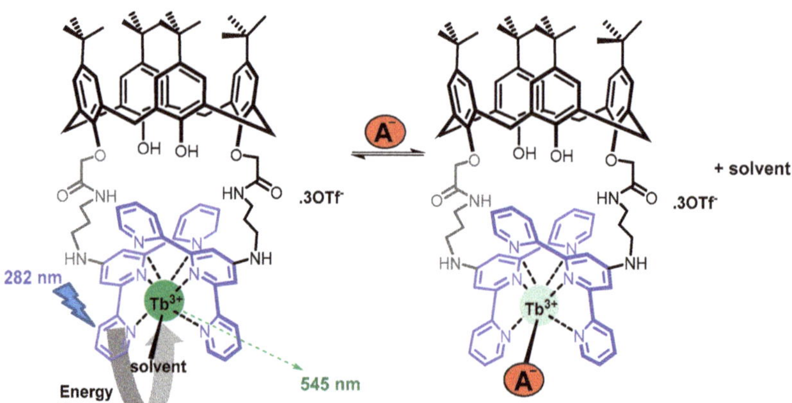

Scheme 1. Sensor Design of [Tb.1]$^{3+}$: Antenna-sensitized lanthanide luminescence is effected when the terpyridine moiety is excited with UV light. There is a change in the photophysical properties of the metal center upon binding due to **(a)** displacement of the solvent molecules with the anion and/or **(b)** anion binding to the linkers (through amide and amine moieties) of the calix[4]arene.

In our sensor design, we postulated that binding of the anions would change the photophysical properties of the complex due to: (1) The coordinatively unsaturated nature of the complexes; anions should bind directly to the metal center leading to displacement of solvent molecules and/or ligands. (2) The linkers of the metal complexes themselves are rich in anion binding moieties (specifically amide and amine moieties); binding to such moieties (via hydrogen bonding) is predicted to change the sensitizing ability of the ligand [32,33]. Either or both changes could form the basis of an appropriate signaling mechanism for our system. This sensor is also ripe with opportunities for diversification (via the antenna, metal center and linkers) for future pattern-based sensing applications [34]. In the following paragraphs, an investigation of this sensor is presented. The investigation begins with the synthesis and characterization of the complexes and is followed by the metal binding and photophysical studies. Finally, insights into the capabilities of the sensor are detailed.

2. Results and Discussion

2.1. Synthesis and Characterization of Ligand and its Lanthanide-Based Complexes

With sensing applications in mind, it was necessary that the synthetic route be facile, amenable to scale-up and with the potential to create a library of ligands from available precursor compounds; amide coupling was chosen. The synthetic route to ligand **1** is shown in Scheme 2 and is summarized as follows. Ligand **1** was obtained by coupling the calix[4]arene dicarboxylic acid **3** [22] and N-[2,2′;6′,2″]terpyridin-4′-yl-propane-1,3-diamine [35] using N-Ethyl-N′-(3-dimethylaminopropyl)carbodiimide hydrochloride (EDCI), 1-Hydroxybenzotriazole hydrate (HOBt) in the presence of N,N-Diisopropylethylamine (DIPEA) in CH_2Cl_2. Subsequent work-up and purification via column-chromatography afforded **1** in 48% percent yield. The structure of ligand **1** was confirmed using ^{1}H and ^{13}C NMR spectroscopy and MS. Key resonances in the ^{1}H NMR spectrum were found in the region of 7–9 ppm due to the terpyridine moieties and the aromatic moieties of the calixarene scaffold. There were also singlets at 4.65 ppm due to the methylene linker as well as at 1.09 and 1.06 ppm due to the *tert*-butyl groups. Of particular importance are the doublet of doublets found at 4.18 and 3.30 ppm due to the methylene bridges indicating that ligand **1** has C_{2v} symmetry and is locked in the cone conformation; this was also supported by ^{13}C NMR data that showed a set of overlapping resonances at 31.3 ppm. The presence of a peak due to the doubly-charged ion $[M+2H]^{2+}$ in the HRMS also supported formation of ligand **1**. Metallation of ligand **1** with a lanthanide(III) triflate salt was effected by stirring equimolar quantities of the salt and the ligand overnight at room temperature in CH_3CN. Precipitation via hexanes of a concentrated solution of **1** in CH_2Cl_2 was used to obtain the lanthanide-based complexes of **1**. The Eu(III) and Tb(III) complexes were isolated in 53% and 46% yield, respectively.

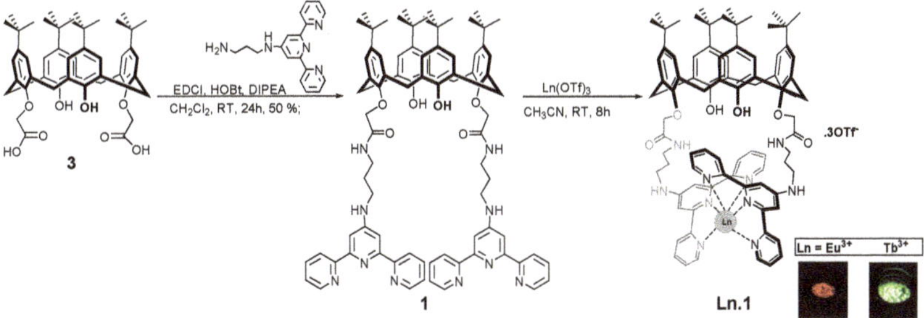

Scheme 2. Synthesis of ligand **1** and lanthanide-based complexes.

2.2. How Well Does Ligand **1** Bind Lanthanide(III) Ions?

The investigation began with ligand **1** as it was our entry point to the sensors and because of its own potential in future sensing applications. Specifically, there was interest in determining how easily metal complexation could be monitored in situ and the strength of the binding of this ligand with lanthanides. We, thus, investigated the binding of ligand **1** to Tb(III) and Eu(III) using UV-vis and fluorescence spectroscopy. For these studies, solutions of ligand **1** (9.7 μM) were titrated with 0–7 equivalents of lanthanide ions (as triflate salts) in CH_3CN.

Fluorescence spectra representing the titration of ligand **1** with $Tb(OTf)_3$ are shown in Figure 1. Prior to the addition of any equivalents of $Tb(OTf)_3$, there is a peak centered at

410 nm due to the ligand. As the titration progresses, this peak decreases in intensity and is slightly red-shifted. There is the concomitant appearance of line spectra in the 450–700 nm region, signaling that the [Tb.**1**](OTf)$_3$ had been formed in situ (see Section 2.3).

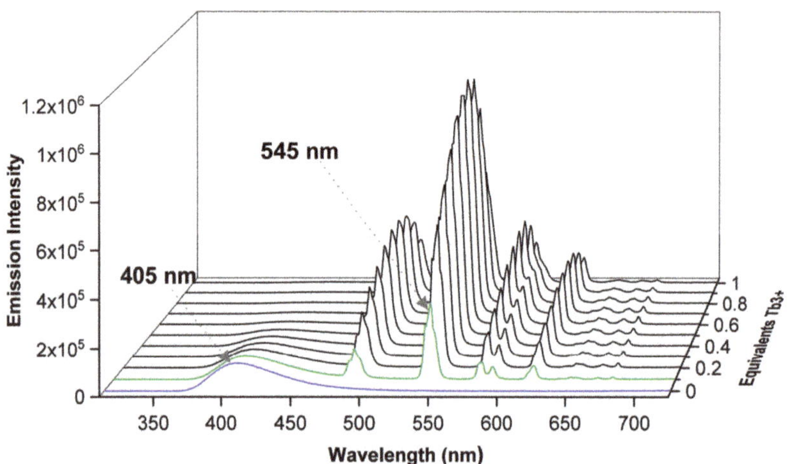

Figure 1. Fluorescence spectra representing the titration of ligand **1** (9.7 µM, acetonitrile) with 0 to 1 equivalents of Tb(OTf)$_3$. Excitation at 280 nm; excitation and emission slit widths are 2 and 1 nm, respectively. Integration time = 0.2 s.

The same titration was repeated and monitored using UV–vis spectroscopy. Shown in Figure 2 is a series of absorption spectra resulting from the titration of ligand **1** with Tb(OTf)$_3$. There is a decrease in the intensity of the peak at 279 nm (slightly red-shifted by ca. 6 nm) and the growth of a shoulder at 325 nm, indicating the complexation of the terpyridine moieties with the metal center. The absorption spectrum at the end of the titration matched that of the Tb.**1**(OTf)$_3$ (see Section 2.3). Binding isotherms obtaining by plotting absorbance vs. Tb(OTf)$_3$ concentration at 279 and 320 nm revealed the presence of an inflection point at 1 equivalent Tb(OTf)$_3$ added, indicated 1:1 ligand: metal binding stoichiometry. These isotherms were subsequently analyzed using a 1:1 ligand: metal model in the nonlinear regression curve-fitting program Hypspec [36] and yielded formation constants of log K = 7.3 ($\pm$0.2) for Tb(III) and log K = 6.6 ($\pm$0.1) for Eu(III), respectively (see supplementary information). These binding constants are within the range of other Eu(III) and Tb(III) complexes based on terpyridine and other pyridine-based ligands [37,38]. For example, the log K for a 1:1 Eu(III) complex based on a tripodal terpyridine ligand was determined to be 7.2 ($\pm$0.3) in methanol [35]. The log K of a 1:1 Eu(III) complex based on a bis-bipyridinehenylphosphine oxide ligand was determined to be 5.8 ($\pm$0.5) in acetonitrile [39].

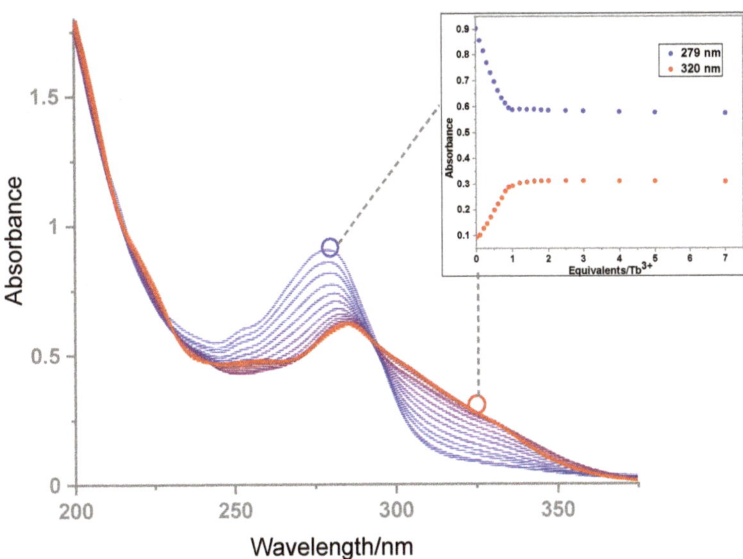

Figure 2. Absorption spectra resulting from the titration of ligand **1** (9.7 µM, acetonitrile) with 0 to 7 equivalents of Tb(OTf)$_3$.

2.3. Photophysical Studies of Ligand 1 and Complexes

Next, the photophysical properties of ligand **1** and its complexes were examined; these data are summarized in Table 1. We begin our discussion of this section by firstly highlighting the absorption, excitation and emission spectra of ligand **1** and its complexes (Figure 3) and the information that was gleaned from these spectra. The spectra of the ligand and its Ln(III) Complexes (9.8 µM) were measured in CH$_3$CN. The absorption spectrum of the ligand (top panel) is dominated by a peak centered at 282 nm (log ε = 4.70 in acetonitrile) due to the $\pi \rightarrow \pi^*$ transition of the terpyridine moieties and the calixarene scaffold; a broad shoulder centered at ca. 296 nm is also present. The emission spectrum of the ligand resulting from excitation at 280 nm shows a single peak centered at 394 nm.

Table 1. Photophysical properties of ligand **1** and lanthanide complexes in CH$_3$CN at room temperature.

Species	Absorption		Emission	
	λmax (nm) (log ε) [a]	λmax (nm)	Φ (%)	τ (ms)
Ligand **1**	282 (4.70)	394	[b]	[b]
[Tb.1]$^{3+}$	289 (4.17), 314 (4.00)	543	4.5(0.2) [c] 4.7(0.2) [d]	0.95(0.02) [e]
[Eu.1]$^{3+}$	288 (4.60), 310 (4.48)	613	0.24(0.05) [c] 0.26(0.04) [d]	1.00(0.04) [f]

[a] log of the extinction coefficient, [b] not measured, [c] measured relative to cesium tris(6-carboxypyridine-2-carboxylato) lanthanide(III), [d] measured relative to tris(bipyridine)ruthenium(II) chloride, [e] $^5D_4 \rightarrow {}^7F_5$ transition, [f] $^5D_0 \rightarrow {}^7F_2$ transition.

The dominant peaks in the absorption spectra of the lanthanide complexes (Figure 3—middle and bottom panels) show a 5 nm bathochromic shift with respect to ligand, indicating that the terpyridine moiety is responsible for complexation of the ligand; the shift is the result of a change in conformation from the *trans, trans* or *cis, trans* form to the *cis, cis* form [40]. We note that absorption for the ligand and complexes is considered to be a highly efficient process as indicated by the high values for molar absorptivity (logε: 4.17–4.70) that were obtained.

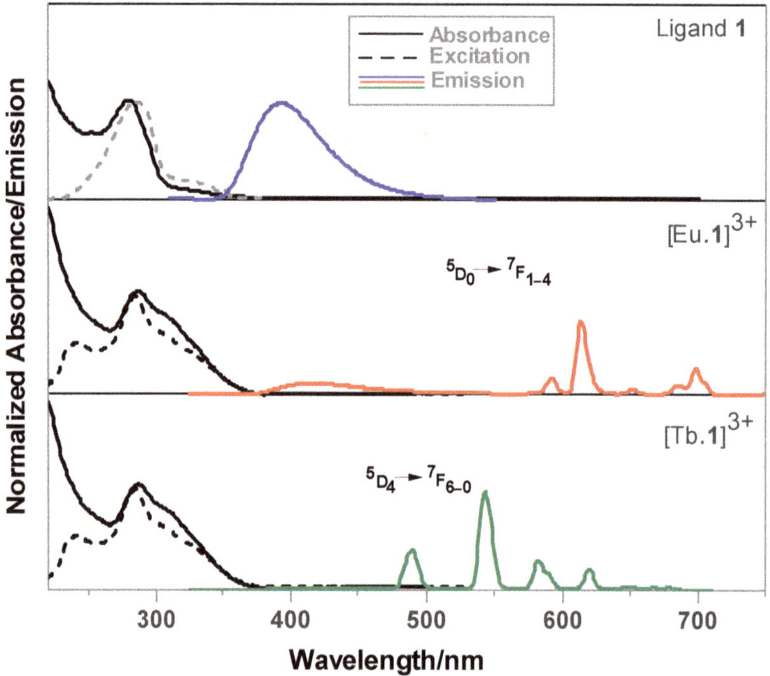

Figure 3. Normalized absorbance, excitation and fluorescence spectra of [Tb.1](OTf)$_3$ (9.9 µM, acetonitrile) shown alongside the normalized ligand absorbance. Fluorescence spectrum conditions: emission: excitation at 280 nm; excitation: emission wavelengths: 394 (ligand), 543 (Tb) and 613 nm (Eu).

The emission spectra of the lanthanide complexes (Figure 3—middle and bottom panels), produced by exciting the complexes in the absorption band of the ligand, resulted in an intense metal-centered luminescence. This luminescence is displayed in the form of line spectra, ranging from 591 to 698 nm for Eu(III), and ranging from 488 to 677 nm for Tb(III). These line spectra serve as spectroscopic signatures of successful lanthanide complexation resulting from terpyridine-based ligand sensitization and hence energy transfer from ligand to metal. In the case of the Eu(III) complexes, energy is transferred from the ligand to the $^5D_0$ excitation state of Eu(III), and there is subsequent deactivation to the ground state multiplets 7F$_J$ (J = 0–4)). For the Tb(III) complexes, the $^5D_4$ excited state and the ground state multiplets are 7F$_J$ (J = 6–0). Similarities in the absorption and excitation spectra of the ligand and complexes are indicative of good "energy matches" between ligand and metal ion that resulted in this successful energy transfer. Blue (for ligand), green (for Tb) and red emission (for Eu) can be detected with the naked eye upon exposure of the ligand and complexes (in the solid state) to UV light (Scheme 2).

The complexes were then defined in terms of the requisite benchmarks of luminescence, i.e., the lifetime and quantum yield of the luminescence of the complexes. Lifetimes (Table 1) were obtained by measuring the luminescence decay at the maximum of the most dominant transitions ($^5D_4 \rightarrow {}^7F_5$ transition at 543 nm for Tb(III) and $^5D_0 \rightarrow {}^7F_2$ transition at 613 nm for Eu(III)). In all cases, the decays were single exponential in nature indicating that the luminescence arose from one excited state. Rate constants were extracted from the single exponential fits and subsequently converted to lifetimes. For both lanthanide complexes, the lifetime measurements were ca. 1 ms—the typical timescale for lifetimes of lanthanide complexes.

The fluorescence quantum yields for the lanthanide complexes were measured in acetonitrile using the relative method [41,42]. In short, this method involves measuring the quantum yield of an analyte relative to a compound with a known fluorescence quantum yield. This is performed by recording the fluorescent emission spectrum of both compounds in solution, integrating this spectrum over the emission range and plotting this value against the absorbance of the solution at the excitation wavelength for multiple solutions at varying, but dilute, concentrations (A < 0.1). A linear trendline is fit to this plot, the gradient of which can be used to calculate the quantum yield of the analyte relative to the standard. This is performed according to the following equation:

$$\phi_x = \phi_{st}\left(\frac{grad_x}{grad_{st}}\right)\left(\frac{\eta_x^2}{\eta_{st}^2}\right)$$

φ_x, η_x and $grad_x$ represent the fluorescence quantum yield, refractive index and gradient of the linear trendline, respectively. The quantum yields were modest: 4.5% for the Tb(III) complex and 0.24% for the Eu(III) complex.

2.4. Anion Binding Studies—Assay Developemt

The Tb complex, [Tb.1](OTf)$_3$, was the metal complex chosen to probe anion binding due to its higher quantum yield—a factor of 20 higher than its Eu counterpart. We selected anions (NO$_3^-$, H$_2$PO$_4^-$, Cl$^-$ and CH$_3$CO$_2^-$-as tetrabutylammonium (TBA) salts) to determine if there were changes in the luminescence of [Tb.1](OTf)$_3$. Luminescence changes were observed upon the addition of these anions when acetonitrile and methanol were used as solvents, but acetonitrile was ultimately chosen as it is easier to carry out studies over a longer period of time due to its slower rate of evaporation compared to methanol. We explored changes in the ground state of the complex (using absorption spectra) and its excited state (using fluorescence spectra). Two regions of the fluorescence spectrum were explored: (1) the π^* to π transition centered at about 400 nm and (2) the Tb excited state to ground state transition (ca. 450–720 nm).

2.5. Binding Studies of the Tb.1(OTf)$_3$ with Dihydrogen Phosphate

To determine the response of our sensor towards anions, [Tb.1](OTf)$_3$ was titrated with solutions of anions as their TBA salts in CH$_3$CN. We first determined if our sensor would exhibit a response towards triflate, the counter anion in our complex. Titration of up to 100 equivalents of TBA triflate did not change the overall response of the sensor by any appreciable amount.

Shown in Figure 4 (top panel—left) is an example of the visual detection of an anion, H$_2$PO$_4^-$, by [Tb.1](OTf)$_3$. The complex changes from fluorescent green to blue in the presence of the anion indicating that the sensor exhibits a "turn-off" response towards H$_2$PO$_4^-$. Fluorescence spectra (Figure 4—top panel) obtained upon the titration of H$_2$PO$_4^-$ with the Tb.1(OTf)$_3$ complex showed a concomitant increase in the π^* to π transition and a decrease in the Tb-centered transition, indicating that the spectra tend towards one that resembles ligand **1**. The binding isotherm that was obtained by plotting the changes at the hypersensitive peak for Tb(III) emission, 545 nm versus equivalents of H$_2$PO$_4^-$, is also shown. No discernable changes were seen in the shapes of the emission spectra from this titration indicating that perhaps the anion was not binding to the metal center or that the probe that was chosen was not one that could effectively monitor the interaction between anion and the first coordination sphere of the metal.

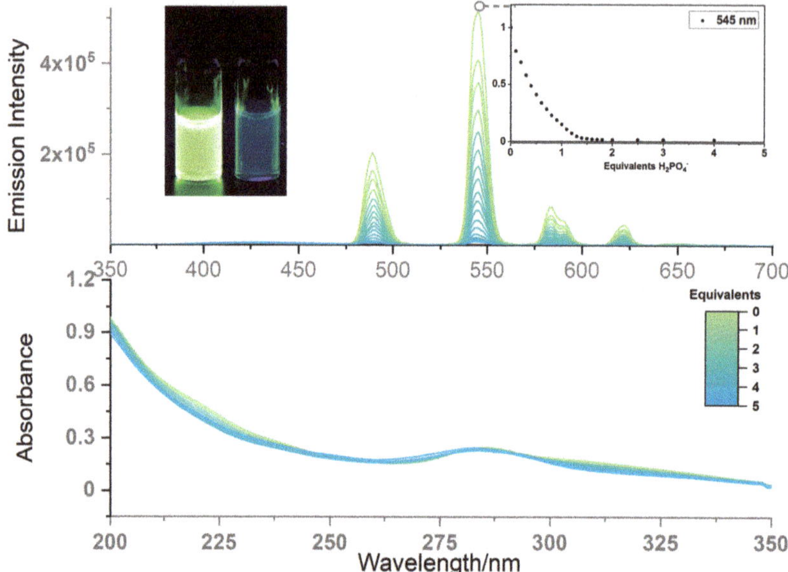

Figure 4. Top panel: 0.9 mM of [Tb.**1**](OTf)$_3$ in the absence (left) and presence (right) of $H_2PO_4^-$ (10 equivalents). Spectra resulting from the titration of [Tb.**1**](OTf)$_3$ (9.9 μM, acetonitrile) with 0 to 7 equivalents of $H_2PO_4^-$. Fluorescence (top panel) and UV-vis (bottom panel). The binding isotherm for the emission intensity at 545 nm in the fluorescence spectrum is shown as an inset in the top panel.

Shown in the bottom panel of Figure 4 is the series of absorption spectra obtained upon titration of [Tb.**1**](OTf)$_3$ with $H_2PO_4^-$. The disappearance of the shoulder at approximately 313 nm and a blue-shift of the peak at about 282 nm were observed—a spectrum at the end that resembled that of ligand **1**.

2.6. The Nature of the Sensor Response Towards Anions

Titration of anions resulted in quenching of the fluorescence of the [Tb.**1**](OTf)$_3$ giving this Tb(III) complex a "turn-off" designation. This "turn-off" response, in our opinion, is due to the following:

(1)   Binding of the anion leads to partial decomplexation of one or both terpyridine moieties from the metal center, thus inhibiting its ability to act as a sensitizer.
(2)   Anion is bound to the sensitizer and or metal center in such a way that it "interferes" with the energy transfer from the triplet state of the ligand to the Tb excited state.

The anions displaced CH_3CN molecules which had limited quenching ability in the first place. Allen and co-workers determined that the luminescence decay rate per inner sphere solvent molecule for Eu(III) complexes was 0 for acetonitrile compared to 0.83 s^{-1} for water and 0.41 s^{-1} for methanol [43]. Depending on the position in the course of the titration, either factor (1) or (2) or a combination thereof could be at play. These factors coupled with the limited quenching ability of CH_3CN could explain the "turn-off" response.

2.7. How Does the Sensor Respond to Anions Other than Dihydrogen Phosphate?

To determine how the sensor would respond towards a variety of anions, we chose: HSO_4^-, NO_3^-, the phosphate derivatives ($H_2PO_4^-$, $HP_2O_7^{3-}$), the halides (Cl^- and F^-) and the carboxylate, $CH_3CO_2^-$—a diverse enough set of anions which allowed us to gain an understanding of how the sensor responds to factors such as charge, basicity, size, shape and the presence of rotamers that could lead to luminescence quenching (Figure 5).

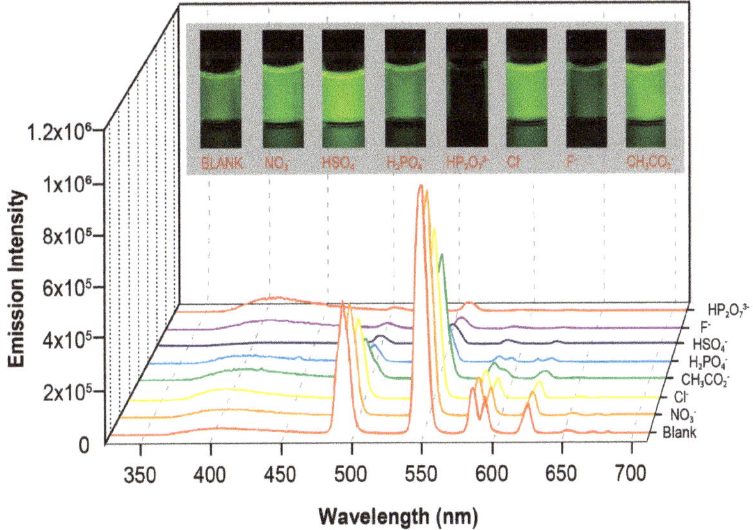

Figure 5. Anions targeted in this study.

Figure 6 shows steady state emission spectra representative of [Tb.**1**](OTf)$_3$ (9.9 μM) in the presence of 1 equivalent of each anion. These spectra were then translated to relative intensity and percent quenching ability (at 545 nm) (Figure 7) by comparing the spectrum in the presence of each anion to a solution of the [Tb.**1**](OTf)$_3$ in the absence of anion. One equivalent was chosen as the readout point as we felt that this point struck a balance between being far enough along in the titration where there was a fair amount of quenching yet, a differential sensor response towards anions could be observed. A series of vials representing the [Tb.**1**](OTf)$_3$ (50 μM) in the presence of 1 equivalent of each anion when exposed to UV light is also shown in Figure 6.

Figure 6. Steady state emission spectra of [Tb.**1**](OTf)$_3$ (9.9 μM) in the presence of 1 equivalent of anions. Pictures show [Tb.**1**](OTf)$_3$ (50 μM) in the presence of 1 equivalent of each anion.

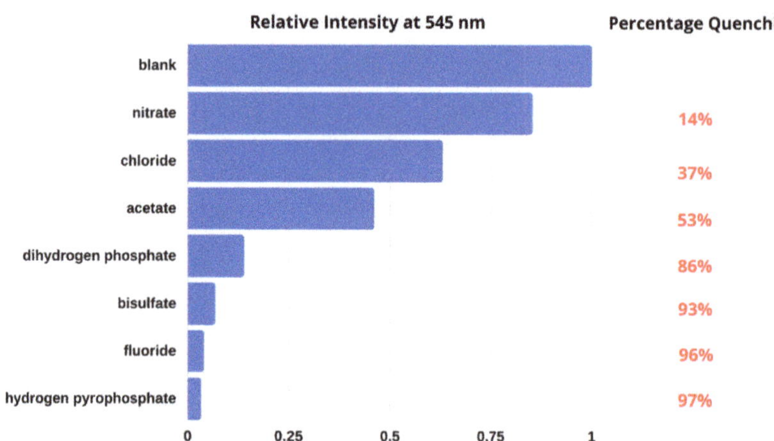

Figure 7. Relative intensity values at 545 nm and the corresponding percent quenching for Tb.**1**(OTf$_3$) in the presence of 1 equivalent of anions.

Two key factors appear to be at work when one examines the overall sensor response: (1) the Lewis (and Bronsted) basicities of the anions [44–46] and (2) the number of OH bonds that the anion possesses. The anions HSO_4^- and $HP_2O_7^{3-}$ each contain one OH bond, whereas $H_2PO_4^-$ contains two OH bonds. These OH bonds are high energy oscillators (3600 cm^{-1}) with the ability to quench the luminescence of the lanthanide sensor via non-radiative decay. This decay is proportional to the number of O-H bonds present and the distance from the metal center [47].

The quenching ability runs parallel to the basicity of the anions, generally, with two exceptions. Anions HSO_4^- and $HP_2O_7^{3-}$ present with greater quenching ability that one would expect if basicity were the only factor at play. We assert that this greater quenching ability is due to an OH bond. The sensor response that was obtained is best considered in three cohorts. The weakest responses were obtained for the weakest bases (NO_3^- and Cl^-), good responses were obtained for $CH_3CO_2^-$ and $H_2PO_4^-$ and the best responses for F^-, HSO_4^- and $HP_2O_7^{3-}$. The best responses are due to basicity (in the case of F^-), "luminescent quenchable" OH bonds (HSO_4^- and $HP_2O_7^{3-}$) and the -3 charge ($HP_2O_7^{3-}$).

Quantitative information regarding the anion binding event was obtained from the titration of [Tb.**1**](OTf)$_3$ (9.9 µM) with anions (0 to 5 equivalents). Binding isotherms (Figure 8) (normalized fluorescence intensity vs. anion concentration plots) were obtained by monitoring changes at the 545 nm peak analyzed using a 1:1 sensor: anion binding model in HypSpec. We, for comparison, show the binding isotherms with all anions at the same concentration but note that for some anions, higher (example: NO_3^-) and lower (example: $HP_2O_7^{3-}$) concentrations were used in experiments so as to obtain the appropriate curvature, a necessity for accurate fitting of the isotherms to nonlinear regression models [48].

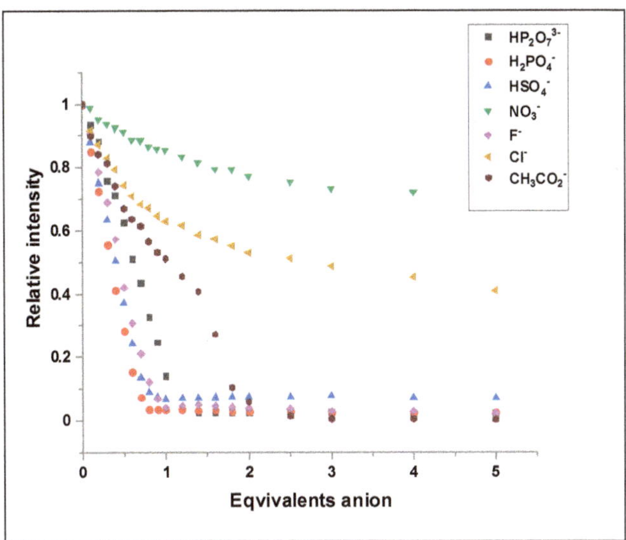

Figure 8. Binding isotherms showing the fluorescence intensity of [Tb.1](OTf)$_3$ at 545 nm vs. equivalents of anion.

The binging constants of [Tb.1](OTf)$_3$ with the anions NO_3^-, Cl^- and $H_2PO_4^-$ were determined to be log K = 5.60 ($\pm$0.16), 6.23 ($\pm$0.33) and 6.50 ($\pm$0.33), respectively (See Figures S13–S15 in the Supplementary Information). Concerning HSO_4^-, F^-, $HP_2O_7^{3-}$ and $CH_3CO_2^-$ anions, their log K values were higher than 7, and hence exceeded the value that can be reliably measured with the program. We note that for $CH_3CO_2^-$, the fit was better to a 1:2 sensor: anion complex, but the log β also exceeded the value that can be reliably measured with the program. For this proof-of-principle study, the goals of which were to construct a lanthanide-based sensor, determine what types of anions that the sensor could respond to and the nature of the sensor response, the aim was to determine not the range but the lower limit of the binding constants of anions. These goals were all successfully achieved.

3. Experimental Section

3.1. Material and Instruments

NMR spectra were recorded on Bruker 400 and 500 MHz spectrophotometers (Bruker, Billerica, MA, USA). Chemicals shifts are referenced to the residual solvent peaks and given in parts per million (ppm) for ^{1}H and ^{13}C NMR Spectra. HRMS were acquired on an Agilent 6530 Q-TOF using electrospray ionization (Agilent, Santa Clara, CA, USA). UV-vis spectra were acquired on a Cary 100-Bio UV-vis spectrophotometer that was purchased from Agilent (Santa Clara, CA, USA) and a PerkinElmer Lambda 35 (PerkinElmer, Waltham, MA, USA). Fluorescence excitation and emission spectra were acquired on a Horiba FluoroMax-4 and a PTI-QM4 steady-state fluorimeter with a 75 W Xenon arc-lamp and a R928 photo multiplier tube. Both fluorimeters were purchased from HORIBA Scientific (Piscataway, NJ, USA). All pH measurements were taken with a Denver Instruments UltraBASIC electronic pH meter (Denver Instruments, Bohemiam, TX, USA). Chemicals were purchased from Sigma-Aldrich, Alfa Aesar or Strem Chemicals Inc. and, unless stated otherwise, were used as received. Deuterated solvents for NMR analysis were purchased from Cambridge Isotope Laboratories Inc. or Sigma-Aldrich. The following solvents were used for chromatography or other analysis and used as received: methanol GR ACS Special Anhydrous (EMD), Dichloromethane $\geq$ 99.9%, ChromAR® for HPLC, for UV spectrophotometry (Macron) and Acetonitrile $\geq$ 99.8% for HPLC-GC (Sigma Aldrich).

Solvents that were used in reactions were treated with a Vacuum Atmospheres Company Solvent Purification System (SPS) (Hawthorne, CA, USA) for drying prior to use or were of the special anhydrous kind and used as received.

3.2. Synthesis of Ligand 1 and Lanthanide Complexes

Ligand **1**. DIPEA (22.8 µL, 1.3×10^{-4} mol) was added to a suspension of compound **3** [6c] (50 mg, 6.6×10^{-5} mol), N-[2,2′;6′,2″]Terpyridin-4′-yl-propane-1,3-diamine [11] (40 mg, 1.3×10^{-4} mol), EDCI (25 mg, 1.3×10^{-4} mol) and HOBt (20 mg, 1.3×10^{-4} mol) in CH_2Cl_2 (30 mL). The reaction mixture was stirred under Ar until TLC analysis showed reaction completion (ca. 8 h). Reaction work-up was effected by diluting the reaction mixture to twice its volume and then washing with DI water and brine (3 times each). The CH_2CL_2 layer was then dried with anhydrous $MgSO_4$.The solvent was removed in vacuo to yield a white residue which was purified via column chromatography (neutral alumina, 5% MeOH in CH_2Cl_2). The product was obtained as a white to cream-colored solid (43 mg, 48.2% yield).^{1}H NMR (400 MHz, CD_3CN, 300 K) δ (in ppm): 8.95 (t, $J = 5.67$ Hz, 2H), 8.57 (d, $J = 4.56$ Hz, 4H), 8.37 (broad s, 4H), 8.18 (s, 2H), 7.87 (t, $J = 7.51$Hz, 4H), 7.60 (broad s, 2H), 7.37 (m, 4H), 7.13 (s, 4H) 6.93 (s, 4H), 4.65 (s, 4H), 4.18–3.30 (dd, $J = 13.9$ Hz, $J = 354.12$ Hz, 8H), 3.68 (m, 4H), 3.43 (m, 4H), 2.07 (m, 4H), 1.09 (s, 18H), 1.06 (s, 18H) ^{13}C NMR (214 MHz, CD_3CN, 300 K) δ (in ppm): 169.44, 150.27, 150.06, 149.72, 144.2, 138.3, 138.7, 134.74, 128.31, 127.09, 126.41, 121.99, 75.72, 41.12, 37.45, 35.05, 34.54, 32.69, 32.04, 31.67, 31.37, 30.40, 30.10, 23.45, 14.44. HRMS (ESI-TOF) m/z: $[M + 2H]^{2+}$ Calcd for $C_{84}H_{95}N_{10}O_6$ 670.37520; Found 670.37740. UV-vis (acetonitrile) λ_{max} [nm ($\times 10^4$ $cm^{-1}M^{-1}$)]: 282 (5.03), 254 (3.67) (sh). Fluorescence (acetonitrile) λ_{max} nm: 400.

[Eu.**1**](OTf)$_3$: Ligand **1** (26.3 mg, 2.0×10^{-5} mol) and Eu(OTf)$_3$ (11.8 mg, 2.0×10^{-5} mol) were added to a round bottom flask in 4 mL of dry CH_3CN. The resulting solution was stirred at room temperature overnight under Ar. The solvent was removed in vacuo, and the resulting residue was dissolved in the minimum amount of CH_2Cl_2. The target complex was precipitated via the addition of hexanes as a cream-colored (20.0 mg isolated, 53%). HRMS (ESI-TOF) m/z: $[M-H-3OTf^-]^{2+}$ Calcd for $C_{84}H_{93}EuN_{10}O_6$ 744.32610; Found 744.32340. (UV-vis (acetonitrile) λmax [nm ($\times 10^4$ cm^{-1} M^{-1})]: 288 (4.0), 310 (3.0) (sh). Fluorescence (acetonitrile) nm: 407, 594, 614, 655, 691, 698.

[Tb.**1**](OTf)$_3$: Ligand **1** (19 mg, 2.0×10^{-5} mol) and Tb(OTf)$_3$ (8.6 mg) were added to a round bottom flask in 4 mL of dry CH_3CN. The resulting solution was stirred at room temperature overnight under Ar. The solvent was removed in vacuo and the resulting residue was dissolved in the minimum amount of CH_2Cl_2. The target complex was precipitated via the addition of hexanes as a white solid (12.7 mg isolated, 46%), respectively. HRMS (ESI-TOF) m/z: $[M-3OTf-H]^{2+}$ Calcd for $C_{84}H_{93}N_{10}O_6Tb$ 748.32860; Found 748.32610. UV-vis (acetonitrile) λ_{max} [nm ($\times 10^4$ cm^{-1} M^{-1})]: 289 (1.5), 314 (01.0) (sh). Fluorescence (acetonitrile) nm: 489, 544, 580, 619, 648, 671, 678.

3.3. Metal Binding Studies

Protocol for UV-Vis and Fluorescence Studies: All solutions were prepared in dry degassed CH_3CN. Solutions of Ligand **1** (9.7 µM) and Ln(OTf)$_3$ (5.8 and 0.58 mM) were prepared. This was carried out so that 5 µL of a solution of Ln(OTf)$_3$ delivered either 0.1 equivalent (0.58 mM) or 1 equivalent (5.8 mM) to the ligand during the titration. After each addition of Ln(OTf)$_3$ to ligand **1**, the cuvette was inverted 2–3 times. After a three-minute equilibration period, the spectrum was taken.

3.4. Anion Binding Studies

Protocol for UV-vis and Fluorescence Studies: Anions (as their TBA salts), except TBAF, were dried under vacuum at 70 °C overnight prior to each titration. TBAF was dried under vacuum at 40 °C for about 4 h. Transferal of solvent for solution preparation was performed under an Ar atmosphere. All solutions were prepared in dry degassed CH_3CN. Solutions of [Tb.**1**](OTf)$_3$ (9.9 µM) and anions (as TBA salts) (6.0 mM and 0.6 mM) were prepared. This was carried out so that 2 μL of a solution of an anion delivered either 0.1 equivalent (0.6 mM) or 1 equivalent (6.6 mM) to [Tb.**1**](OTf)$_3$ during the titration. After each addition of anion solution to [Tb.**1**](OTf)$_3$, the cuvette was inverted 2–3 times. After a three-minute equilibration period, the spectrum was taken.

Protocol for Visual Detection Studies: All solutions were prepared in dry degassed CH$_3$CN. A solution of [Tb.**1**](OTf)$_3$ (500 μM) was prepared and added to vials. Solutions of anions (9 mM) were prepared. This was carried out so that 5 μL of a solution of an anion delivered 1 equivalent with respect to [Tb.**1**](OTf)$_3$.

4. Conclusions

The development of lanthanide complexes based on a terpyridine-functionalized calix[4]arene ligand chelated with Eu(III) or Tb(III) has been reported. These complexes were synthesized in two steps and characterized using ^{1}H NMR, ^{13}C NMR and MS. The photophysical properties were also studied; high molar absorptivity values, modest quantum yields and luminescence lifetimes on the ms timescale were obtained. The anion sensing ability of the Tb complex was evaluated due to its higher quantum yield (in acetonitrile solution), and it was found via visual detection, UV-vis and fluorescence studies to be responsive towards a range of anions. Large binding constants (ranging from log K = 5.6 to >7) were obtained from fluorescence titrations and were indicative of strong binding between the sensor and anions. Developing a sensor array based on the diversification of this system is currently being investigated in our laboratory.

Supplementary Materials: The following material is available online: Experimental details of quantum yield and lifetime measurements; NMR spectra (^{1}H, ^{13}C, COSY, HMQC, HMBC) of ligand **1**; UV-Vis and Fluorescence spectra of titrations of ligand **1** with Eu(OTf)$_3$; UV-Vis and Fluorescence spectra of the titrations of [Tb.**1**](OTf)$_3$ with anions and data fits from HypSpec 2014 models (PDF).

Author Contributions: N.Y.E.: conceptualization and design of the work; supervision of the research project; acquisition, analysis and interpretation of data; writing, reviewing and editing. D.M.S.: design of quantum yield studies; acquisition, analysis and interpretation of data (synthesis, visual detection and quantum yield studies); writing of quantum yield studies. I.R.G.-D.: analysis, interpretation and writing of lifetime measurements. J.L.S.: acquisition and analysis of the synthesis of metal complexes. All authors have read and agreed to the published version of the manuscript.

Funding: This research received no external funding.

Data Availability Statement: Data contained within the article and Supplementary Materials are available on request from the authors.

Acknowledgments: N.Y.E. would like to acknowledge undergraduate researchers Mackenzie Schanzlin and Levi Burner for preliminary experiments and Professor Micheal Orleski (Misericordia-Physics Department) who took some of the pictures appearing in this manuscript. She would also like to thank the Misericordia University Faculty and Summer Research Grants Program for financial support of this work. She is grateful to Professors Gong Chen and Eric Anslyn as she has spent some summers doing research in their laboratories. She thanks Professors Marco Bonizzoni and Peter Gans for helpful discussions. N.Y.E. acknowledges OriginLab (technical support) for assistance with some of the figures appearing in the manuscript. She also acknowledges the late Professor Owen Morgan for mentorship.

Conflicts of Interest: The authors declare no conflict of interest.

Sample Availability: Samples of the compounds are not available from the authors.

References

1. Sessler, J.L.; Gale, P.; Cho, W.; Stoddart, J.F.; Rowan, S.J.; Aida, T.; Rowan, A.E. *Anion Receptor Chemistry*; The Royal Society of Chemistry: London, UK, 2006; pp. 1–26.
2. Begum, R.A.; Kang, S.O.; Day, V.W.; Bowman-James, K. Structural Aspects of Anion Coordination Chemistry. *Anion Coord. Chem.* **2011**, 141–225. [CrossRef]
3. Kubik, S. Receptors for Biologically Relevant Anions. *Anion Coord. Chem.* **2011**, 363–464. [CrossRef]
4. Gale, P.A.; Caltagirone, C. Anion Sensors. *Chemosensors* **2011**, 395–427. [CrossRef]

5. Wenzel, M.; Hiscock, J.R.; Gale, P.A. Anion Receptor Chemistry: Highlights from 2010. *Chem. Soc. Rev.* **2012**, *41*, 480–520. [CrossRef]

6. Busschaert, N.; Caltagirone, C.; Van Rossom, W.; Gale, P.A. Applications of Supramolecular Anion Recognition. *Chem. Rev.* **2015**, *115*, 8038–8155. [CrossRef] [PubMed]

7. Wang, X.; Chang, H.; Xie, J.; Zhao, B.; Liu, B.; Xu, S.; Pei, W.; Ren, N.; Huang, L.; Huang, W. Recent Developments in Lanthanide-Based Luminescent Probes. *Coord. Chem. Rev.* **2014**, 201–212. [CrossRef]

8. Gunnlaugsson, T.; Pope, S.J. Lanthanide Ion Complexes as Chemosensors. In *Luminescence of Lanthanide Ions in Coordination Compounds and Nanomaterials*; Wiley: Hoboken, NJ, USA, 2014; pp. 231–268.

9. Spangler, C.; Schäferling, M. Luminescent Chemical and Physical Sensors Based on Lanthanide Complexes. In *Lanthanide Luminescence Springer Series on Fluorescence (Methods and Applications)*; Hänninen, P., Härmä, H., Eds.; Springer: Berlin/Heidelberg, Germany, 2010; Volume 7. [CrossRef]

10. Lee, M.H.; Kim, J.S.; Sessler, J.L. Small Molecule-Based Ratiometric Fluorescence Probes for Cations, Anions, and Biomolecules. *Chem. Soc. Rev.* **2015**, *44*, 4185–4191. [CrossRef] [PubMed]

11. Aletti, A.B.; Gillen, D.M.; Gunnlaugsson, T. Luminescent/colorimetric Probes and (Chemo-) Sensors for Detecting Anions Based on Transition and Lanthanide Ion receptor/binding Complexes. *Coord. Chem. Rev.* **2018**, *354*, 98–120. [CrossRef]

12. Shinoda, S.; Tsukube, H. Luminescent Lanthanide Complexes as Analytical Tools in Anion Sensing, pH Indication and Protein Recognition. *Analyst* **2011**, *136*, 431–435. [CrossRef]

13. Hewitt, S.H.; Macey, G.; Mailhot, R.; Elsegood, M.R.J.; Duarte, F.; Kenwright, A.M.; Butler, S.J. Tuning the Anion Binding Properties of Lanthanide Receptors to Discriminate Nucleoside Phosphates in a Sensing Array. *Chem. Sci.* **2020**, *11*, 3619–3628. [CrossRef]

14. Bünzli, J.C.G.; Eliseeva, S.V. Basics of Lanthanide Photophysics. In *Lanthanide Luminescence. Springer Series on Fluorescence (Methods and Applications)*; Hänninen, P., Härmä, H., Eds.; Springer: Berlin/Heidelberg, Germany, 2010; Volume 7. [CrossRef]

15. dos Santos, C.M.G.; Harte, A.J.; Quinn, S.J.; Gunnlaugsson, T. Recent Developments in the Field of Supramolecular Lanthanide Luminescent Sensors and Self-Assemblies. *Coord. Chem. Rev.* **2008**, *252*, 2512–2527. [CrossRef]

16. Butler, S.J.; Parker, D. Anion Binding in Water at Lanthanide Centres: From Structure and Selectivity to Signalling and Sensing. *Chem. Soc. Rev.* **2013**, *42*, 1652–1666. [CrossRef] [PubMed]

17. Shunmugam, R.; Tew, G.N. Terpyridine-Lanthanide Complexes Respond to Fluorophosphate Containing Nerve Gas G-Agent Surrogates. *Chem. Eur. J.* **2008**, *14*, 5409–5412. [CrossRef] [PubMed]

18. Gupta, K.; Patra, A.K. A Luminescent pH-Responsive Ternary Europium(III) Complex of β-Diketonates and Terpyridine Derivatives as Sensitizing Antennae—Photophysical Aspects, Anion Sensing, and Biological Interactions. *Eur. J. Inorg. Chem.* **2018**, *2018*, 1882–1890. [CrossRef]

19. Ghosh, S.; Abbas, Z.; Dasari, S.; Patra, A.K. Luminescent Eu^{3+} and Tb^{3+} complexes of 4-aminophenyl terpyridine (ptpy): Photophysical aspects, DNA and serum protein binding properties. *J. Lumin* **2017**, *187*, 46–52. [CrossRef]

20. Hau, F.K.; Lo, H.; Yam, V.W. Synthesis and Photophysical Studies of Calixarene-Based Alkynylplatinum(II) Terpyridine Complexes with various Receptor Sites for Colorimetric and Luminescence Sensing of Anions. *Chem. Eur. J.* **2016**, *22*, 3738–3749. [CrossRef]

21. Li, L.; Du, L.; Sun, J.; Yan, C. Synthesis, Crystal Structure and Fluorescent Sensor Ability of Bis-Terpyridinyl-Calix[4]Arene Derivatives. *Chem. Res. Chin. Univ.* **2013**, *29*, 874–878. [CrossRef]

22. Molard, Y.; Parrot-Lopez, H. Molecular Scaffolds for Di-Metallic Complexation: The Synthesis, Characterisation and Complexation Properties of Tetrakis-Terpyridinyl-Calix[4]Arene. *Tetrahedron Lett.* **2002**, *43*, 6355–6358. [CrossRef]

23. Muravev, A.A.; Agarkov, A.S.; Galieva, F.B.; Yakupov, A.T.; Bazanova, O.B.; Rizvanov, I.K.; Shokurov, A.V.; Zaitseva, A.V.; Selektor, S.L.; Solovieva, S.E.; et al. New Terpyridine Derivatives of Thiacalix[4]Arenes in Solution and at the Water-Air Interface. *Russ. Chem. Bull.* **2020**, *69*, 339–350. [CrossRef]

24. Baldini, L.; Sansone, F.; Casnati, A.; Ungaro, R. Calixarenes in Molecular Recognition. In *Supramolecular Chemistry*; Wiley: Hoboken, NJ, USA, 2012.

25. Kumar, R.; Sharma, A.; Singh, H.; Suating, P.; Kim, H.S.; Sunwoo, K.; Shim, I.; Gibb, B.C.; Kim, J.S. Revisiting Fluorescent Calixarenes: From Molecular Sensors to Smart Materials. *Chem. Rev.* **2019**, *119*, 9657–9721. [CrossRef]

26. Jennings, L.B.; Shuvaev, S.; Fox, M.A.; Pal, R.; Parker, D. Selective Signalling of Glyphosate in Water using Europium Luminescence. *Dalton Trans.* **2018**, *47*, 16145–16154. [CrossRef] [PubMed]

27. Caffrey, D.F.; Gunnlaugsson, T. Displacement Assay Detection by a Dimeric Lanthanide Luminescent Ternary Tb(Iii)–cyclen Complex: High Selectivity for Phosphate and Nitrate Anions. *Dalton Trans.* **2014**, *43*, 17964–17970. [CrossRef] [PubMed]

28. Butler, S.J. Quantitative Determination of Fluoride in Pure Water using Luminescent Europium Complexes. *Chem. Commun.* **2015**, *51*, 10879–10882. [CrossRef] [PubMed]

29. Shuvaev, S.; Starck, M.; Parker, D. Responsive, Water-Soluble Europium(III) Luminescent Probes. *Chem. Eur. J.* **2017**, *23*, 9974–9989. [CrossRef]

30. Dickins, R.S.; Aime, S.; Batsanov, A.S.; Beeby, A.; Botta, M.; Bruce, J.I.; Howard, J.A.K.; Love, C.S.; Parker, D.; Peacock, R.D.; et al. Structural, Luminescence, and NMR Studies of the Reversible Binding of Acetate, Lactate, Citrate, and Selected Amino Acids to Chiral Diaqua Ytterbium, Gadolinium, and Europium Complexes. *J. Am. Chem. Soc.* **2002**, *124*, 12697–12705. [CrossRef]

31. Edwards, N.Y.; Liu, F.; Chen, G. Experimental and Computational Studies of Anion Recognition by Pyridine-Functionalized Calixarenes. *Supramol. Chem.* **2013**, *25*, 481–489. [CrossRef]

32. dos Santos, C.M.G.; Gunnlaugsson, T. The Recognition of Anions using Delayed Lanthanide Luminescence: The use of Tb(Iii) Based Urea Functionalised Cyclen Complexes. *Dalton Trans.* **2009**, 4712–4721. [CrossRef]

33. dos Santos, C.M.G.; Fernandez, P.B.; Plush, S.E.; Leonard, J.P.; Gunnlaugsson, T. Lanthanide Luminescent Anion Sensing: Evidence of Multiple Anion Recognition through Hydrogen Bonding and Metal Ion Coordination. *Chem. Commun.* **2007**, 3389–3391. [CrossRef]

34. Umali, A.P.; Anslyn, E.V. A general approach to differential sensing using synthetic molecular receptors. *Curr. Opin. Chem. Biol.* **2010**, *14*, 685–692. [CrossRef]

35. Kotova, O.; Daly, R.; dos Santos, C.M.G.; Boese, M.; Kruger, P.E.; Boland, J.J.; Gunnlaugsson, T. Europium-Directed Self-Assembly of a Luminescent Supramolecular Gel from a Tripodal Terpyridine-Based Ligand. *Angew. Chem. Int. Ed.* **2012**, *51*, 7208–7212. [CrossRef]

36. Gans, P.; Sabatini, A.; Vacca, A. Determination of Equilibrium Constants from Spectrophotometric [Sic] Data obtained from Solutions of Known pH: The Program pHab. *Ann. Chim.* **1999**, *89*, 45–49.

37. Marie, C.; Miguirditchian, M.; Guillaumont, D.; Tosseng, A.; Berthon, C.; Guilbaud, P.; Duvail, M.; Bisson, J.; Guillaneux, D.; Pipelier, M.; et al. Complexation of Lanthanides(III), Americium(III), and Uranium(VI) with Bitopic N,O Ligands: An Experimental and Theoretical Study. *Inorg. Chem.* **2011**, *50*, 6557–6566. [CrossRef] [PubMed]

38. Canard, G.; Koeller, S.; Bernardinelli, G.; Piguet, C. Effective Concentration as a Tool for Quantitatively Addressing Preorganization in Multicomponent Assemblies: Application to the Selective Complexation of Lanthanide Cations. *J. Am. Chem. Soc.* **2008**, *130*, 1025–1040. [CrossRef] [PubMed]

39. Montalti, M.; Prodi, L.; Zaccheroni, N.; Charbonniere, L.; Douce, L.; Ziessel, R. A Luminescent Anion Sensor Based on a Europium Hybrid Complex. *J. Am. Chem. Soc.* **2001**, *123*, 12694–12695. [CrossRef] [PubMed]

40. Nakamoto, K. Ultraviolet Spectra and Structures of 2,2′-Bipyridine and 2,2′,2″-Terpyridine in Aqueous Solution. *J. Phys. Chem.* **1960**, *64*, 1420–1425. [CrossRef]

41. Crosby, G.A.; Demas, J.N. Measurement of Photoluminescence Quantum Yields. Review. *J. Phys. Chem.* **1971**, *75*, 991–1024. [CrossRef]

42. Recording Fluorescence Quantum Yields. Available online: http://www.horiba.com/fileadmin/uploads/Scientific/Documents/Fluorescence/quantumyieldstrad.pdf (accessed on 15 September 2020).

43. Dissanayake, P.; Mei, Y.; Allen, M.J. Luminescence-Decay as an Easy-to-use Tool for the Study of Lanthanide-Containing Catalysts in Aqueous Solutions. *ACS Catal.* **2011**, *1*, 1203–1212. [CrossRef]

44. Bordwell pKa Table. Available online: https://organicchemistrydata.org/hansreich/resources/pka/#pka_dmso_compilation (accessed on 15 September 2020).

45. Muckerman, J.T.; Skone, J.H.; Ning, M.; Wasada-Tsutsui, Y. Toward the accurate calculation of pKa values in water and acetonitrile. *Biochim. Biophys. Acta (BBA) Bioenerg.* **2013**, *1827*, 882–891. [CrossRef]

46. Oliveri, I.P.; Di Bella, S. Lewis Basicity of Relevant Monoanions in a Non-Protogenic Organic Solvent using a Zinc(II) Schiff-Base Complex as a Reference Lewis Acid. *Dalton Trans.* **2017**, *46*, 11608–11614. [CrossRef]

47. Lo, W.; Wong, W.; Law, G. Friend Or Foe? the Role of Solvents in Non-Triplet, Intraligand Charge Transfer Sensitization of Lanthanide(III) Luminescence. *RSC Adv.* **2016**, *6*, 74100–74109. [CrossRef]

48. Anslyn, E.V.; Dougherty, D.A. *Modern Physical Organic Chemistry*; University Science: Sausalito, CA, USA, 2006; pp. 216–218.

Article

Isolation of a Nitromethane Anion in the Calix-Shaped Inorganic Cage

Yuji Kikukawa [1,2,*]**, Hiromasa Kitajima** [1]**, Sho Kuwajima** [1] **and Yoshihito Hayashi** [1,*]

1   Department of Chemistry, Graduate School of Natural Science and Technology,
    Kanazawa University Kakuma, Kanazawa 920-1192, Japan; h.kitajima8923@stu.kanazawa-u.ac.jp (H.K.);
    skuwaji@stu.kanazawa-u.ac.jp (S.K.)
2   Japan Science and Technology Agency, Precursory Research for Embryonic Science and Technology,
    4-1-8 Honcho, Kawaguchi 332-0012, Japan
*   Correspondence: kikukawa1@se.kanazawa-u.ac.jp (Y.K.); hayashi@se.kanazawa-u.ac.jp (Y.H.);
    Tel.: +81-76-264-5691 (Y.H.)

Academic Editor: Mario Berberan-Santos and Paula M. Marcos
Received: 30 October 2020; Accepted: 27 November 2020; Published: 1 December 2020

Abstract: A calix-shaped polyoxometalate, $[V_{12}O_{32}]^{4-}$ (**V12**), stabilizes an anion moiety in its central cavity. This molecule-sized container has the potential to control the reactivity of an anion. The highly-reactive cyanate is smoothly trapped by **V12** to form $[V_{12}O_{32}(CN)]^{5-}$. In the CH_3NO_2 solution, cyanate abstracts protons from CH_3NO_2, and the resultant $CH_2NO_2^-$ is stabilized in **V12** to form $[V_{12}O_{32}(CH_2NO_2)]^{5-}$ (**V12(CH$_2$NO$_2$)**). A crystallographic analysis revealed the double-bond characteristic short bond distance of 1.248 Å between the carbon and nitrogen atoms in the nitromethane anion in **V12**. ^1H and ^{13}C NMR studies showed that the nitromethane anion in **V12** must not be exchanged with the nitromethane solvent. Thus, the **V12** container restrains the reactivity of anionic species.

Keywords: calixarene-like structure; polyoxometalates; nitronate; crystal structure; host–guest chemistry; anion receptor

---

1. Introduction

Calixarene possesses a rigid conformation with a broader hydrophobic upper rim, a narrow hydrophilic lower rim, and a central annulus. Due to this attractive architecture, the host–guest chemistry of calixarene has been developed [1–4]. It can include various chemical moieties of neutral molecules, cations, and anions through hydrophobic, cation–π, anion–π, and hydrogen-bonding interactions [5]. The host property is finely tuned by the functional modification of the upper rim and/or the lower rim. Calixarene is utilized in various fields of science, such as catalysis, sensors, functional materials, analytical chemistry, electrochemistry, photochemistry, biochemistry, and pharmacy [6–11]. The cavity diameters, considering the van der Waals radii, around the upper rim of calix [4] arenes and calix [6] arenes are 3.8 Å and 5.0 Å, respectively [12].

Polyoxometalates are a large family of metal-oxide cluster anions. They show several unique properties related to their well-defined structures [13–15]. Tungsten- and molybdenum-based polyoxometalates adopt lacunary structures by removing constituent elements. The lacunary type polyoxometalates act as inorganic multidentate ligands to stabilize several metal or metal-oxide cores. On the other hand, vanadium-based polyoxometalates formed by the condensation of VO_5 square pyramids stabilize anionic moiety at the center of their clusters [16,17]. Up to now, various kinds of anions have been included in the vanadium-based polyoxometalates. Most anion-including polyoxometalates are sphere structures. [18] Although each electrophilicity directed to the base of the VO_5 square pyramid is weak, the electrophilicity is strengthened by the condensation of the VO_5 pyramids oriented to the center of a sphere to stabilize an anion. The distance between the central anion and its nearest neighbor vanadium is far from bonding distances, showing that the central anion is floating in the container. Among polyoxometalates composed of VO_5 square pyramids, a 'calix'-shaped dodecavanadate $[V_{12}O_{32}]^{4-}$ (**V12**) is known [19–30]. The broader upper rim consists of eight edge-shared VO_5 pyramids, with a 4.4 Å cavity entrance and a narrow lower rim consisting of four vertex-shared VO_5 pyramids (Figure 1). This attractive structure was first reported by Day et al. in 1989 [19]. In the initial stage of the **V12** chemistry, the utility is limited to the guest exchange among nitriles because of the strong affinity between the nitrile group and the **V12** container [20–22]. Recently, we have developed the host–guest chemistry of **V12** [23–29]. By avoiding the usage of the nitrile group in the synthetic procedure by controlling the oxidation state of the vanadium sources, **V12** with guest-exchangeability to various electron-rich groups can be prepared [25]. In addition, the guest-removal from **V12** is successfully accomplished [28]. Importantly, during the guest removal, even in the solid state, one of the VO_5 square pyramids of lower rim is flipped, and an oxygen atom is located at the center of the **V12** to fill the cavity. The structure is retrieved by the exposure to the guest vapor, such as acetonitrile, nitromethane, dichloromethane, 1,2-dichloroethane, bromomethane, CO_2, or Br_2. Thus, **V12** can be categorized as one of the guest-responsible Supramolecular Association–Dissociation Switches [3]. In the case of Br_2, Br_2 inserted into **V12** is polarized due to the unique charge distribution of the inside of **V12**, and the peak of the Br–Br vibration is detected in IR at 185 cm^{-1} [29]. Furthermore, the stabilization of several kinds of anions in **V12** has been reported [25–27]. The most interesting example is NO^--including **V12** [30]. While a standard chemistry text book introduces the electron structure of an anionic nitrogen monoxide, the isolation of anionic NO^- is very rare due to its stability and short life under normal experimental conditions.

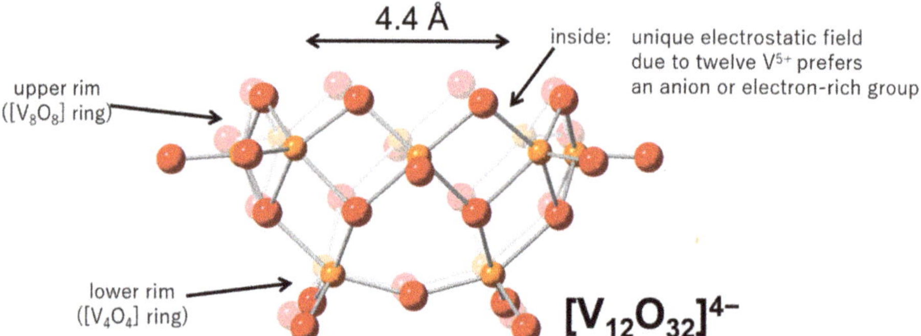

Figure 1. Schematic representation of a dodecavanadate. The red and orange spheres represent oxygen and vanadium atoms, respectively.

Cyanide shows high nucleophilicity and acts as a ligand for several kinds of metal cations. The high nucleophilicity also makes it act as efficient Brønsted and Lewis base catalysts [31–34]. The discrete cyanide catalyzes the cyanosilylation of aldehydes, deprotonation, deacetylation, and Michael addition reactions. In this report, the host–guest chemistry of **V12** is applied to the reaction inhibitor. The addition of **V12** during the catalytic reaction with cyanide quenches the reaction to form cyanide-including **V12**. Successively, the guest exchange reaction proceeds to form nitromethane-anion-including **V12** ($[V_{12}O_{32}(CH_3NO_2)]^{4-}$, **V12(CH₂NO₂)**) in nitromethane. The crystal structure and ^{1}H and ^{13}C NMR spectra are also discussed.

2. Results and Discussion

2.1. Reactivity of Cyanide and the Effect of the Addition of V12

Tetraethylammmonium cyanide {Et$_4$N}CN is commercially available and stable. Bare cyanide in {Et$_4$N}CN shows higher reactivity than that in metal cyanide complexes. In the presence of 1 mol% of {Et$_4$N}CN, the reaction of acetophenone and trimethylsilyl cyanide gave the corresponding cyanohydrin trimethylsilyl ether with a 93% yield in 5 min (Figure 2). The turnover frequency reaches 19 min^{-1}. This value is the highest level among the previous reports [35–38]. By the addition of **V12(NM)** into the reaction solution, **V12(NM)** was easily decomposed due to the formation of electrophilic trimethylsilyl cations and/or the successive formation of nucleophilic cyanide.

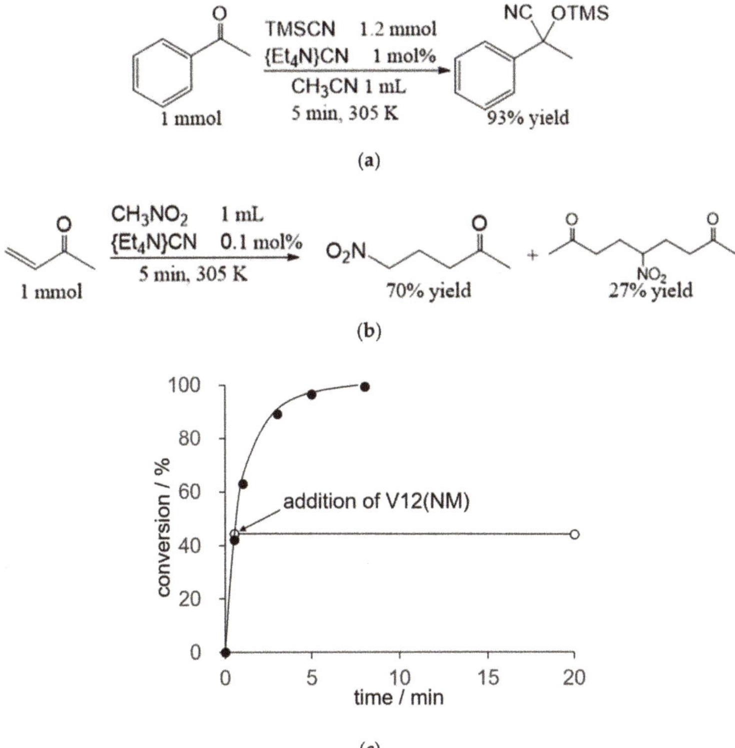

Figure 2. Catalytic performance of {Et$_4$N}CN for (**a**) cyanosilylation, (**b**) Michael addition, and (**c**) time course profiles of Michael addition and the effect of the addition of **V12(NM)**. One equivalent of **V12(NM)** respective to {Et$_4$N}CN was added 30 s after the reaction started.

Cyanide also catalyzes Michael addition [34]. In the presence of 0.1 mol% of {Et$_4$N}CN, the conversion of methyl vinyl ketone in nitromethane reached 97% in 5 min, to give 5-nitro-2-pentanone with a 70% yield and 5-nitro-2,8-nonanedione with a 27% yield (Figure 2). This reaction proceeded as follows. Nitromethane was deprotonated by the cyanide catalyst, and the reactive nitronate attacked the beta carbon to give the products. After 30 sec, about 40% of the methyl vinyl ketone was converted under the catalytic condition. By the addition of one equivalent of **V12(NM)** with respect to {Et$_4$N}CN into the reaction solution after 30 sec, the reaction immediately stopped (Figure 2). Even if ten times the amounts of the catalyst and **V12(NM)** were used, the reaction did not proceed.

The ^{51}V NMR of the reaction solution was measured. The results described below were beyond our imagination. From our previous report, it was assumed that the reaction was quenched by the incorporation of cyanide into the **V12**. However, this was not all. After 30 min from the addition of **V12(NM)** into the reaction solution, the ^{51}V NMR showed t three signals at −564, −572, and −578 ppm, which are different signals from those of cyanido-including **V12** ($[V_{12}O_{32}(CN)]^{5-}$, **V12(CN)**) (Figure S1). This spectrum is also different from that of **V12(NM)**. The spectrum most closely resembles that of acetate-including **V12** ($([V_{12}O_{32}(CH_3CO_2)]^{5-}$, **V12(OAc)**)), with three signals at −567 (sharp), −578 (broad), and −585 ppm (broad) (Figure S2).

The ^{51}V NMR spectrum was monitored without the addition of methyl vinyl ketone (Figure S3). Compound **V12(NM)** in nitromethane showed three signals at −591, −596, and −606 ppm. Through the addition of {Et$_4$N}CN, three signals at −578, −586, and −599 ppm due to **V12(CN)** were observed. With time, the intensity of the signals of **V12(CN)** decreased, and that of the three signals at −564, −572, and −578 ppm increased. The cold-spray ionized mass (CSI-MS) spectrum of the nitromethane solution of **V12(CN)** showed peaks at 1930 of $\{(Et_4N)_6[V_{12}O_{32}(CN)]\}^+$ (Figure S4). With time, this peak intensity decreased and the intensity of the peak at 1964 assignable to $\{(Et_4N)_6[V_{12}O_{32}(CH_2NO_2)]\}^+$ increased. Considering these results and the reaction mechanism, by the addition of **V12(NM)**, cyanide was actually trapped in **V12** to form **V12(CN)**; the successive deprotonation of nitromethane proceeded, and $CH_2NO_2^-$ was stabilized in **V12** to form $[V_{12}O_{32}(CH_2NO_2)]^{5-}$ (**V12(CH$_2$NO$_2$)**). Thus, the reaction of methyl vinyl ketone and CH_3NO_2 with {Et$_4$N}CN catalyst stopped at the step of the formation of $CH_2NO_2^-$ in **V12**.

2.2. Crystal Strucuture and Charactorization

Fortunately, we can obtain crystals suitable for the X-ray structure analysis by the cation exchange from Et$_4$N$^+$ to Me$_4$N$^+$ (Table S1, Figure 3). The anion structure exhibits the typical **V12** structure with the guest moiety of $CH_2NO_2^-$ in the concave. This is the first report on the crystal structure of nitromethane anions, as far as we know. Four CH_3NO_2 as crystalline solvents and five {Me$_4$N}$^+$ were determined, supporting the theory that the moiety in **V12** is a −1 charged anion. These results agreed well with the elemental and thermogravimetric analyses. In the case of neutral CH_3NO_2 as a guest, one of the oxygen atoms of the nitro group is inserted into the cavity, and the other oxygen is located at the same level of the entrance oxygen atoms of **V12** (Figure S5). On the other hand, two oxygen atoms of $CH_2NO_2^-$ were packed into the concave of **V12**. The shortest distance between the nearest vanadium atom and an oxygen atom of $CH_2NO_2^-$ is 2.538(6) Å, showing that $CH_2NO_2^-$ is not directly bonded to vanadium centers. Although the visual aspect of **V12(CH$_2$NO$_2$)** is similar to that of **V12(OAc)**, each bond distance in the guest is different. $CH_2NO_2^-$ possesses a 1.255(11) Å of C−N bond, and 1.325(8) Å and 1.313(8) Å of N−O bonds, while ^-OAc possesses a 1.506(5) Å of C−C bond, and 1.257(4) Å and 1.260(4) Å of C−O bonds. Generally, a nitromethane anion exhibits resonance structures: anion charge locates on a carbon atom with a single C−N bond, and anion charge locates on oxygen atoms with a double C=N bond. The 1.255(11) Å bond distance of the C−N of $CH_2NO_2^-$ shows the composition of a double bond between carbon and nitrogen atoms. During the above mentioned catalytic reaction, the formation of more reactive $^-CH_2-NO_2$ was restrained by the inclusion of $CH_2NO_2^-$ in **V12**, and the reaction stopped.

The ^{51}V NMR spectrum of **V12(CH$_2$NO$_2$)** in nitromethane is maintained for more than 60 min. In order to determine the ^{13}C NMR for the nitromethane anion in **V12**, ^{13}C-enriched nitromethane-anion-including **V12** was prepared. The $^{51}VNMR$ spectrum showed isotope shift (Figure S1) [39]. The ^{13}C NMR spectrum showed a peak at 109.3 ppm of nitromethane anions, in addition to peaks at 52.2 and 6.5 ppm of {Et$_4$N}$^+$. In an off resonance-decoupled ^{13}C, a peak at 109.2 ppm was tripled, showing that two protons are attached to the carbon (Figure S6) [40,41]. The peak intensity of the nitromethane anion is maintained for more than 60 min, suggesting that the included ^{13}C-nitromethane anion is not exchanged with the ^{12}C-one derived from the nitromethane solvent. In the ^1H NMR spectra in CD_3NO_2, a peak at 5.27 ppm due to the nitromethane anion was detected

(Figure S7). The peak intensity is retained for more than 60 min, showing that the proton exchange reaction between the included nitromethane anion and nitromethane solvent does not proceed.

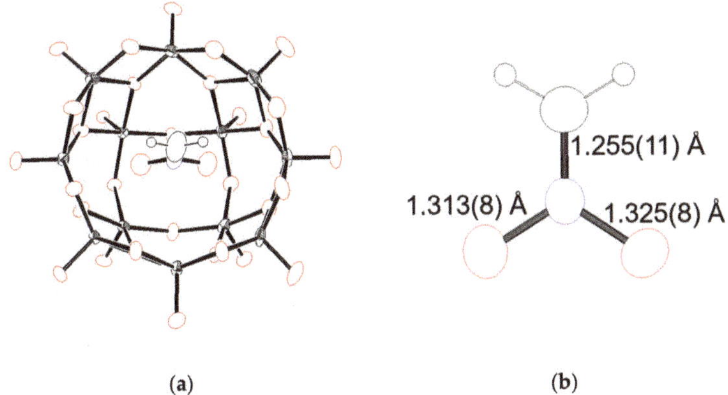

Figure 3. Ortep representations of (**a**) an anion section of **V12(CH₂NO₂)** and (**b**) a $CH_2NO_2^-$ fragment in the cavity of **V12**. The black octant shading and spheres represent vanadium and hydrogen atoms, respectively. The red, blue, and black boundaries represent oxygen, nitrogen, and carbon atoms, respectively.

3. Materials and Methods

All of the reagents were obtained from commercial suppliers and were used without further purification unless otherwise noted. The solvents for the NMR measurements of the authentic **V12(CH₂NO₂)** were dehydrated by molecular sieve 4A. The **V12(NM)** and **V12(CN)** were synthesized following the literature [26].

For the synthesis of $\{Et_4N\}_5[V_{12}O_{32}(CH_2NO_2)]$ and $\{Me_4N\}_5[V_{12}O_{32}(CH_2NO_2)]$, the tetraethyl salt of **V12(CN)** (100 mg, 0.056 mmol) was dissolved in 5 mL of CH_3NO_2 and stirred for 1 h. The addition of an excess amount of diethyl ether with vigorous stirring gave a brown powder of $\{Et_4N\}_5[V_{12}O_{32}(CH_2NO_2)]$. The precipitates were collected by filtration and dried (85 mg, 83% yield). ^{51}V NMR (in CD_3NO_2); δ = −564, −572, and −578 ppm. ^1H NMR (in CD_3NO_2); δ = 5.28 (s), 3.40 (q), and 1.36 (t) ppm. ^{13}C NMR (in CD_3NO_2); δ = 109.2, 52.2 and 6.5 ppm. The peak due to CH_2NO_2 was only observed by using a ^{13}C-enriched carbon source. ^{13}C-enriched $[V_{12}O_{32}(^{13}CH_2NO_2)]^{5-}$ was obtained by the dissolution of **V12(CN)** into $^{13}CH_3NO_2$. IR (Attenuated Total Reflection (ATR) without ATR correction): 2983, 2949, 2882, 1927, 1642, 1547, 1481, 1455, 1392, 1392, 1305, 1254, 1173, 1102, 1056, 1035, 973, 902, 853, 831, 750, 699, 628, 543, and 511 cm^{-1} (Figure S8). For the crystallographic analysis, tetramethylammonium salt was prepared. The tetraethylammonium salt of **V12(CH2NO2)** (91.7 mg, 0.05 mmol) was dissolved in 2 mL nitromethane. To this solution, 3 mL nitromethane solution of $\{Me_4N\}ClO_4$ (43.4 mg, 0.25 mmol) was added. The immediately-formed precipitates were removed by filtration and stood at 5 °C for 2 days. Elemental analysis calcd for $\{Me_4N\}_5[V_{12}O_{32}(CH_2NO_2)]\cdot 4CH_3NO_2$ ($C_{25}H_{74}N_{10}O_{42}V_{12}$): C 16.70%, H 4.15%, N 7.79%; found: C 16.85%, H 4.17%, N 7.32%. The Thermogravimetric analysis data showed 14% mass-decreasing (the desorption of four CH_3NO_2 as crystalline solvent) by 180 °C.

The catalytic reaction was carried out as follows. Into a screw-capped test tube, 1 mmol substrate, 1 mL solvent, $\{Et_4N\}CN$ (catalyst) 1 mol% for cyanosililation and 0.1 mol% for Michael addition and naphthalene (internal standard) were added and stirred at 800 rpm at 305 K.

The IR spectra were measured with the ATR method (Zn/Se prism) on a JASCO FT/IR-4200 spectrometer. The ^1H, ^{13}C and ^{51}V NMR spectra were recorded on a JEOL JNM-LA400. The thermogravimetry data were collected on a Rigaku Thermo plus EVO2 instrument with a temperature sweep rate of 10 °C/min under N_2 flow (200 mL/min). The elemental analyses of C, H, and N were performed by the Research Institute for Instrumental Analysis at Kanazawa University. The GC analyses were performed on a Shimadzu GC-2014 with a flame ionization detector (FID) equipped with a InertCap Pure-WAX or ZB-WAXplus capillary column (internal diameter = 0.25 mm, length = 30 m).

The diffraction measurements of **V12(CH$_2$NO$_2$)** were performed on a Bruker D8 VENTURE diffractometer with graphite monochromated Cu Kα radiation (λ = 1.54178 Å). The data reduction and absorption correction were carried out using the APEX3 program [42]. The structural analyses were performed using APEX3 and WinGX [43]. The structures were refined by SHELXL-2013 [44]. The non-hydrogen atoms were refined anisotropically. The hydrogen atoms were positioned geometrically and refined using a riding model. The atoms in one of the tetramethylammoinium cations are restrained with a SIMU command. CCDC 2,041,581 contains the supplementary crystallographic data for this paper. These data can be obtained, free of charge, from The Cambridge Crystallographic Data Centre.

4. Conclusions

Calix-shaped dodecavanadate **V12** acts as an efficient trap for the reactive anion species. By dissolving cyanide-including **V12** in nitromethane, nitromethane is activated and the nitromethane anion—the reaction intermediate—is stabilized in **V12**, which enables the X-ray crystallographic analysis.

Supplementary Materials: Figure S1: ^{51}V NMR spectra of the reaction mixture of methyl vinyl ketone (1 mmol), CH$_3$NO$_2$ (1 mL), {Et$_4$N}CN (10 µmol), and **V12(NM)** (10 µmol) after 20 min., Figure S2: ^{51}V NMR spectra of (a) [V$_{12}$O$_{32}$(^{13}CH$_2$NO$_2$)]$^{5-}$, (b) **V12(CH$_2$NO$_2$)**, (c) **V12(OAc)**, (d) **V12(NM)** and (e) **V12(CN)** in nitromethane, Figure S3: ^{51}V NMR spectra of the dehydrated-nitromethane solution of (a) **V12(NM)**, and {Et$_4$N}CN (10 µmol), and **V12(NM)** (10 µmol) after (b) 1 min, (c) 5 min, (d) 10 min and (e) 20 min, Figure S4: CSI-MS spectra (positive mode) of the nitromethane solution of **V12(CN)** (a) just after dissolution, (b) after 10 min and (c) after 30 min, Figure S5: Anion structures of (a) **V12(CH$_2$NO$_2$)**, (b) **V12(OAc)**, (c) **V12(NM)** and (d) **V12(CN)**, Figure S6: (a) Decoupling and (b) off resonance-decoupled ^{13}C NMR spectra of **V12(CH$_2$NO$_2$)** in CD$_3$NO$_2$, Figure S7: ^{1}H NMR spectrum of **V12(CH$_2$NO$_2$)** in CD$_3$NO$_2$, Figure S8: IR spectrum of **V12(CH$_2$NO$_2$)** (ATR without ATR correction), Table S1: Crystallographic data for **V12(CH$_2$NO$_2$)**.

Author Contributions: Conceptualization, Y.K. and Y.H.; methodology, Y.K.; catalytic reaction, Y.K. and H.K.; NMR measurement, H.K.; CSI-MS measurement, Y.K.; crystallization, S.K.; crystal structure determination, Y.K. and S.K.; writing, Y.K. and Y.H.; visualization, Y.K. and H.K.; funding acquisition, Y.K. All authors have read and agreed to the published version of the manuscript.

Funding: This research was funded by JST PRESTO, grant number JPMJPR1655"; Nippon Sheet Glass Foundation for Materials Science and Engineering, JSPS KAKENHI, grant number JP18K14239; and the Core-to-Core Program, and Kanazawa University SAKIGAKE project.

Acknowledgments: The authors thank K. Yamaguchi and K. Suzuki (University of Tokyo) and their co-workers for their help with the CSI-MS measurements.

Conflicts of Interest: The authors declare no conflict of interest.

References

1. Ikeda, A.; Shinkai, S. Novel Cavity Design Using Calix[*n*]arene Skeletons: Toward MolecularRecognition and Metal Binding. *Chem. Rev.* **1997**, *97*, 1713–1734. [CrossRef] [PubMed]

2. Rebek, J., Jr. Host-guest chemistry of calixarene caplixarene capsules. *Chem. Commun.* **2000**, 637–643. [CrossRef]

3. Blanco-Gómez, A.; Cortón, P.; Neira, L.B.I.; Pazos, E.; Peinador, C.; Garía, M.D. Controlled binding of organic guests by stimuli-responsive macrocycles. *Chem. Soc. Rev.* **2020**, *49*, 3834–3862. [CrossRef] [PubMed]

4. Islam, M.; Georghiou, P.E.; Rahman, S.; Yamato, T. Calix[3]arene-Analogous Metacyclophanes: Synthesis, Structures and Properties with Infinite Potential. *Molecules* **2020**, *25*, 4202. [CrossRef]

5. Ortolan, A.O.; Øestrøm, I.; Caramori, G.; Parreira, R.L.T.; Muñoz-Castro, A.; Bickelhaupt, F.M. Anion Recognition by Organometallic Calixarenes: Analytsis from Relativistic DFT Calculations. *Organometallics* **2018**, *37*, 2167–2176. [CrossRef]

6. Guo, D.-S.; Liu, Y. Supramolecular Chemistry of *p*-Sulfonatocalix[n]arenes and Its Biological Applications. *Acc. Chem. Res.* **2014**, *47*, 1925–1934. [CrossRef]

7. Diamond, D.; McKervey, M.A. Calixarene-based Sensing Agents. *Chem. Soc. Rev.* **1996**, *25*, 15–24. [CrossRef]

8. Kumar, R.; Sharma, A.; Singh, H.; Suating, P.; Kim, H.S.; Sunwoo, K.; Shim, I.; Gibb, B.C.; Kim, J.S. Revisiting Fluorescent Calixarenes: From Molecular Sensors to Smart Materials. *Chem. Rev.* **2019**, *119*, 9657–9721. [CrossRef]

9. Ma, X.; Zhao, Y. Biomedical Applications of Supramolecular Systems Based on Host-Guest Interactions. *Chem. Rev.* **2015**, *115*, 7794–7839. [CrossRef]

10. Homden, D.M.; Redshaw, C. The Use of Calixarenes in Metal-Based Catalysis. *Chem. Rev.* **2008**, *108*, 5086–5130. [CrossRef]

11. Ludwing, R. Calixarenes in analytical and separation chemistry. *Fresenius J. Anal. Chem.* **2000**, *367*, 103–128. [CrossRef] [PubMed]

12. Shinkai, S.; Araki, K.; Manabe, O. Does the Calixarene Cavity recognize the Size of Guest Molecules? On the 'Hole-size Selectivity' in Water-soluble Calixarenes. *J. Chem. Soc. Chem. Commun.* **1988**, 187–189. [CrossRef]

13. Hill, C.L. Themed issue on Polyoxometalates. *Chem. Rev.* **1998**, *98*, 1–390. [CrossRef] [PubMed]

14. Cronin, L.; Müller, A. Themed issue on Polyoxometalate cluster science. *Chem. Soc. Rev.* **2012**, *41*, 7325–7648.

15. Misra, A.; Kozma, K.; Streb, C.; Nyman, M. Beyond Charge Balance: Counter-Cations in Polyoxometalate Chemistry. *Angew. Chem. Int. Ed.* **2020**, *59*, 596–612. [CrossRef] [PubMed]

16. Hayashi, Y. Hetero and Lacunary Polyoxovanadate chemistry: Synthesis, reactivity and structural aspects. *Coord. Chem. Rev.* **2001**, *255*, 2270–2280. [CrossRef]

17. Streb, C. *Polyoxometalate-Based Assemblies and Functional Materials*; Song, Y.-F., Ed.; Springer: Cham, Switzerland, 2018; pp. 31–47.

18. Rehder, D. *Bioinorganic Vanadium Chemistry*; John Wiley & Sons, Ltd.: Hoboken, NJ, USA, 2008.

19. Day, V.W.; Klemperer, W.G.; Yaghi, O.M. Synthesis and characterization of a soluble oxide inclusion complex, [CH3CN.cntnd.(V12O324-)]. *J. Am. Chem. Soc.* **1989**, *111*, 5959–5961. [CrossRef]

20. Klemperer, W.G.; Marquart, T.A.; Yaghi, O.M. Shape-selective binding of nitriles to the inorganic cavitand vanadate, $V_{12}O_{32}{}^{4-}$. *Mater. Chem. Phys.* **1991**, *29*, 97–104. [CrossRef]

21. Rohmer, M.-M.; Benard, M. An Interpretation of the Structure of the Inclusion Complexes [RCN⊂$(V_{12}O_{32})^{4-}$] (R = CH3, C6H5) from Electrostatic Potentials. *J. Am. Chem. Soc.* **1994**, *116*, 6959–6960. [CrossRef]

22. Rohmer, M.-M.; Devemy, J.; Wiest, R.; Benard, M. Ab Initio Modeling of the Endohedral Reactivity of Polyoxometallates: 1. Host-Guest Interactions in [RCN⊂$(V_{12}O_{32})^{4-}$] (R = H, CH3, C6H5). *J. Am. Chem. Soc.* **1996**, *118*, 13007–13014. [CrossRef]

23. Kurata, T.; Hayashi, Y.; Isobe, K. Synthesis and characterization of chloride-incorporated dodecavanadate from dicopper complex of macrocyclic octadecavanadate. *Chem. Lett.* **2010**, *39*, 708–709. [CrossRef]

24. Inoue, Y.; Kikukawa, Y.; Kuwajima, S.; Hayashi, Y. A chloride capturing system via proton-induced structure transformation between opened- and closed-forms of dodecavanadates. *Dalton Trans.* **2016**, *45*, 7563–7569. [CrossRef] [PubMed]

25. Kuwajima, S.; Ikinobu, Y.; Watanabe, D.; Kikukawa, Y.; Hayashi, Y.; Yagasaki, A. A Bowl-Type Dodecavanadate as a Halide Receptor. *ACS Omega* **2017**, *2*, 268–275. [CrossRef] [PubMed]

26. Kuwajima, S.; Kikukawa, Y.; Hayashi, Y. Small-Molecule Anion Recognition by a Shape-Responsive Bowl-Type Dodecavanadate. *Chem. Asian J. Chem.* **2017**, *12*, 1909–1914. [CrossRef]

27. Kuwajima, S.; Arai, Y.; Kitajima, H.; Kikukawa, Y.; Hayashi, Y. Synthesis and structural characterization of tube-type tetradecavanadates. *Acta Crystallogr. C* **2018**, *74*, 1295–1299.

28. Kikukawa, Y.; Seto, K.; Uchida, S.; Kuwajima, S.; Hayashi, Y. Solid-State Umbrella-type Inversion of a VO_5 Square-Pyramidal Unit in a Bowl-type Dodecavanadate Induced by Insertion and Elimination of a Guest Molecule. *Angew. Chem. Int. Ed.* **2018**, *57*, 16051–16055. [CrossRef] [PubMed]

29. Kikukawa, Y.; Seto, K.; Watanabe, D.; Kitajima, H.; Katayama, M.; Yamashita, S.; Inada, Y.; Hayashi, Y. Induced Fitting and Polarization of a Bromine Molecule in an Electrophilic Inorganic Molecular Cavity and Its Bromination Reactivity. *Angew. Chem. Int. Ed.* **2020**, *59*, 14399–14403. [CrossRef]

30. Kawanami, N.; Ozeki, T.; Yagasaki, A. NO-Anion Trapped in a Molecular Oxide Bowl. *J. Am. Chem. Soc.* **2000**, *122*, 1239–1240. [CrossRef]

31. North, M.; Omedes-Pujol, M.; Yong, C. Kinetics and mechanism of the racemic addition of trimethylsilyl cyanide to aldehydes catalysed by Lewis bases. *Org. Biomol. Chem.* **2012**, *10*, 4289–4298. [CrossRef]

32. Holmes, B.T.; Arthur, S.W. Aliphatic thioacetate deprotection using catalytic tetrabutylammonium cyanide. *Tetrahedron* **2005**, *61*, 12339–12342. [CrossRef]

33. Park, H.J.; Lee, S.S. Catalytic Deacetylation of p-Nitrophenyl Thioacetate by Cyanide Ion and Its Sensor Applications. *Anal. Sci.* **2019**, *35*, 589–593. [CrossRef] [PubMed]

34. Miyashita, A.; Numata, A.; Suzuki, Y.; Iwamoto, K.; Higashino, T. Olefin-Insertion Raction between the Carbonyls of Benzils; Formation of 1,4-Diketones by Michael Addition Catalyzed by Cyanide Ion. *Chem. Lett.* **1997**, *24*, 697–698. [CrossRef]

35. Ullah, B.; Chen, J.; Zhang, Z.; Xing, H.; Yang, Q.; Bao, Z.; Ren, Q. 1-Ethyl-3-methylimidazolium acetate as a highly efficient organocatalyst for cyanosilylation of carbonyl compounds with trimethylsilyl cyanide. *Sci. Rep.* **2017**, *7*, 42699. [CrossRef] [PubMed]

36. Huang, X.; Chen, L.; Ren, F.; Yang, C.; Li, J.; Shi, K.; Gou, X.; Wang, W. Lewis Acid Rather than Brønsted Acid Sites of Montmorillonite K10 Act as a Powerful and Reusable Green Heterogeneous Catalyst for Rapid Cyanosilylation of Ketones. *Synlett* **2017**, *28*, 439–444. [CrossRef]

37. Wang, W.; Wang, X.; Zhou, S.; Xu, X.; Du, J.; Zhang, L.; Mu, X.; Wei, Y.; Zhu, X.; Wang, S. Syntheses, Structures, and Catalytic Activities of the Anionic Heterobimetallic Rare-Earth Metal Complexes Supported by Pyrrolyl-Substituted 1,2-Diimino Ligands. *Inorg. Chem.* **2018**, *57*, 10390–10400. [CrossRef]

38. Rawat, S.; Bhandari, M.; Prashanth, B.; Singh, S. Three Coordinated Organoaluminum Cation for Rapid and Selective Cyanosilylation of Carbonyls under Solvent-Free Conditions. *ChemCatChem* **2020**, *12*, 2407–2411. [CrossRef]

39. Jameson, C.J.; Rehder, D.; Hoch, M. Isotope and Temperature Dependence of Transition-Metal Shielding in Complexes of the Type M(XY)$_6$. *J. Am. Chem. Soc.* **1987**, *109*, 2589–2594. [CrossRef]

40. Griswold, A.A.; Starcher, P.S. The Nuclear Magnetic Resonance Spectra of aci-Nitro Anions. *J. Org. Chem.* **1965**, *30*, 1687–1690. [CrossRef]

41. Engelke, R.; Earl, W.L.; Rohlfing, C.M. Production of the Ntromethane Acl Ion by UV Irradiation: Its Effect on Detonation Sensitivity. *J. Phys. Chem.* **1986**, *90*, 545–547. [CrossRef]

42. *APEX3, SAINT, and SADABS*; Bruker AXS Inc.: Madison, WI, USA, 2015.

43. Farrugia, L.J. WinGX suite for small-molecule single-crystal crystallography. *J. Appl. Crystallogr.* **1999**, *32*, 837–838. [CrossRef]

44. Gruene, T.; Hahn, H.W.; Luebben, A.V.; Sheldrick, G.M. Refinement of macromolecular structures against neutron data with SHELXL2013. *J. Acta Cryst.* **2014**, *47*, 462–466.

Sample Availability: Samples of the compounds are available from the authors.

Publisher's Note: MDPI stays neutral with regard to jurisdictional claims in published maps and institutional affiliations.

Article

Study on the Influence of Chirality in the Threading of Calix[6]arene Hosts with Dialkylammonium Axles

Carmen Talotta [1], Gerardo Concilio [1], Paolo Della Sala [1], Carmine Gaeta [1],
Christoph A. Schalley [2,3,*] and Placido Neri [1,*]

[1] Dipartimento di Chimica e Biologia "A. Zambelli", Università di Salerno, Via Giovanni Paolo II 132, I-84084 Fisciano, Italy; ctalotta@unisa.it (C.T.); gconcilio@unisa.it (G.C.); pdellasala@unisa.it (P.D.S.); cgaeta@unisa.it (C.G.)

[2] Institut für Chemie und Biochemie, Freie Universität, Arnimallee 20, 14195 Berlin, Germany

[3] School of Life Sciences, Northwestern Polytechnical University, Xi'an 710072, China

* Correspondence: c.schalley@fu-berlin.de (C.A.S.); neri@unisa.it (P.N.)

Academic Editors: Mario Berberan-Santos and Paula M. Marcos
Received: 19 October 2020; Accepted: 12 November 2020; Published: 15 November 2020

Abstract: The influence of chirality in calixarene threading has been studied by exploiting the "superweak anion approach". In particular, the formation of chiral pseudo[2]rotaxanes bearing a classical stereogenic center in their axle and/or wheel components has been considered. Two kind of pseudo[2]rotaxane stereoadducts, the "*endo*-chiral" and "*exo*-chiral" ones, having the stereogenic center of a cationic axle inside or outside, respectively, the calix-cavity of a chiral calixarene were preferentially formed with specifically designed chiral axles by a fine exploitation of the so-called "*endo*-alkyl rule" and a newly defined "*endo*-α-methyl-benzyl rule" (*threading of a hexaalkoxycalix[6]arene with a directional (α-methyl-benzyl)benzylammonium axle occurs with an* endo-α-methyl-benzyl *preference*). The obtained pseudorotaxanes were studied in solution by 1D and 2D NMR, and in the gas-phase by means of the enantiomer-labeled (EL) mass spectrometry method, by combining enantiopure hosts with pseudoracemates of one deuterated and one unlabeled chiral axle enantiomer. In both instances, there was not a clear enantiodiscrimination in the threading process with the studied host/guest systems. Possible rationales are given to explain the scarce reciprocal influence between the guest and host chiral centers.

Keywords: calixarenes; threading; chirality; barfate salts; pseudorotaxane; chiral axles; chiral wheels.

---

1. Introduction

Over the past two decades, there has been a great scientific interest for the synthesis of mechanomolecules [1], such as rotaxanes and catenanes [2–4]. Mechanomolecules have found many applications in very different topics, such as nanoelectronics [5–7], molecular machines [2,3,8–10], and catalysis [11–13]. The mechanical bond [1], as a building element of rotaxane and catenane architectures, is usually obtained by template-directed synthesis [14] based on the threading of a linear molecule (axle) through a macrocyclic component (wheel). The most convenient synthetic routes are the threading-followed-by-stoppering, the clipping, and the slipping methods [1,14] that exploit weak intermolecular interactions (hydrogen bond, halogen bond, π-stacking, cation-π, or metal coordination) to assemble the molecular components.

The peculiar stereochemical features of mechanomolecules are one of the most fascinating aspects of these architectures. In particular, the chirality within such supramolecular architectures [15,16] is of special interest, because it is relevant for enantioselective recognition and sensing [17], asymmetric catalysis [13,18–20], and unidirectional molecular motors [21–23]. The simplest way to obtain chiral rotaxanes and catenanes is to introduce a classical stereogenic element in one their components to give a chiral axle and/or a chiral wheel (Figure 1A) [16,24,25]. Another more sophisticated approach is given by "mechanical chirality" [26–30] (Figure 1B) in which chirality arises from the combination of achiral "directional" elements spatially restricted by the mechanical bond. Of course, the first approach is more convenient from the synthetic point of view, because a difficult resolution step, on an appropriate preparation scale, can be avoided by resorting to suitable enantiopure moieties available from the chiral pool.

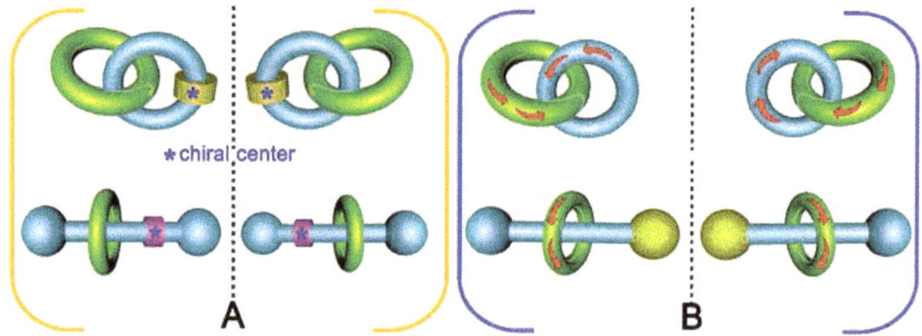

Figure 1. Chiral catenanes and rotaxanes. Chirality arising from (**A**) classical chiral elements and (**B**) "mechanical chirality" with achiral "directional" elements.

Over the past ten years, our group has shown that the threading of scarcely preorganized calix[6]arene macrocycles (e.g., **1**, Scheme 1) with dialkylammonium axles (e.g., **2$^+$**) occurs in CDCl$_3$ when the ammonium linear system is coupled with the weakly coordinating barfate anion tetrakis[3,5-bis(tri-fluoromethyl)phenyl]borate [B(Ar^F)$_4$]$^-$ "superweak anion" [31–35]. This method was defined, in short, as the "superweak anion approach" [31–34]. Interestingly, we also observed that the threading of alkylbenzylammonium cations may result in two different stereoisomeric pseudorotaxanes [32], in which the alkyl or the benzyl moiety is hosted inside the calixarene cavity. These have been termed as "*endo*-alkyl" or "*endo*-benzyl" isomers, respectively (Scheme 1). The *endo*-alkyl "orientational mechanostereoisomer" is usually preferred [32], thus leading to the definition of the so-called "*endo*-alkyl rule": *threading of a directional alkylbenzylammonium axle through a hexaalkoxycalix[6]arene occurs with an* endo-*alkyl preference* [36].

A full exploitation of the "*endo*-alkyl rule" within the "superweak anion approach" has led to several interesting examples of stereoprogrammed calixarene-based pseudorotaxanes, to their integrative self-sorting [36], and to the synthesis of the corresponding rotaxane and catenane mechanomolecules [37,38].

The threading of a tertiary ammonium axle [39] has led to the first examples of dissymmetric calixarene-based pseudo[2]rotaxanes (Scheme 1, lower right) obtained by combination of two achiral components. In detail, this peculiar chirality was generated by the structural directionality of calix[6]arene macrocycle which led to differentiate the two benzyl units of the prochiral tertiary ammonium axle [39]. Interestingly, this represents an example of a dissymmetric pseudorotaxane in which an atomic stereogenic center is generated by the threading of an axle with a directional wheel.

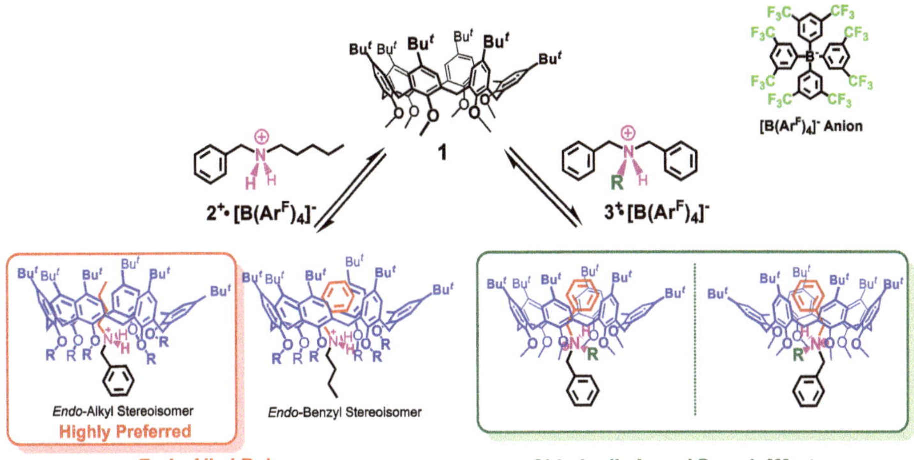

Scheme 1. Structures of calix[6]arene host **1**, alkylammonium axles **2⁺** and **3⁺**, BArF anion, examples of classical and chiral pseudo [2] rotaxane.

On the basis of the above background and considering the great potential of chiral calixarene-based MIMs in different fields such as enantioselective recognition, sensing, and catalysis, we decided to synthesize chiral calix-wheels by introducing chiral pendant groups on the calix[6]arene scaffold (**4–7**, Chart 1) and to study their threading with chiral secondary dialkylammonium cations (**9⁺–10⁺**) by exploiting the superweak anion approach.

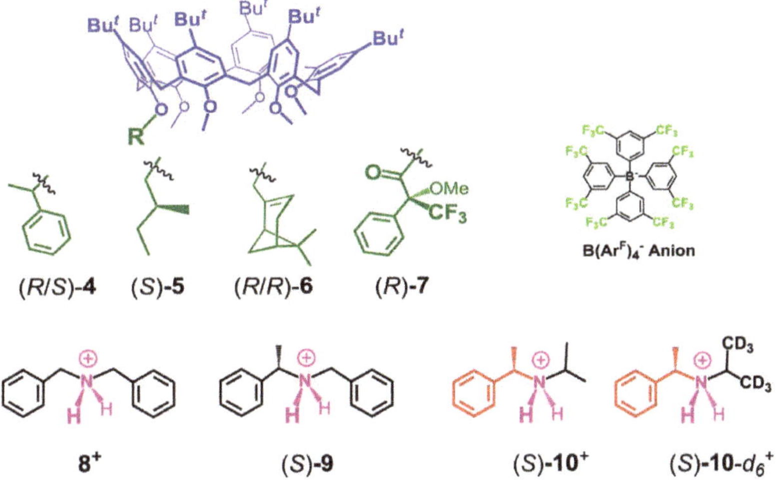

Chart 1. Structures of chiral calix[6]arene hosts **4–7**, alkylammonium axles **8⁺**, **9⁺**, **10⁺**, **10-d_6⁺**, and [B(Ar^F)_4]⁻ anion.

2. Results and Discussion

2.1. Synthesis of Chiral Calixarenes

Pentamethoxy-calix[6]arene-mono-ol **14** [40–42] was the ideal precursor for the synthesis of chiral calixarene derivatives (*R/S*)-**4**, (*S*)-**5**, (*R*)-**6**, and (*R*)-**7** (Scheme 2) obtainable by the introduction (alkylation or esterification) of a chiral pendant moiety on its hydroxyl group located at the lower rim.

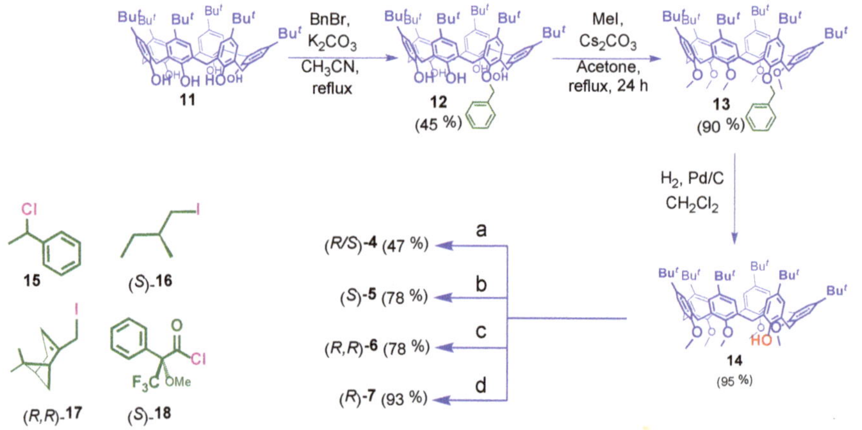

Scheme 2. Synthesis of chiral calixarene derivatives **4**–**7**. (**a**) **15**, NaH, *dry*-DMF 80 °C, 12 h. (**b**) **16**, NaH, *dry*-DMF, 80 °C, 12 h. (**c**) **17**, NaH, *dry*-DMF, 80 °C, 12 h. (**d**) **18**, DMAP, *dry*-NEt$_3$, *dry*-DMF, 70 °C, 12 h.

Derivative **14** was synthesized by following the reaction sequence reported in Scheme 2 [40–42]. In particular, native *p-tert*-butylcalix[6]arene **11** was monobenzylated using benzyl bromide and K$_2$CO$_3$ to give calix[6]arene **12** in 45% yield [42]. Derivative **12** was exhaustively methylated by treatment with MeI in acetone using Cs$_2$CO$_3$ as a base to give calix[6]arene **13** in 90% yield [42]. The removal of the benzyl group was easily accomplished by hydrogenolysis to give mono-ol **14** [42].

The treatment of **14** with α-methylbenzylbromide in the presence of NaH leads to the racemic compound (*R/S*)-**4** in 47% yield.

The ^{1}H NMR spectrum (400 MHz, CDCl$_3$) of (*R/S*)-**4** at 298 K (Figure 2b) shows six ^{t}Bu singlets at 0.89, 0.96, 0.98, 1.27, 1.28, and 1.32 ppm (1:1:1:1:1:1 ratio) while the ArCH$_2$Ar groups gave rise to six sharp AX systems at 4.62/3.59 (*J* = 13.4 Hz), 4.18/3.55 (*J* = 14.0 Hz), 4.24/3.66 (*J* = 14.0 Hz), 4.11/3.75 (*J* = 13.6 Hz), 4.09/3.79 (*J* = 13.9 Hz), 4.10/2.97 (*J* = 13.9 Hz). The C*H*(CH$_3$) group shows a resonance at 5.02 ppm (quadruplet, *J* = 6.1 Hz, 1H), while the methyl group resonates at 1.66 ppm as a doublet (*J* = 6.1 Hz). In addition, aromatic signals attributable to the benzyl group at the lower rim resonate at 6.75–7.53 ppm.

Chiral calixarenes in enantiopure form (*S*)-**5**, (*R,R*)-**6**, and (*R*)-**7** (Scheme 2) were obtained by a similar functionalization of the OH group of calix[6]arene-mono-ol **14** with enantiopure chiral reagents. All four compounds were fully characterized by NMR and MS (Figure 2e) spectroscopy (see the experimental section and SI).

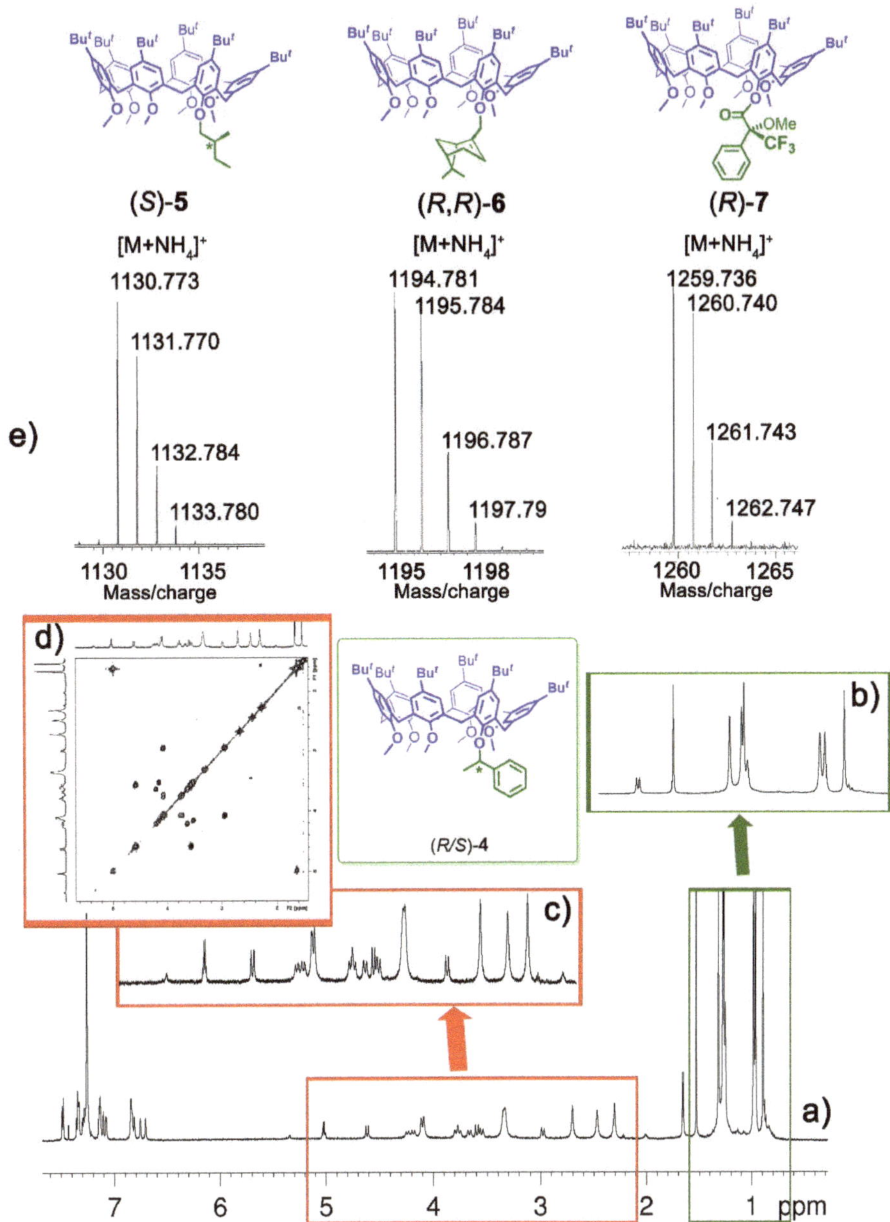

Figure 2. ^{1}H NMR spectrum (CDCl$_3$, 600 MHz) of derivative (*R/S*)-**4** at 298 K (**a**). Different portions of its ^{1}H NMR (**b**,**c**) and COSY−45 (**d**) spectrum. Partial ESI mass spectra of compounds (*S*)-**5**, (*R*,*R*)-**6**, and (*R*)-**7** (**e**).

2.2. Syntheses of Chiral Axles

With the aim to observe an enantiodiscrimination effect in the formation of pseudo[2]rotaxane, the new linear chiral derivatives (*S*)-**9**$^+$·[B(Ar^F)$_4$]$^-$, (*S*)-**10**$^+$·[B(Ar^F)$_4$]$^-$, and (*S*)-**10**-$d_6$$^+$·[B(Ar^F)$_4$]$^-$

were designed. The synthesis of (*S*)-**9$^+$**·[B(Ar^F)$_4$]$^-$ is outlined in Scheme 3A. A mixture of (*S*)-α-methyl-benzylamine (*S*)-**15** and benzaldehyde **19** in chloroform was stirred for 2 h at room temperature and then directly reduced with NaBH$_4$. The secondary amine was treated with HCl/MeOH to give chloride salt **21**. Finally, a counterion exchange with NaBAr^F led to (*S*)-**9$^+$**·[B(Ar^F)$_4$]$^-$ salt. In a similar way, (*S*)-**10$^+$**·[B(Ar^F)$_4$]$^-$ was synthesized from (*S*)-α-methyl-benzylamine (*S*)-**15** and acetone **22** at reflux for 18 h and then reduced with NaBH$_4$. HCl (37%) was then used to obtain the chloride salt which was exchanged with Na[B(Ar^F)$_4$]$^-$ (Scheme 3B).

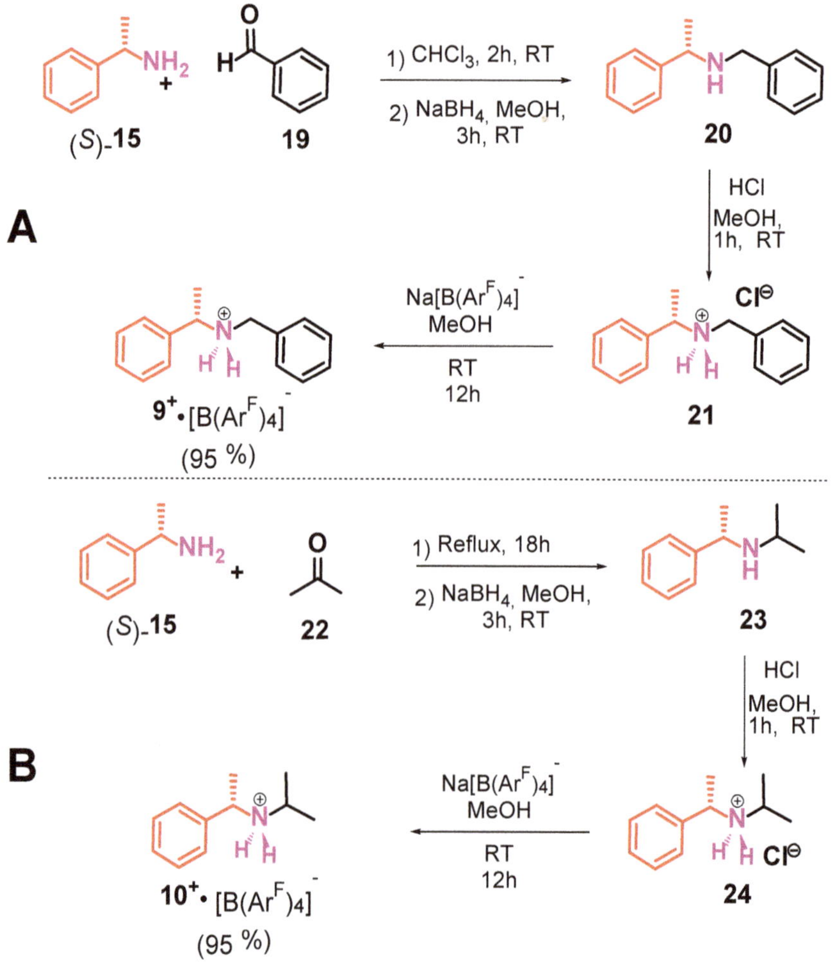

Scheme 3. Synthesis of enantiopure derivatives (*S*)-**9$^+$**·[B(Ar^F)4]$^-$ (**A**) and (*S*)-**10$^+$**·[B(Ar^F)4]$^-$ (**B**).

The synthesis of (*S*)-**10**-$d_6$$^+$·B(Ar^F)$_4$$^-$ (Scheme 4) was obtained from the amine-acetone coupling, by using Ti(O*i*Pr)$_4$ as a catalyst (see experimental section for further details). The acidification with HCl (37%) gave the ammonium chloride and the usual salt exchange gave the required axle. In a similar way, the corresponding (*R*)-axles were also made in order to have access to the pseudoracemates.

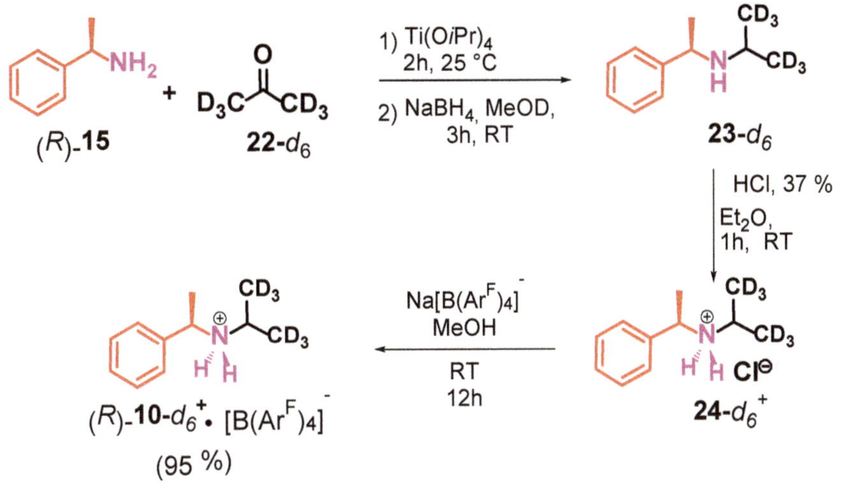

Scheme 4. Synthesis of enantiopure derivative (*R*)-**10**-d_6^+·[B(Ar^F)4]$^-$.

2.3. NMR Threading Studies of Chiral Calixarene (R/S)-4

2.3.1. Threading of Racemic Calix[6]arene-wheel (*R/S*)-**4** with Dibenzylammonium Axles **8**$^+$ and **9**$^+$

Through-the-annulus threading with linear axles was initially studied by using NMR spectroscopy. As an initial study, we decided to investigate the complexing ability of racemic (*R/S*)-**4** towards achiral dibenzylammonium cation **8**$^+$ (Scheme 5).

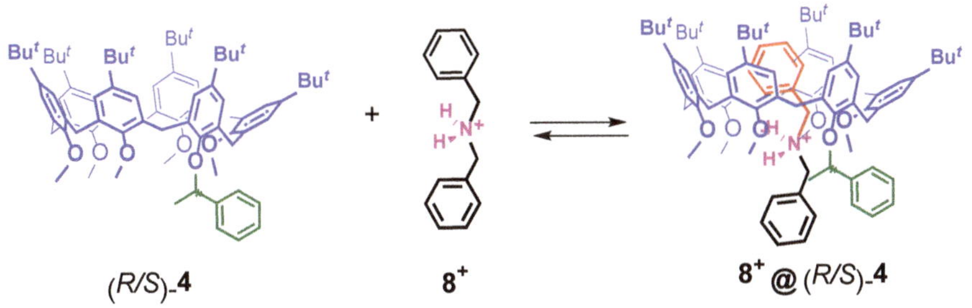

Scheme 5. Formation of the **8**$^+$@(*R/S*)-**4** pseudo[2]rotaxanes.

The ^{1}H NMR spectrum of a 1:1 mixture of (*R/S*)-**4** and **8**$^+$·B(Ar^F)$_4$$^-$ salt in CDCl$_3$ at 25 °C (Figure 3) showed, immediately after mixing, important changes that confirmed the formation of the pseudorotaxane [32] **8**$^+$@(*R/S*)-**4** (Scheme 5) by a threading equilibrium slow on the NMR timescale. In details, shielded aromatic signals attributable to the benzylic unit (*ortho*-BnH, *meta*-BnH, and *para*-BnH) hosted inside the calixarene cavity, were detected respectively at 4.55, 5.29, and 5.95 ppm, while the benzylic-CH$_2$ was found at 5.11 ppm (Figure 3b,c) [32].

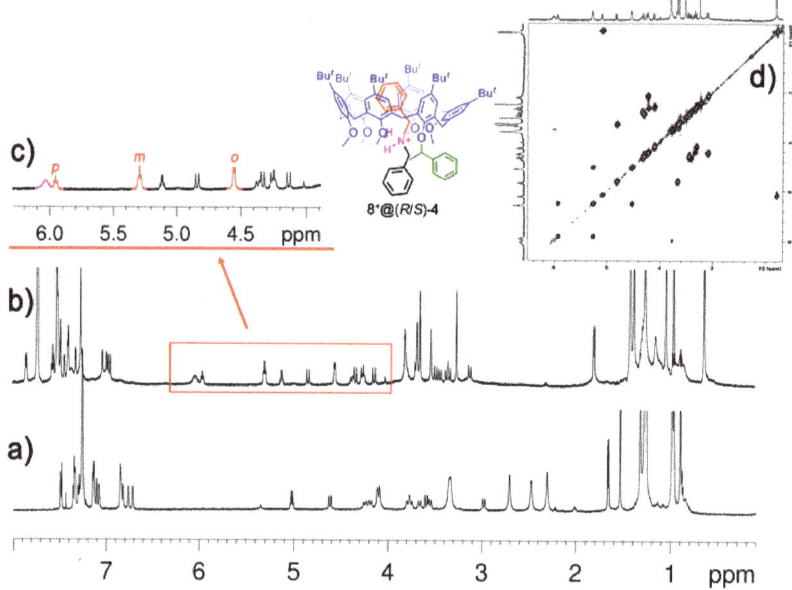

Figure 3. Significant portions of the ^{1}H NMR spectra (600 MHz, CDCl$_3$, 298 K) of: (**a**) (*R/S*)-**4**; (**b**) 1:1 mixture (3 mM) of (*R/S*)-**4** and **8$^+$**·B(Ar^F)$_4^-$ after mixing. Insets: (**c**) expansion of the ^{1}H NMR and (**d**) COSY portions of the mixture; the BnH protons of the shielded *endo*-cavity benzylammonium moiety are marked in red.

The spectrum remained unchanged after 12 h at 55 °C, thus demonstrating that the system had reached the equilibrium immediately after the mixing. A 2D COSY spectrum (Figure 3d) allowed the assignment of all the relevant resonances of **8$^+$**@(*R/S*)-**4** racemic pseudo[2]rotaxane.

DFT calculations at the B97D3/SVP/SVPFIT level of theory were performed on pseudorotaxane **8$^+$**@(*R*)-**4**. A close inspection of the optimized pseudo[2]rotaxane structure (Figure 4) reveals the presence of typical H-bonds between the ammonium group of the dibenzylic **8$^+$** axle and the oxygen atoms at the calix[6]arene lower rim (average: N···O distance of 3.6 Å and N-H···O angle of 158.7°). In addition, C-H···π interactions between the H-atoms of the axle and the calixarene aromatic walls, (C-H···π^{centroid} distance of 2.9 Å and C-H···π^{centroid} angle of 131.1°) [43] play a crucial role in the stabilization of the supramolecular structure.

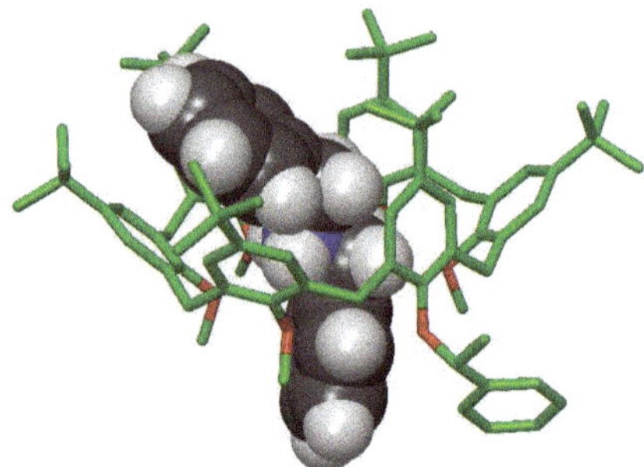

Figure 4. DFT-optimized structures of **8$^+$**@(*R*)-**4** pseudo[2]rotaxane at the B97D3/SVP/SVPFIT level of theory.

With these results in hand we turned our attention to the study of the threading of calixarene (*R/S*)-**4** with the enantiopure dibenzylammonium axle (*S*)-**9**$^+$·B(Ar^F)$_4$$^-$ (Scheme 6). In this specific case, the formation of a pseudo[2]rotaxane could lead to four different stereoisomers: two "*endo*-chiral" stereoadducts with the stereogenic center of cation **9**$^+$ inside the calix-cavity for both calixarene enantiomers (being the calixarene in racemic form) and two "*exo*-chiral" ones with the stereogenic center outside the cavity (Scheme 6). The ^{1}H NMR spectrum of an equimolar (3 mM) solution of (*R/S*)-**4** and (*S*)-**9**$^+$·B(Ar^F)$_4$$^-$ (Figures 5 and 6) showed the formation of only "*endo*-chiral" stereoisomers. In fact, we can observe a diagnostic doublet at negative values (in red in Figure 5b) attributable to the methyl group of the α-methylbenzyl moiety shielded by the aromatic walls.

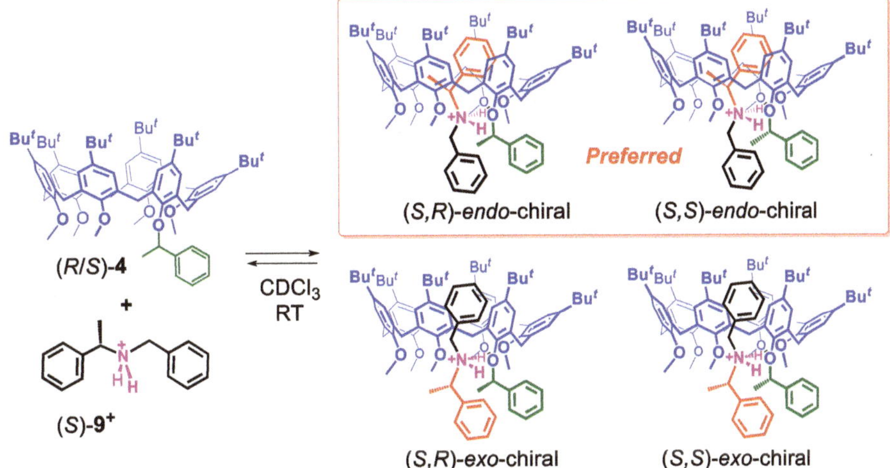

Scheme 6. Stereoisomeric pseudo[2]rotaxanes obtainable by threading of racemic (*R/S*)-**4** with enantiopure axle (*S*)-**9**$^+$.

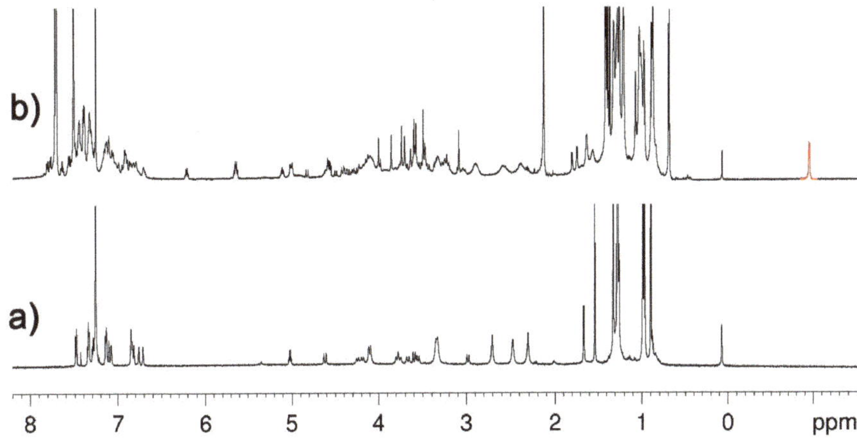

Figure 5. ^{1}H NMR spectra (600 MHz, CDCl$_3$, 298 K) of: (**a**) (*R/S*)-**4**. (**b**) 1:1 mixture (3 mM) of (*R/S*)-**4** and (*S*)-**9**$^+$·B(Ar^F)$_4$$^-$. Marked in red the doublet attributable to the methyl group of the axle shielded inside the calixarene cavity.

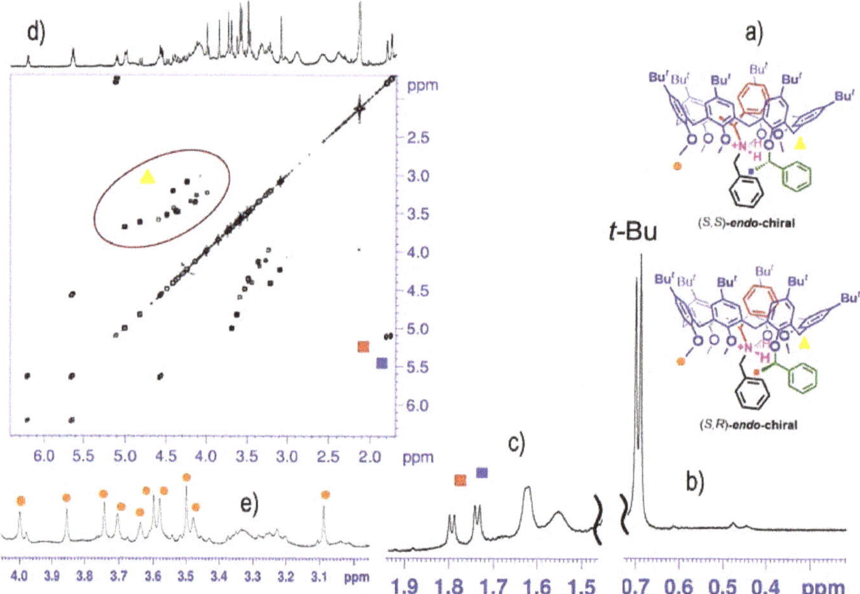

Figure 6. (**a**) Chemical drawing of the two possible *endo*-chiral (*S*)-**9**⁺@(*R*)-**4** and (*S*)-**9**⁺@(*S*)-**4** pseudorotaxanes. (**b,c,e**) Expansions of the ^{1}H NMR spectrum (600 MHz, CDCl$_3$, 298 K) showing the signals of the two (*S*)-**9**⁺@(*R*)-**4** and (*S*)-**9**⁺@(*S*)-**4** diastereoisomers, in equal ratio. (**d**) Expansion of the COSY spectrum of the 1:1 mixture (3 mM) of (*R/S*)-**4** and (*S*)-**9**⁺·B(Ar^F)$_4^-$, marked the 12 ^2J-correlations attributable to the methylene groups of the two stereoisomers.

In addition, shielded *endo*-cavity benzyl resonances were present in the 4.8–6.5 ppm range (Figure 5b). A 2D-COSY spectrum (Figure 6d and Supplementary Materials) showed all the basic correlations necessary to confirm the formation of the "*endo*-chiral" (*S*)-**9**⁺@(*R*)-**4** and (*S*)-**9**⁺@(*S*)-**4** pseudo[2]rotaxane diastereoisomers.

Note that this result represents an important extension of the above mentioned "*endo*-alkyl rule", because it demonstrates that the calix-cavity prefers to host the α-methyl-benzyl moiety with respect to the simple unsubstituted benzyl group. Evidently, the α-methyl group gives rise to additional stabilizing interactions with the aromatic walls which direct the preferential formation of the "*endo*-chiral" diastereoisomers - an information of pivotal importance in the design of new calixarene-based MIMs.

On this basis, a new stereoselectivity rule (named as "*endo*-α-methyl-benzyl rule") for the threading of calixarene macrocycles was defined: *threading of a hexaalkoxycalix[6]arene with a directional (α-methyl-benzyl)benzylammonium axle occurs with an* endo-α-methyl-benzyl *preference*.

DFT calculations (vide infra) confirmed that the α-methyl group gives rise to additional stabilizing C-H···π interactions with the calixarene aromatic walls which leads to the preferential formation of the "*endo*-chiral" diastereoisomer.

Furthermore, a quantitative evaluation about the number and integrals of the signals (Figure 6) in the methoxy (Figure 6e) and *tert*-butyl (Figure 6b) regions of the spectrum, indicated that the two (*S*)-**9**⁺@(*R*)-**4** and (*S*)-**9**⁺@(*S*)-**4** pseudo[2]rotaxane diastereoisomers were formed in equal ratio. This means that no enantiodiscrimination was observed in their formation probably because the two chiral centers were too remote from each other to exert a mutual influence. In accord with this experimental observation, DFT calculations at the B97D3/SVP/SVPFIT level of theory indicated a very slight energy difference between the two (*S*)-**9**$^+$@(*R*)-**4** and (*S*)-**9**$^+$@(*S*)-**4** *endo*-chiral pseudorotaxane stereoisomers (Figure 7).

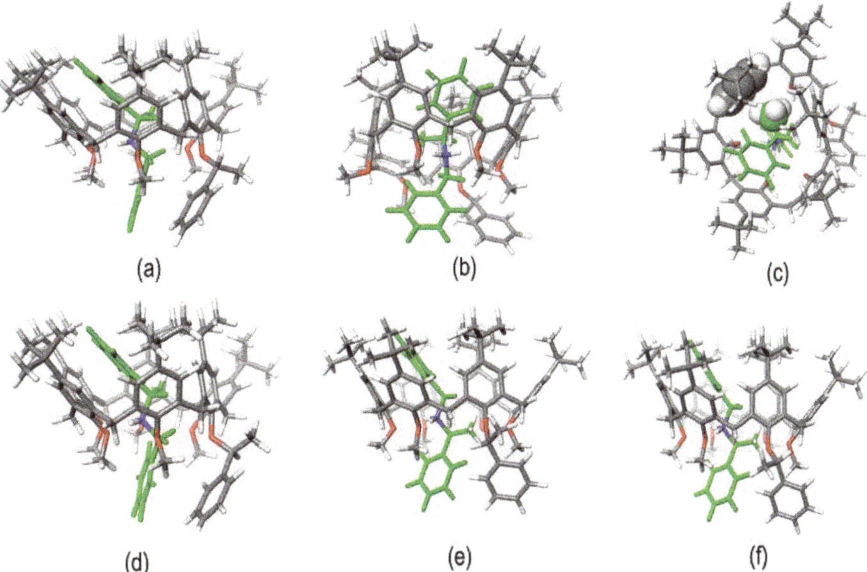

Figure 7. DFT optimized structures at the B97D3/SVP/SVPFIT level of theory of the stereoisomeric *exo*-chiral and *endo*-chiral **9**$^+$@**4** pseudorotaxanes. (**a**) *endo*-chiral (*S*)-**9**$^+$@(*S*)-**4** pseudorotaxane. (**b**) Particular of the $^+$NH^{axle}...OR^{calix} H-bonding interactions in (*S*)-**9**$^+$@(*S*)-**4**. (**c**) Particular of the C-H···π interactions in (*S*)-**9**$^+$@(*S*)-**4**. (**d**) *endo*-chiral (*S*)-**9**$^+$@(*R*)-**4** pseudorotaxane. (**e**) *exo*-chiral (*S*)-**9**$^+$@(*R*)-**4** pseudorotaxane. (**f**) *exo*-chiral (*S*)-**9**$^+$@(*S*)-**4** pseudorotaxane.

Single point energy DFT calculations (Figure 7), evidenced a significant energy difference between the *endo*-chiral and *exo*-chiral stereoisomers in Scheme 6 ($\Delta E = E_{exo\text{-}chiral} - E_{endo\text{-}chiral} > 3.0$ kcal/mol), that corroborated the experimental results. A close inspection of the DFT optimized structure of (*S*)-**9**$^+$@(*R*)-**4** *endo*-chiral pseudorotaxane (Figure 7a,b) evidenced the presence of two classical $^+$NH^{axle}...OR^{calix} H-bonding interactions with a N···O mean distance of 2.85 Å and a N-H···O mean angle of 160° (SI). In addition, stabilizing C-H···π interactions (Figure 7c and SI) were found for the (*S*)-**9**$^+$@(*S*)-**4** *endo*-chiral pseudo[2]rotaxane between the α-methyl group of the axle of (*S*)-**9**$^+$ and the calixarene aromatic walls with a C-H···$\pi^{centroid}$ distance of 2.49 Å (SI).

The DFT optimized structure of (*S*)-**9**$^+$@(*R*)-**4** *exo*-chiral pseudorotaxane, showed that the α-methyl group of axle (*S*)-**9**$^+$ lies on the same plane of the oxygen-calixarene atoms (Figure 7e). Thus, because of the steric hindrance between the methyl group of the axle and the calixarene OR groups, a slight distortion of the geometrical H-bonding parameters was found. In particular, the $^+$NH^{axle}...OR^{calix} interactions showed a longer N···O mean distance of 3.02 Å and a smaller N-H···O mean angle of 145° (SI). These values are indicative of weaker H-bonding interactions between the axle (*S*)-**9**$^+$ and the wheel (*R*)-**4** in (*S*)-**9**$^+$@(*R*)-**4** *exo*-chiral pseudorotaxane.

2.3.2. Threading of Racemic Calix[6]arene-wheel (*R/S*)-**4** with Isopropylbenzylammonium Axle (*S*)-**10**$^+$

We envisioned that to improve the enantiodiscrimination, the distance between the two chiral centers should be reduced. This can be done by exploiting the "*endo*-alkyl rule" with an alkylbenzylammonium axle bearing the chiral center on the benzyl side. For this reason, we designed and synthesized derivative (*S*)-**10**$^+$·[B(Ar^F)$_4$]$^-$ (see above).

Again, the combination of (*S*)-**10**⁺·[B(Ar^F)₄][−] and racemic wheel (*R/S*)-**4** may (Scheme 7) could result in the formation of four stereoisomers: two "*endo*-chiral" and two "*exo*-chiral" stereoadducts with the isopropyl moiety outside or inside the calix cavity, respectively.

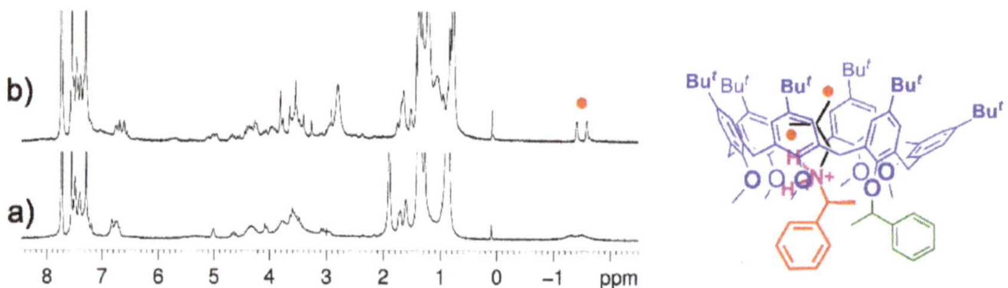

Scheme 7. Formation of the two "*exo*-chiral" (*S*)-**10**⁺@(*R/S*)-**4** pseudo[2]rotaxane diastereoisomers.

The ¹H NMR spectrum of a 1:1 mixture (3 mM) of (*R/S*)-**4** with (*S*)-**10**⁺·[B(Ar^F)₄][−] salt in CDCl₃ at 25 °C (Figure 8) showed, immediately after mixing, the sharpening of all signals and the peculiar presence two doublets for the diastereotopic methyl protons of the *endo*-isopropyl group. This means that the "*exo*-chiral" orientation is preferred leading to (*S*)-**10**⁺@(*R/S*)-**4** pseudo[2]rotaxane diastereoisomers in accordance with the *endo*-alkyl rule.

Figure 8. ¹H NMR spectra (CDCl₃, 400 MHz, 298 K) of equimolar solution (3 mM) of (*R/S*)-**4** and (*S*)-**10**⁺·[B(Ar^F)₄][−] at: (**a**) 298 K and (**b**) 243 K.

In order to establish if any enantiodiscrimination is occurring in the formation of (*S*)-**10**⁺@(*R/S*)-**4** stereo-adducts we carefully examined different regions of its ¹H NMR spectrum, but no conclusive results were obtained because of the large number of partially overlapping signals. 2D NMR techniques were also used with no conclusive results.

2.4. MS Experiments of the Threading of Chiral Calixarene (S)-5, (R)-6, and (R)-7

To obtain more straightforward insight, enantiomer-labeled (EL) mass spectrometry was used. A fundamental feature of this method is the use of a pseudoracemate, i.e., a 1:1 mixture of one deuterated and the opposite, non-labeled enantiomer. The two possible diastereomeric pseudorotaxanes would then have different masses and are, thus, distinguishable. A 1/1 mixture of labelled and unlabelled guest enantiomers was mixed with 0.5 equivalent of the target chiral host. It was considered the competitive equilibrium system through Equations (1) and (2).

$$H + G_S^+ \underset{\rightleftharpoons}{\overset{K_S}{}} (HG_S)^+ \tag{1}$$

$$H + G_R^+ \underset{\rightleftharpoons}{\overset{K_R}{}} (HG_{R-d6})^+ \tag{2}$$

Therefore, the peak intensity ratio, $I_S/I_{R\text{-}d6} = I[(HG_S)^+]/I[(HG_{R\text{-}d6})^+]$, of the diastereomeric host-guest complex ions, was expected to become a measure of the enantio-discrimination ability of the host toward the two enantiomers of the chiral guest.

- $I_S/I_{R\text{-}d6} > 1$ means that a given chiral host binds more strongly the (*S*)-enantiomer (the larger the $I_S/I_{R\text{-}d6}$ ratio value, and higher the degree of chiral discrimination by the host).
- $I_S/I_{R\text{-}d6} < 1$ means that a given chiral host binds more strongly the (*R*)-labelled guest.
- $I_S/I_{R\text{-}d6} = 1.0 \pm 0.1$ means that a given chiral host cannot differentiate the chirality of a given guest.

2.4.1. Concentration Effect

The solution concentration can affect the ionization process and, hence, the intensity ratio $I[(HG_S)^+]/I[(HG_{R\text{-}d6})^+]$. Therefore, the first step was to choose the best concentration value to use. It was seen that the peak intensities were quite low up to a 300 µM solution and the intensity ratio did not change increasing the concentration over 300 µM. Therefore, 300 µM solutions were used in determining isotopic and chiral recognition effects.

2.4.2. Isotopic Effects

Of course, isotope effects may also operate in addition to any stereochemical effect. However, this can easily be tested by a control experiment with either the oppositely labeled pseudoracemate of axles or by repeating the same experiment with the other enantiomer of the calixarene. To test, whether isotope effects play a significant role, a control experiment in Figure 9 with a 1:1 mixture of (*S*)-**5** and labeled and unlabeled (*R*)-**10**$^+$·[B(Ar^F)$_4$]$^-$ was done. All intensity ratios are close to 1.0 (Figure 9) so that an isotope effect can be ruled out.

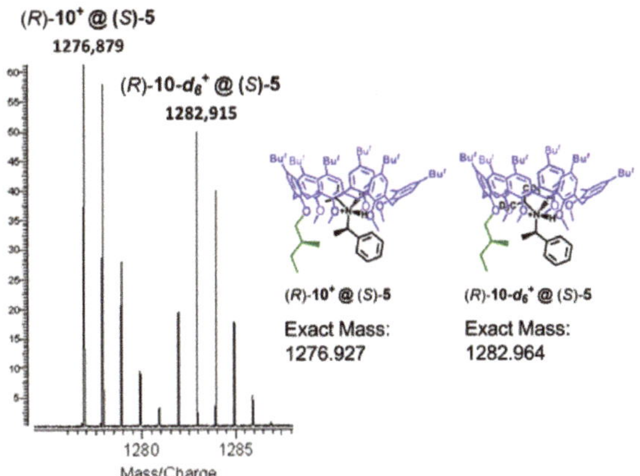

Figure 9. Significant portion of the mass spectrum of a 1:2:2 mixture (CH$_2$Cl$_2$, 300 µM) of derivatives (*S*)-**5**, (*R*)-**10**$^+$·[B(Ar^F)$_4$]$^-$, and (*R*)-**10**-$d_6$$^+$·[B(Ar^F)$_4$]$^-$, respectively.

The $[I_R/I_{R\text{-}d6}]$ values for the MS spectra of compounds (*S*)-**5**, (*R*)-**6**, and (*R*)-**7** (Figure 9 and Figures S21–S23) showed a slight isotopic effect (see experimental section for the exact values and the calculation procedures). According to the literature, the directionality of isotope effects (IEs) is difficult to predict from system to system. The literature supports observation of deuterium IEs from both solution and gas phase [44]. Therefore, the observed effects in our systems might be due to decreased van der Waals interactions between the guest's deuterated moiety and the host; or a preferential ionization of one diastereomeric complex over the other one; or a different gas-phase behavior between the deuterated and non-deuterated guests.

2.4.3. Chiral Discrimination

In order to determine a chiral discrimination effect of a 1:1 mixture of a pair of labelled (R)-enantiomer, (R)-**10**-$d_6{}^+$·[B(Ar^F)$_4$]$^-$, and unlabelled (S)-enantiomer guest, (S)-**10**$^+$·[B(Ar^F)$_4$]$^-$, was used (Figure 10) with 0.5 equivalent of the corresponding host and the $I_S/I_{R\text{-}d6}$ ratio value was measured by inspection of the mass spectrum. The resulting mass spectra for the investigation of the isotopic effect are reported below (Figure 10) and in SI (Figures S24–S26).

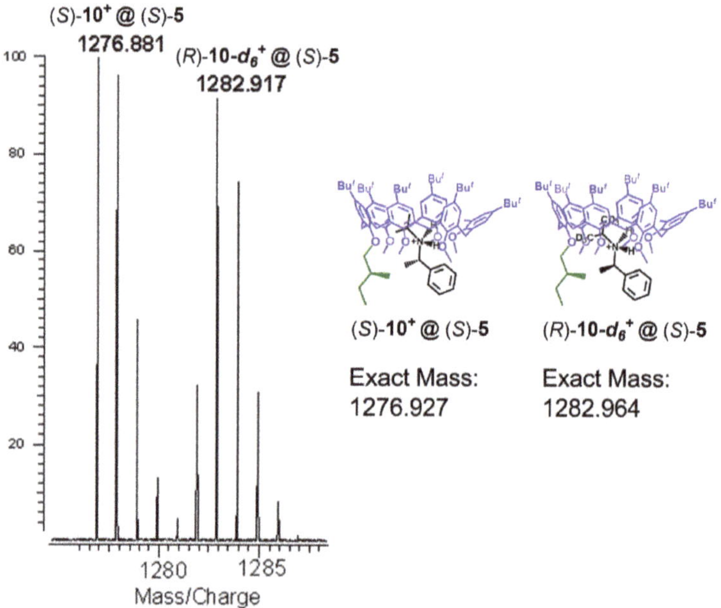

Figure 10. Significant portion of the mass spectrum of a 1:2:2 mixture (CH$_2$Cl$_2$, 300 µM) of derivatives (S)-**5**, (S)-**10**$^+$·[B(Ar^F)$_4$]$^-$, and (R)-**10**-$d_6{}^+$·[B(Ar^F)$_4$]$^-$.

For the hosts (S)-**5**, (R)-**6**, and (R)-**7** the $I_S/I_{R\text{-}d6}$ were slightly different from the unit. However, their values were almost identical to those regarding the isotopic effect and, therefore, there was not a clear enantiodiscrimination with the hosts studied (see experimental section for the exact values and the calculation procedures). Hence, the mass spectrometric experiments indicate that no significant differentiation of the two diastereomeric complexes occurs.

2.5. Possible Rationales

The above results clearly indicated that, independently by the used technique, the enantiodiscrimination ability of the studied hosts are negligible. This unexpected result could be explained by the scarce reciprocal influence between the guest and host chiral centers. In particular, in the case of the "*endo*-chiral" complexes above defined, it is clear that from the point of view of the guest the cavity appears as "too-symmetric" due to the high conformational mobility of the aromatic walls, which do not give any real steric restriction towards the guest geometry. In analogy with chiral crown-ethers [45–47], this is a lack of "central-cavity discrimination".

In the case of the "*exo*-chiral" complexes the two host/guest chiral centers appear to be close enough to affect each other. The observed lack of influence could be then ascribed to their high reciprocal freedom of movements associated to their scarce supramolecular interactions, which is a lack of "lateral discrimination" [45–47]. On the basis of these considerations, we can expect that all the factors able to rigidify the cavity in a fixed asymmetric geometry should improve the "central-cavity discrimination", while the "lateral discrimination" should be improved by the presence of stereoelectronic complementary interacting groups.

3. Conclusions

In this work we report an initial study on the influence of chirality in calixarene threading by exploiting the "superweak anion approach" to give chiral pseudo[2]rotaxanes bearing a classical stereogenic element in their axle and/or wheel components. In a first instance, "*endo*-chiral" pseudo[2]rotaxane stereoadducts, having the stereogenic center of a cationic axle inside the calix-cavity of a chiral calixarene, were studied. Successively, the "*exo*-chiral" ones, with the stereogenic center outside the calix-cavity, were also considered. Both "*endo*-chiral" and "*exo*-chiral" pseudo[2]rotaxane stereoadducts were preferentially formed with specifically designed chiral axles by a fine exploitation of the so-called "*endo*-alkyl rule" and a newly defined "*endo*-α-methyl-benzyl rule" [*threading of a hexaalkoxycalix[6]arene with a directional (α-methyl-benzyl)benzylammonium axle occurs with an endo-α-methyl-benzyl preference*]. In both instances, the use of 1D and 2D NMR techniques, to establish if any enantiodiscrimination is occurring, led to no conclusive results because of the large number of partially overlapping signals. The pseudorotaxanes were then studied in the gas-phase by means of mass spectrometry, using the enantiomer-labeled guest method. This required the synthesis of enantiopure host/guest couples, including a pair protiated/deuterated chiral axles. Also in this instance, there was not a clear enantiodiscrimination in the threading process with the studied hosts. Possible rationales were given to explain the scarce reciprocal influence between the guest and host chiral centers, which can be useful for future studies in the chiral threading of classical calixarenes and new macrocyclic arenes.

4. Experimental Section

ORD spectra were recorded on a JASCO J600 (Tokyo, Japan) spectropolarimeter at room temperature, in acetonitrile as solvent. During the measurement, the instrument was thoroughly purged with nitrogen. Mass spectra were recorded with a Finnigan Mat 711 (Waltham, MA, USA) (EI, 80 eV, 8 kV), an Agilent 6210 ESI-TOF, and an Agilent QFT-7 FTICR (Santa Clara, CA, USA) mass spectrometer with Micromass Z-Spray ESI source, sample cone 25V, HV 2500 V. Flash chromatography was performed on Merck (Darmstadt, Germany) silica gel (60, 40–63 μm). All chemicals were reagent grade and were used without further purification. Anhydrous solvents were purchased from Aldrich (Darmstadt, Germany). When necessary compounds were dried in vacuo over $CaCl_2$. Reaction temperatures were measured externally. Reactions were monitored by TLC on Merck silica gel plates (0.25 mm) and visualized by UV light, or by spraying with H_2SO_4-$Ce(SO_4)_2$ or phosphomolybdic acid. 1D NMR spectra were recorded on a Bruker (Billerica, MA, USA) Avance-400 spectrometer [400 (1H) and 100 MHz (13C)], Bruker Avance-300 spectrometer [300 (1H) and 75 MHz (13C)] and Bruker Avance-250 spectrometer [250 (1H) and 63 MHz (13C)]; chemical shifts are reported relative to the residual solvent peak ($CHCl_3$: δ 7.26, $CDCl_3$: δ 77.23; CD_3OH: δ 3.31, CD_3OD: δ 49.0). Derivatives **11**, **12**, and **13** [18,19] were synthesized according to literature procedures.

Synthesis of derivative (*R/S*)-**4**. NaH (0.072 g, 3.0 mmol) was added to a solution of derivative **14** (0.31 g, 0.30 mmol) in dry DMF (15 mL) and stirred for 1 h at 0 °C. The mixture was allowed to cool at 25 °C, then benzylbromide (0.17 g, 1.00 mmol) was added. The resulting mixture was stirred at 80 °C for 12 h under nitrogen atmosphere, then the solvent was removed under reduced pressure and the mixture was partitioned between CH_2Cl_2 and H_2O. The organic layer was washed with 1N HCl (30 mL), brine (30 mL), and dried over Na_2SO_4. The crude product was purified by column chromatography (SiO_2; CH_2Cl_2) to give (0.16 g, 0.14 mmol) of (*R/S*)-**4** as a white solid. Mp: 190–193 °C dec. **ESI(+) MS:** *m/z* =1147.81 (MH$^+$); ^{1}**H NMR** (600 MHz, $CDCl_3$, 298 K): δ 7.47–6.71 (overlapped, 17H, ArH_{calix} + ArH$_{Bn}$), 5.02 (q, *J* = 6.1, 1H, C*H*), 4.62 and 3.59 (AX, *J* = 13.4 Hz, 2H, ArCH_2Ar), 4.18 and 3.55 (AX, *J* = 14.0 Hz, ArCH_2Ar, 2H), 4.24 and 3.66 (AX, *J* = 14.0 Hz, ArCH_2Ar, 2H), 4.11 and 3.75

(AX, J = 13.6 Hz, ArCH_2Ar, 2H), 4.09 and 3.79 (AX, J = 13.9 Hz, ArCH_2Ar, 2H), 4.10 and 2.97 (AX, J = 13.9 Hz, ArCH_2Ar, 2H), 3.34–3.33 (s, overlapped, OCH_3, 6H), 2.70 (s, OCH_3, 3H), 2.47 (s, OCH_3, 3H), 2.30 (s, OCH_3, 3H), 1.66 (d, J = 6.1, 3H, CH_{3Bn}), 1.32 (s, C(CH_3)$_3$, 9H), 1.28 and 1.27 (s, C(CH_3)$_3$, 9 H, each one), 0.98, 0.96, 0.89 (s, C(CH_3)$_3$, 9H, each one).^{13}**C NMR** (100 MHz, CDCl$_3$, 298 K): δ 154.5, 154.47, 154.43, 153.6, 151.3, 145.8, 145.7, 145.4, 143.3, 134.2, 134.0, 133.8, 133.6, 133.6, 133.5, 133.4, 133.4, 128.4, 127.8, 127.1, 126.7, 124.9, 124.3, 124.1, 81.0, 77.4, 60.3, 60.2, 60.0, 59.9, 34.3, 34.3, 34.2, 34.1, 32.1, 31.7, 31.4, 31.3, 30.8, 30.6, 29.9, 29.5, 22.9.

Synthesis of derivative (*S*)-5. NaH (0.024 g, 1.0 mmol) was added to a solution of derivative **14** (0.10 g, 0.10 mmol) in dry DMF (15 mL) and stirred for 1 h. The mixture was allowed to cool at room and (*S*)-(+)-1-iodo-2-methylbutane (0.99 g 0.5 mmol) was added. The resulting mixture was kept at 80 °C for 12 h under a nitrogen atmosphere, then the solvent was removed under reduced pressure and the mixture was partitioned between CH$_2$Cl$_2$ and H$_2$O. The organic layer was washed with 1N HCl (30 mL), brine (30 mL), and dried over Na$_2$SO$_4$. The crude product was purified by column chromatography (SiO$_2$; CH$_2$Cl$_2$) to give derivative (*S*)-5 as a white solid (0.87 g, 0.078 mmol, 78%). Mp: 200–203 °C. [α]25D = +45 (*c* 2.5 CHCl$_3$). **ESI(+) MS:** *m/z* =1130.773 [M + NH$_4$]$^+$; ^{1}**H NMR** (300 MHz, TCDE, 423 K): δ 7.31 (br s, ArH, 2H), 7.17 (br s, ArH, 4H), 6.99 (brs, ArH, 2H), 6.95 (brs, ArH, 2H), 6.92 (br s, ArH, 2H), 4.04 (br s, ArCH_2Ar, 4H), 4.00 (br s, ArCH_2Ar, 8H), 3.88–3.74 (m, OCH_2CH(CH$_3$)CH$_2$CH$_3$, 2H), 3.37 (s, OCH_3, 6H), 3.00 (s, OCH_3, 3H), 2.79 (s, OCH_3, 6H), 2.07 (m, OCH$_2$CH(CH$_3$)CH$_2$CH$_3$, 1H), 1.77 (m, OCH$_2$CH(CH$_3$)CH_2CH$_3$, 2H), 1.38 (s, C(CH_3)$_3$, 18H), 1.33 (s, C(CH_3)$_3$, 9H), 1.24–0.97 (overlapped, OCH$_2$CH(CH_3)CH$_2$CH_3, 6H), 1.15 (s, C(CH_3)$_3$, 18H), 1.05 (s, C(CH_3)$_3$, 9H); ^{13}**C NMR** (75 MHz, CDCl$_3$, 298 K): δ 154.4, 153.6, 152.2, 145.8, 145.6, 133.9, 133.7, 133.6 (2), 133.5, 133.3, 127.6, 127.0, 125.0, 124.2, 60.2, 60.1, 60.0, 36.2, 34.3, 34.2, 32.1, 31.7, 31.4, 30.6, 30.3, 29.9, 29.5, 26.4, 22.9, 16.9, 14.3, 11.7.

Synthesis of derivative(*R,R*)-6. NaH (0.05 g, 2.0 mmol) was added to a solution of derivative **14** (0.20 g, 0.20 mmol) in dry DMF (10 mL) and stirred for 1 h. The mixture was allowed to cool at room and (*R,R*)-myrtenyl iodide (0.27 g, 1.00 mmol) was added. The resulting mixture was stirred at 80 °C for 12 h under a nitrogen atmosphere, then the solvent was removed under reduced pressure and the mixture was partitioned between CH$_2$Cl$_2$ and H$_2$O. The organic layer was washed with 1N HCl (30 mL), brine (30 mL), and dried over Na$_2$SO$_4$. The crude product was purified by column chromatography (SiO$_2$; CH$_2$Cl$_2$) to give derivative (*R*)-6 as a white solid (0.16 g, 0.14 mmol, 75%). Mp: 190–194 °C dec. [α]25D = −15 (*c* 2.5 CHCl$_3$). **ESI(+) MS:** *m/z* = 1194.781 [M + NH$_4$]$^+$, ^{1}**H NMR** (300 MHz, TCDE, 353 K): δ 7.28 (br s, ArH, 2H), 7.17–7.15 (overlapped, ArH, 4H), 6.90 (br s, ArH, 2H), 6.87 (br s, ArH, 2H), 6.84 (br s, ArH, 2H), 5.79 (br s, C=CH, 1H), 4.84 and 3.51 (AX, J = 14.1 Hz, ArCH_2Ar, 4H), 4.30 (br s OCH_2, 2H), 4.27 and 3.66 (AX, J = 15.2 Hz, ArCH_2Ar, 4H), 4.16 and 3.78 (AX, J = 14.9 Hz, ArCH_2Ar, 2H), 3.38 (s, OCH_3, 3H), 3.36 (s, OCH_3, 3H), 2.77 (s, OCH_3, 3H), 2.63 (s, OCH_3, 3H), 2.58 (s, OCH_3, 3H), 2.53–2.20 (4H), 1.40 (s, CH$_3$, 3H), 1.38 (s, C(CH_3)$_3$, 18H), 1.32 (s, C(CH_3)$_3$, 9H), 1.05 (s, C(CH_3)$_3$, 18H), 1.00 (s, CH$_3$, 3H), 0.97 (s, C(CH_3)$_3$, 9H), 0.92 (m, 2H).^{13}**C NMR** (75 MHz, CDCl$_3$, 298 K) 154.9, 154.2, 152.8, 146.3, 146.2, 145.4, 134.5, 134.4, 134.2, 134.1, 133.8, 127.5, 125.6, 124.8, 120.0, 77.9, 75.9, 60.5, 44.3, 41.6, 38.9, 35.3, 34.9, 34.7, 32.4, 32.2, 31.2, 32.0, 31.9, 31.1, 30.9, 27.0, 25.9, 21.9.

Synthesis of derivative (*R*)-7. Derivative **14** (0.10 g, 0.10 mmol), DMAP (3.7 mg, 0.030 mmol), and triethylamine (1.0 mL) were mixed with (*S*)-Mosher's acid chloride (47 mg, 0.20 mmol) in dry DMF (5.0 mL) and the reaction mixture was stirred at 70 °C for 12 h. The reaction mixture was cooled to room temperature and CH$_2$Cl$_2$ (7 mL) was added. The organic phase was washed with an aqueous solution of HCL (1 N) (3 × 10 mL) and successively with a saturated solution of NaHCO$_3$. The organic phase was dried over Na$_2$SO$_4$, filtered and dried. The crude product was purified by column chromatography (CH$_2$Cl$_2$: CH$_3$OH, 98:2, *v/v*) to give derivative (*R*)-7 as white solid (0.12 g, 0.093 mmol, 93 %). Mp: >210 °C dec. [α]25D = −22 (*c* 2.5 CHCl$_3$). **ESI(+) MS:** *m/z* =1259.736 [M + NH$_4$]$^+$; ^{1}**H NMR** (300 MHz, CDCl$_3$, 353 K): δ 7.82 (br m, ArH, 2H), 7.47 (br m, ArH, 3H), 7.24–7.17 (overlapped, ArH, 6H), 6.88–6.83 (overlapped, ArH, 6H), 4.35–4.12 (overlapped, ArCH_2Ar, 5H), 3.95 (d, ArCH_2Ar, J = 15.4 Hz, 1H), 3.77–3.33 (overlapped, ArCH_2Ar, 6H), 3.78 (s, OCH_3, 3H), 3.52 (s, OCH_3, 3H) 3.40 (s, OCH_3, 3H), 2.70

(s, OCH$_3$, 3H), 2.57 (s, OCH$_3$, 3H), 2.47 (s, OCH$_3$, 3H), 1.39 (s, C(CH$_3$)$_3$, 9H), 1.36 (s, C(CH$_3$)$_3$, 9H), 1.33 (s, C(CH$_3$)$_3$, 9H), 1.03 (s, C(CH$_3$)$_3$, 9H), 1.02 (s, C(CH$_3$)$_3$, 9H), 0.94 (s, C(CH$_3$)$_3$, 9H); ^{13}**C NMR** (75 MHz, CDCl$_3$, 353 K): δ 165.9, 155.3, 155.1, 154.2, 149.5, 146.6, 146.4, 144.1, 134.2, 134.0, 133.9, 133.6, 133.5, 132.8, 132.3, 130.6, 129.3, 128.7, 128.3, 128.2, 128.0, 127.8, 127.7, 125.6, 125.3, 125.2, 125.1, 60.7, 60.6, 56.5, 34.8, 34.7, 32.2, 32.0, 31.8, 31.3, 31.1.

Synthesis of derivative (*S*)-**9**$^+$·[B(Ar^F)$_4$]$^-$. (*S*)-α-Methylbenzylamine (0.020 mol) was added to benzaldehyde (0.020 mol) in dry CHCl$_3$ (2 mL) and the reaction mixture was stirred at room temperature for 2 h to give the imine intermediate in a quantitative yield.

The resulting imine (0.020 mol) was dissolved in dry MeOH (20 mL) under nitrogen atmosphere and NaBH$_4$ (0.20 mol) was added at 0 °C and the mixture was allowed to warm at room temperature and stirred for 3 h. The solvent was removed under reduced pressure and the residue partitioned between AcOEt (30 mL) and an aqueous saturated solution of NaHCO$_3$ (30 mL). The organic layer was dried over MgSO$_4$ and the solvent was removed under reduced pressure, to give derivative **20** as a yellow viscous liquid. The compound was used for the next step without further purification. The crude product (0.010 mol) was dissolved in Et$_2$O (20 mL) at room temperature and an aqueous solution of HCl (37% *w/w*, 0.02 mol) was added dropwise. The mixture was stirred for 1 h, until the formation of a white precipitate. The solid was collected by filtration, purified by crystallization with acetonitrile and dried under vacuum, to give derivative **21** as a white solid. Derivative **21** was dissolved in dry MeOH (0.2 M), then NaB(ArF)$_4$ (1.1 equiv) was added and the mixture was kept under stirring overnight in the dark. The solvent was removed and deionized water was added, obtaining a brown precipitate that was filtered off and dried under vacuum to give derivatives **9**$^+$·[B(Ar^F)$_4$]$^-$. Derivative **9**$^+$·[B(Ar^F)$_4$]$^-$: (0.090 g, 0.22 mmol, 95%). [α]25D = −12 (*c* 2.0 MeOH). Mp: >170 °C dec. **ESI(+) MS**: *m/z* = 212.15 (M+). ^{1}**H NMR** (400 MHz, CD$_3$OD, 298 K): δ 1.65 (d, *J* = 7.1, 3H, CH$_3$), 3.85 and 4.05 (AB, *J* = 13.1, 2H), 4.36 (q, *J* = 7.1, 1H), 7.32–7.42 (overlapped, 6H, ArH), 7.56–7.59 (overlapped, 6H, ArH); ^{13}**C NMR** (100 MHz, CD$_3$OD, 298 K) δ 18.3, 49.3, 58.3, 117.1, 120.3, 123.0, 125.7, 127.2, 128.4, 128.6, 128.9, 129.2, 129.3, 129.4, 129.5, 130.9, 134.4, 136.0, 160.8, 161.2, 161.7, 162.2.

Synthesis of derivative (*R*)- *or* (*S*)-**10**$^+$·[B(Ar^F)$_4$]$^-$. (*R*)- or (*S*)-α-Methylbenzylamine (0.08 mol) was dissolved in acetone (60 mL) and the mixture was stirred at reflux for 18 h. The reaction mixture was then cooled to room temperature and the excess of ketone was removed under reduced pressure.

The resulting imine (0.080 mol) was dissolved in dry MeOH (20 mL) under nitrogen atmosphere and NaBH$_4$ (0.080 mol) was added at 0 °C, then the mixture was allowed to warm at room temperature and stirred for 3 h. The solvent was removed under reduced pressure and the residue partitioned between AcOEt (30 mL) and an aqueous saturated solution of NaHCO$_3$ (30 mL). The organic layer was dried over MgSO$_4$ and the solvent was removed under reduced pressure, to give derivative (*R*)- *or* (*S*)-**23** as a yellow viscous liquid. The compound was used for the next step without further purification. The crude product (0.010 mol) was dissolved in Et$_2$O (20 mL) at room temperature and an aqueous solution of HCl (37% *w/w*, 0.08 mol) was added dropwise. The mixture was kept under stirring for 1 h, until the formation of a white precipitate. The solid was collected by filtration, purified by crystallization with *n*-hexane/MeOH and dried under vacuum, to give derivative (*R*)- *or* (*S*)-**24** as a white solid. Derivative (*R*)- *or* (*S*)-**24** was dissolved in dry MeOH, then NaB(ArF)$_4$ (1.1 equiv) was added and the mixture was kept under stirring overnight in the dark. The solvent was removed and deionized water was added, obtaining a brown precipitate that was filtered off and dried under vacuum to give derivative (*R*)- *or* (*S*)-**10**$^+$·[B(Ar^F)$_4$]$^-$.

Derivative (*S*)-**10**$^+$·[B(Ar^F)$_4$]$^-$: (0.087 g, 0.085 mmol, 95%). Mp: >150 °C dec. [α]25D = −22 (*c* 2.0 MeOH). **ESI(+) MS**: *m/z* = 164.14 (M+). ^{1}**H NMR** (400 MHz, CDCl$_3$, 298 K): δ 1.24 and 1.29 (d, *J* = 6.1 Hz, CH$_3$, 3H each one), 1.66 (d, *J* = 7.1 Hz, CH$_3$, 3H), 4.32 (m, C*H*(CH$_3$)$_2$, 1H), 4.41 (q, *J* = 7.1, 1H, H), 7.22 (br, ArH, 2H), 7.46–7.53 (overlapped, ArH, 7H), 7.69 (br s, ArH, 8H); ^{13}**C NMR** (100 MHz, CD$_3$OD, 298 K) δ 19.1, 19.4, 20.0, 50.4, 57.8, 117.8, 120.7, 123.4, 126.1, 126.5, 128.6, 128.8, 129.0, 129.3, 129.6, 130.6, 131.4, 135.0, 161.1, 161.6, 162.1, 162.6.

Synthesis of derivative (R)-**10**-$d_6^+\cdot[B(Ar^F)_4]^-$. (R)-α-Methylbenzylamine (0.12 g, 0.0010 mol), Ti(i-OPr)$_4$ (0.28 g, 0.001 mol) and deuterated acetone (0.0030 mol) were mixed together and stirred for 2 h at room temperature.

The resulting mixture was diluted with MeOD (10 mL) under nitrogen atmosphere and NaBH$_4$ (0.0020 mol) was added at 0 °C, then the mixture was allowed to warm to room temperature and stirred for 3 h.

The solvent was removed under reduced pressure and the residue partitioned between AcOEt (30 mL) and an aqueous saturated solution of NaHCO$_3$ (30 mL). The organic layer was dried over MgSO$_4$ and the solvent was removed under reduced pressure, to give derivative (R)-**23**-d_6 as a yellow viscous liquid. The compound was used for the next step without further purification. The crude product (0.0010 mol) was dissolved in Et$_2$O (20 mL) at room temperature and an aqueous solution of HCl (37% w/w, 0.003 mol) was added dropwise. The mixture was kept under stirring for 1 h, until the formation of a white precipitate. The solid was collected by filtration, purified by crystallization with n-hexane/MeOH and dried under vacuum, to give derivative (R)-**24**-d_6 as a white solid. Derivative (R)-**24**-d_6 was dissolved in dry MeOH (0.2 M), then NaB(ArF)$_4$ (1.1 eq) was added and the mixture was kept under stirring overnight in the dark. The solvent was removed and deionized water was added, obtaining a brown precipitate that was filtered off and dried under vacuum to give derivatives (R)-**10**-$d_6^+\cdot[B(Ar^F)_4]^-$.

Derivative (R)-**10**-$d_6^+\cdot[B(Ar^F)_4]^-$: (0.98 g, 0.00095 mol, 95%). Mp: >160 °C dec. **ESI(+) MS**: m/z = 170.18 (M+). **^{1}H NMR** (400 MHz, CD$_3$OD, 298 K): δ 1.64 (d, J = 7.1 Hz, 3H), 3.09 (s, 1H), 4.49 (q, J = 7.1 Hz, 1H), 7.47–7.59 (overlapped, 17H, ArH). **^{13}CNMR** (63 MHz, CDCl$_3$, 298 K): δ 149.6, 149.4, 148.4, 148.3, 146.7, 144.6, 143.7, 143.1, 136.6, 132.7, 129.3, 128.6, 127.6, 127.4, 127.2, 127.0, 126.9, 126.4, 126.2, 125.9, 125.6, 78.1, 34.5, 34.2, 34.1, 33.5, 32.9, 31.8.

4.1. General Procedure for MS Experiments (Isotopic Effect)

4.1.1. Sample Preparation

Calixarene derivatives (1.9 × 10^{-3} mmol) were dissolved in 0.5 mL of CHCl$_3$ (3.8 × 10^{-3} M solution). Then, the appropriate barfate salts, (R)-**10**$^+\cdot[B(Ar^F)_4]^-$ (3.8 × 10^{-3} mmol, 7.6 × 10^{-3} M) and (R)-**10**-$d_6^+\cdot[B(Ar^F)_4]^-$ (3.8 × 10^{-3} mmol, 7.6 × 10^{-3} M) were added and the mixture was stirred for 15 min. Then, the solution was diluted to a concentration of 300 µM with CH$_2$Cl$_2$ before sample injection.

4.1.2. MS Conditions

Sample concentration 300 µM; flow rate 2–4 µL/min; sample cone: 25 V; HV 2500 V; source temperature and temperature of desolvation gas were kept constant at 40 °C, no nebulizer gas was used for the experiments.

4.1.3. $I_R/I_{R\text{-dn}}$ Evaluation

Due to incomplete labeling of the acetone-d_6 used to prepare the axles, a small contribution of the d_5-labeled axle is present. As this can only be the same enantiomer as the corresponding d6-labeled isotopologue, the intensities of both were added.

This operation is valid assuming that no significant differences and isotope effect occur between the partially deuterated compounds (d_5) and the fully deuterated one (d_6).

4.2. General Procedure for MS Experiments (Chiral Recognition Effect)

Calixarene derivatives (1.9 × 10^{-3} mmol) were dissolved in 0.5 mL of CHCl$_3$ (3.8 × 10^{-3} M solution). Then, the appropriate barfate salts of (S)-**10**$^+$ (3.8 × 10^{-3} mmol, 7.6 × 10^{-3} M) and (R)-**10**-d_6^+ (3.8 × 10^{-3} mmol, 7.6 × 10^{-3} M) were added and the mixture was stirred for 15 min. The solution was diluted at the desired concentration with CH$_2$Cl$_2$ just before the MS analysis.

MS conditions: sample concentration 300 μM; flow rate 2–4 μL/min; sample cone: 25 V; HV 2500 V; source temperature and temperature of desolvation gas were kept constant at 40 °C, no nebulizer gas was used for the experiments.

$I_S/I_{R\text{-dn}}$ **evaluation**: The $I_S/I_{R\text{-dn}}$ ratios were determined as described above.

4.3. DFT Calculations

Input structure files for DFT calculations were obtained by molecular modeling with YASARA 20.7.4 program [48]. The force field parameters were generated with the AutoSMILES utility, which employs semiempirical AM1 geometry optimization and assignment of charges, followed by the assignment of the AM1BCC atom and bond types with refinement using the RESP charges, and finally the assignments of general AMBER force field atom types. DFT optimized structures were obtained by calculations at the B97D3/SVP/SVPFIT level of theory implemented in Gaussian 16 suite of programs [49].

Supplementary Materials: The following are available online. 1D and 2D, ^{1}H and ^{13}C NMR spectra of compounds **4–7**; ESI-MS spectra of derivatives **5–7**; 1D and 2D NMR spectra of pseudorotaxanes **4@8$^+$**, **4@9$^+$**, and **4@10$^+$**; MS experiments of threading of chiral calixarenes **5–7**; cartesian coordinates of the DFT-optimized structure of pseudorotaxanes.

Author Contributions: C.T.: investigation, acquisition and analysis of data, editing. G.C.: investigation, acquisition and analysis of data, editing. P.D.S.: performance of DFT calculations, analysis, and writing. C.G.: analysis and interpretation of NMR data, writing, and review. C.A.S.: conceptualization, analysis, interpretation, writing, review, and editing. P.N.: conceptualization, supervision, acquisition, interpretation of data, writing, review and editing. All authors have read and agreed to the published version of the manuscript.

Funding: This research was funded by Deutsche Forschungsgemeinschaft grant number CRC 765 and by University of Salerno grant number FARB 2014-2018.

Acknowledgments: Financial support from the Deutsche Forschungsgemeinschaft (CRC 765 and core facility BioSupraMol) is gratefully acknowledged. This work was supported by the University of Salerno (FARB and PhD fundings). We thank Andreas Springer for help with the mass spectrometric measurements.

Conflicts of Interest: The authors declare no conflict of interest.

References

1. Bruns, C.J.; Stoddart, J.F. *The Nature of the Mechanical Bond: From Molecules to Machines*, 1st ed.; John Wiley & Sons: New York, NY, USA, 2017.

2. Feringa, B.L. The Art of Building Small: From Molecular Switches to Motors. *Angew. Chem. Int. Ed.* **2017**, *56*, 11060–11078. [CrossRef] [PubMed]

3. Sauvage, J.-P. From Chemical Topology to Molecular Machines. *Angew. Chem. Int. Ed.* **2017**, *56*, 11080–11093. [CrossRef] [PubMed]

4. Stoddart, J.F. Mechanically Interlocked Molecules (MIMs)-Molecular Shuttles, Switches, and Machines. *Angew. Chem. Int. Ed.* **2017**, *56*, 11094–11125. [CrossRef]

5. Yu, H.; Luo, Y.; Beverly, K.; Stoddart, J.F.; Tseng, H.; Heath, J.R. The Molecule–Electrode Interface in Single-Molecule Transistors. *Angew. Chem. Int. Ed.* **2003**, *42*, 5706–5711. [CrossRef] [PubMed]

6. Coskun, A.; Spruell, J.M.; Barin, G.; Dichtel, W.R.; Flood, A.H.; Botrosghi, Y.Y.; Stoddart, J.F. High hopes: Can molecular electronics realise its potential? *Chem. Soc. Rev.* **2012**, *41*, 4827–4859. [CrossRef]

7. Mendes, P.M.; Flood, A.H.; Stoddart, J.F. Nanoelectronic devices from self-organized molecular switches. *Appl. Phys. A* **2005**, *80*, 1197–1209. [CrossRef]

8. Kassem, S.; Lee, A.T.L.; Leigh, D.A.; Marcos, V.; Palmer, L.I.; Pisano, S. Stereodivergent synthesis with a programmable molecular machine. *Nature* **2017**, *549*, 374–378. [CrossRef]

9. Erbas-Cakmak, S.; Leigh, D.A.; McTernan, C.T.; Nussbaumer, A.L. Artificial molecular machines. *Chem. Rev.* **2015**, *115*, 10081–10206. [CrossRef]

10. Balzani, V.; Credi, A.; Venturi, M. *Molecular Devices and Machines*, 2nd ed.; Wiley-VCH: Weinheim, Germany, 2008.

11. Beswick, J.; Blanco, V.; De Bo, G.; Leigh, D.A.; Lewandowska, U.; Lewandowski, B.; Mishiro, K. Selecting reactions and reactants using a switchable rotaxane organocatalyst with two different active sites. *Chem. Sci.* **2015**, *6*, 140–143. [CrossRef]

12. Blanco, V.; Leigh, D.A.; Marcos, V. Artificial switchable catalysts. *Chem. Soc. Rev.* **2015**, *44*, 5341–5370. [CrossRef]

13. Blanco, V.; Leigh, D.A.; Lewandowska, U.; Lewandowski, B.; Marcos, V. Goldberg Active Template Synthesis of a [2]Rotaxane Ligand for Asymmetric Transition-Metal Catalysis. *J. Am. Chem. Soc.* **2014**, *136*, 15775–15780. [CrossRef] [PubMed]

14. *Molecular Catenanes, Rotaxanes and Knots: A Journey Through the World of Molecular Topology*; Sauvage, J.P.; Dietrich-Buchecker, C. (Eds.) Wiley-VCH: Weinheim, Germany, 1999.

15. Jamieson, E.M.G.; Modicom, F.; Goldup, S.M. Chirality in rotaxanes and catenanes. *Chem. Soc. Rev.* **2018**, *47*, 5266–5311. [CrossRef] [PubMed]

16. Evans, N.H. Chiral Catenanes and Rotaxanes: Fundamentals and Emerging Applications. *Chem. Eur. J.* **2018**, *24*, 3101–3112. [CrossRef] [PubMed]

17. Lim, J.Y.C.; Marques, I.; Félix, V.; Beer, P.D. Enantioselective Anion Recognition by Chiral Halogen-Bonding [2]Rotaxanes. *J. Am. Chem. Soc.* **2017**, *139*, 12228–12239. [CrossRef]

18. Heard, A.W.; Goldup, S.M. Synthesis of a Mechanically Planar Chiral Rotaxane Ligand for Enantioselective Catalysis. *Chem* **2020**, *6*, 994–1006. [CrossRef]

19. Blanco, V.; Leigh, D.A.; Marcos, V.; Morales-Serna, J.A.; Nussbaumer, A.L. A Switchable [2]Rotaxane Asymmetric Organocatalyst That Utilizes an Acyclic Chiral Secondary Amine. *J. Am. Chem. Soc.* **2014**, *136*, 4905–4908. [CrossRef]

20. Cakmak, Y.; Erbas-Cakmak, S.; Leigh, D.A. Asymmetric Catalysis with a Mechanically Point-Chiral Rotaxane. *J. Am. Chem. Soc.* **2016**, *138*, 1749–1751. [CrossRef]

21. Leigh, D.A.; Wong, J.K.Y.; Dehez, F.; Zerbetto, F. Unidirectional rotation in a mechanically interlocked molecular rotor. *Nature* **2003**, *424*, 174–179. [CrossRef]

22. Lewandowski, B.; De Bo, G.W.; Ward, J.; Papmeyer, M.; Kuschel, S.; Aldegunde, M.J.; Gramlich, P.M.E.; Heckmann, D.; Goldup, S.M.; D'Souza, D.M.; et al. Sequence-specific peptide synthesis by an artificial small-molecule machine. *Science* **2013**, *339*, 189–193. [CrossRef]

23. Wilson, M.R.; Solà, J.; Carlone, A.; Goldup, S.M.; Lebrasseur, N.; Leigh, D.A. An autonomous chemically fuelled small-molecule motor. *Nature* **2016**, *534*, 235–240. [CrossRef]

24. Ashton, P.R.; Iriepa, I.; Reddington, M.V.; Spencer, N.; Slawin, A.M.Z.; Stoddart, J.F.; Williams, D.J. An optically-active [2]catenane made to order. *Tetrahedron Lett.* **1994**, *35*, 4835–4838. [CrossRef]

25. Ashton, P.R.; Heiss, A.M.; Pasini, D.; Raymo, F.M.; Shipway, A.N.; Stoddart, J.F.; Spencer, N. Diastereoselective Self-Assembly of [2]Catenanes. *Eur. J. Org. Chem.* **1999**, 995–1004. [CrossRef]

26. Yamamoto, C.; Okamoto, Y.; Schmidt, T.; Jäger, R.; Vögtle, F. Enantiomeric resolution of cycloenantiomeric rotaxane, topologically chiral catenane, and pretzelshaped molecules: Observation of pronounced circular dichroism. *J. Am. Chem. Soc.* **1997**, *119*, 10547–10548. [CrossRef]

27. Kaida, Y.; Okamoto, Y.; Chambron, J.-C.; Mitchell, D.K.; Sauvage, J.-P. The separation of optically active copper (I) catenates. *Tetrahedron Lett.* **1993**, *34*, 1019–1022. [CrossRef]

28. Bordoli, J.R.; Goldup, S.M. An Efficient Approach to Mechanically Planar Chiral Rotaxanes. *J. Am. Chem. Soc.* **2014**, *136*, 4817–4820. [CrossRef] [PubMed]

29. Mochizuki, Y.; Ikeyatsu, K.; Mutoh, Y.; Hosoya, S.; Saito, S. Synthesis of mechanically planar chiral rac-[2]rotaxanes by partitioning of an achiral [2]rotaxane: Stereoinversion induced by shuttling. *Org. Lett.* **2017**, *19*, 4347–4350. [CrossRef]

30. Goldup, S.M. Mechanical chirality: A chiral catalyst with a ring to it. *Nat. Chem.* **2016**, *8*, 404–406. [CrossRef]

31. Gaeta, C.; Talotta, C.; De Rosa, M.; Soriente, A.; Neri, P. *Calixarenes and Beyond*; Neri, P., Sessler, J.L., Wang, M.-X., Eds.; Springer: Dordrecht, The Netherlands, 2016; pp. 783–809.

32. Gaeta, C.; Troisi, F.; Neri, P. endo-Cavity Complexation and Through-the-Annulus Threading of Large Calixarenes Induced by Very Loose Alkylammonium Ion Pairs. *Org. Lett.* **2010**, *12*, 2092–2095. [CrossRef]

33. Gaeta, C.; Talotta, C.; Margarucci, L.; Casapullo, A.; Neri, P. Through-the-Annulus Threading of the Larger Calix[8]arene Macrocycle. *J. Org. Chem.* **2013**, *78*, 7627–7638. [CrossRef]

34. Gaeta, C.; Talotta, C.; Farina, F.; Camalli, M.; Campi, G.; Neri, P. Conformational Features and Recognition Properties of a Conformationally Blocked Calix[7]arene Derivative. *Chem. Eur. J.* **2012**, *18*, 1219–1230. [CrossRef]

35. Arduini, A.; Orlandini, G.; Secchi, A.; Credi, A.; Silvi, S.; Venturi, M. *Calixarenes and Beyond*; Neri, P., Sessler, J.L., Wang, M.-X., Eds.; Springer: Dordrecht, The Netherlands, 2016; pp. 761–781.

36. Talotta, C.; Gaeta, C.; Qi, Z.; Schalley, C.A.; Neri, P. Pseudorotaxanes with Self-Sorted Sequence and Stereochemical Orientation. *Angew. Chem. Int. Ed.* **2013**, *52*, 7437–7441. [CrossRef] [PubMed]

37. Talotta, C.; Gaeta, C.; Neri, P. Stereoprogrammed Direct Synthesis of Calixarene-Based [3]Rotaxanes. *Org. Lett.* **2012**, *14*, 3104–3107. [CrossRef] [PubMed]

38. Gaeta, C.; Talotta, C.; Mirra, S.; Margarucci, L.; Casapullo, A.; Neri, P. Catenation of Calixarene Annulus. *Org. Lett.* **2013**, *15*, 116–119. [CrossRef] [PubMed]

39. Talotta, C.; De Simone, N.A.; Gaeta, C.; Neri, P. Calix[6]arene threading with weakly interacting tertiary ammonium axles: Generation of chiral pseudorotaxane architectures. *Org. Lett.* **2015**, *17*, 1006–1009. [CrossRef] [PubMed]

40. Janssen, R.G.; Verboom, W.; Reinhoudt, D.N.; Casnati, A.; Freriks, M.; Pochini, A.; Ugozzoli, F.; Ungaro, R.; Nieto, P.M.; Carramolino, M.; et al. Procedures for the Selective Alkylation of Calix[6]arenes at the Lower Rim. *Synthesis* **1993**, *4*, 380–386. [CrossRef]

41. De Mendoza, J.; Carramolino, M.; Cuevas, F.; Manule Nieto, P.; Reinhoudt, D.N.; Verboom, W.; Ungaro, R.; Casnati, A. Selective Functionalization of Calix[6]arenes at the Upper Rim. *Synthesis* **1994**, *1*, 47–50. [CrossRef]

42. De Rosa, M.; Soriente, A.; Concilio, G.; Talotta, C.; Gaeta, C.; Neri, P. Nucleophilic Functionalization of the Calix [6] arene Para- and Meta-Position via p-Bromodienone Route. *J. Org. Chem.* **2015**, *80*, 7295–7300. [CrossRef]

43. Suezawa, H.; Ishihara, S.; Umezawa, Y.; Tsuboyama, S.; Nishio, M. The Aromatic CH/π Hydrogen Bond as an Important Factor in Determining the Relative Stability of Diastereomeric Salts Relevant to Enantiomeric Resolution–A Crystallographic Database Study. *Eur. J. Org. Chem.* **2004**, *2004*, 4816–4822. [CrossRef]

44. Schug, K.A.; Maier, N.M.; Lindner, W. Deuterium isotope effects observed during competitive binding chiral recognition electrospray ionization—mass spectrometry of cinchona alkaloid-based systems. *J. Mass Spectrom.* **2006**, *41*, 157–161. [CrossRef]

45. Behr, J.P.; Lehn, J.-M.; Moras, D.; Thierry, J.C. Chiral and functionalized face-discriminated and side-discriminated macrocyclic polyethers. Syntheses and crystal structures. *J. Am. Chem. Soc.* **1981**, *103*, 701–703. [CrossRef]

46. Bako, P.; Fenichel, L.; Toke, L. The complexing ability of crown ethers incorporating glucose. *J. Inclusion Phenom. Mol. Recognit. Chem.* **1993**, *16*, 17–23. [CrossRef]

47. Jung, Y.E.; Song, B.M.; Chang, S.K. Molecular recognition of alkyl- and arylakyl-amines in dischloromethane and chloroform by calix[4]-crown ethers. *J. Chem. Soc. Perkin Trans.* **1995**, *2*, 2031–2034. [CrossRef]

48. Krieger, E.; Vriend, G. YASARA View—Molecular graphics for all devices—From smartphones to workstations. *Bioinformatics* **2014**, *30*, 2981–2982. [CrossRef] [PubMed]

49. Frisch, M.J.; Trucks, G.W.; Schlegel, H.B.; Scuseria, G.E.; Robb, M.A.; Cheeseman, J.R.; Scalmani, G.; Barone, V.; Mennucci, B.; Petersson, G.A.; et al. *Gaussian 16, revision A.03*; Gaussian Inc.: Wallingford, CT, USA, 2016.

Sample Availability: Samples of the compounds are not available from the authors.

Publisher's Note: MDPI stays neutral with regard to jurisdictional claims in published maps and institutional affiliations.

 *molecules*

Article

Dihomooxacalix[4]arene-Based Fluorescent Receptors for Anion and Organic Ion Pair Recognition

Alexandre S. Miranda [1,2], Paula M. Marcos [1,3,*], José R. Ascenso [4], Mário N. Berberan-Santos [2], Rachel Schurhammer [5], Neal Hickey [6] and Silvano Geremia [6]

[1] Centro de Química Estrutural, Faculdade de Ciências da Universidade de Lisboa, Edifício C8, 1749-016 Lisboa, Portugal; miranda.m.alexandre@gmail.com

[2] IBB–Institute for Bioengineering and Biosciences, Instituto Superior Técnico, Universidade de Lisboa, 1049-001 Lisboa, Portugal; berberan@tecnico.ulisboa.pt

[3] Faculdade de Farmácia da Universidade de Lisboa, Av. Prof. Gama Pinto, 1649-003 Lisboa, Portugal

[4] Instituto Superior Técnico, CQE, Complexo I, Av. Rovisco Pais, 1049-001 Lisboa, Portugal; jose.ascenso@ist.utl.pt

[5] Laboratoire de Modélisation et Simulations Moléculaires, Université de Strasbourg, UMR 7140, F-67000 Strasbourg, France; rschurhammer@unistra.fr

[6] Department of Chemical and Pharmaceutical Sciences, Centre of Excellence in Biocrystallography, University of Trieste, via L. Giorgieri 1, 34127 Trieste, Italy; nhickey@units.it (N.H.); sgeremia@units.it (S.G.)

* Correspondence: pmmarcos@fc.ul.pt

Academic Editor: Renata Riva

Received: 11 September 2020; Accepted: 8 October 2020; Published: 14 October 2020

Abstract: Fluorescent dihomooxacalix[4]arene-based receptors **5a–5c**, bearing two naphthyl(thio)ureido groups at the lower rim via a butyl spacer, were synthesised and obtained in the cone conformation in solution. The X-ray crystal structures of 1,3- (**5a**) and 3,4-dinaphthylurea (**5b**) derivatives are reported. Their binding properties towards several anions of different geometries were assessed by ^{1}H-NMR, UV-Vis absorption and fluorescence titrations. Structural and energetic insights of the naphthylurea **5a** and **5b** complexes were also obtained using quantum mechanical calculations. The data showed that all receptors follow the same trend, the association constants increase with the anion basicity, and the strongest complexes were obtained with F$^-$, followed by the oxoanions AcO$^-$ and BzO$^-$. Proximal urea **5b** is a better anion receptor compared to distal urea **5a**, and both are more efficient than thiourea **5c**. Compounds **5a** and **5b** were also investigated as heteroditopic receptors for biologically relevant alkylammonium salts, such as the neurotransmitter γ-aminobutyric acid (GABA·HCl) and the betaine deoxycarnitine·HCl. Chiral recognition towards the guest *sec*-butylamine·HCl was also tested, and a 5:2 selectivity for (*R*)-*sec*-BuNH$_3^+$·Cl$^-$ towards (*P*) or (*M*) enantiomers of the inherently chiral receptor 5a was shown. Based on DFT calculations, the complex [(*S*)-*sec*-BuNH$_3^+$·Cl$^-$/(*M*)-5a] was indicated as the more stable.

Keywords: dihomooxacalix[4]arenes; naphthyl(thio)urea anion receptors; alkylammonium hydrochlorides; ditopic receptors; chiral recognition; NMR studies; UV-Vis absorption studies; fluorescence studies; X-ray diffraction; DFT calculations

---

1. Introduction

Calixarenes are among the most versatile macrocyclic compounds studied in supramolecular chemistry owing to their structural features [1,2]. They can be functionalized at the upper and lower rims, and they possess a pre-organized cavity available in different sizes and conformations. As a result, these compounds have been largely exploited as ion receptors.

Fluorescence spectroscopy, due to its high sensitivity, has been used for ion binding determination [3,4]. Fluorogenic moieties, such as naphthalene, anthracene and pyrene are among the most incorporated in the calixarene framework, leading to the development of fluorescent probes for anion and also ion-pair recognition. Examples of such fluorescent calix[4] arene [5–10], calix[5] arene [11] and calix[6]arene [12–14] receptors have been reported in the literature.

Anions play important roles in many biological and chemical systems, and also in the environment [15,16]. Synthetic anion receptors, namely calixarenes containing amide or (thio)urea groups interact exclusively through H-bonding with the anions. The NH groups provide strong and directional hydrogen bonds, resulting in well preorganized receptors. Some of these hosts can also act as ditopic receptors, simultaneously binding both ions of a given salt [17,18]. These receptors combine different binding sites in the same molecule, such as hydrogen bonds and oxygen donor atoms, besides an aromatic cavity that can establish π-CH interactions with the counter cation.

As part of our on-going interest on the host-guest properties of substituted dihomooxacalix[4]arenes (calix[4]arene analogues in which one CH_2 bridge is replaced by one CH_2OCH_2 group) with (thio)urea units [19–23], we have extended our research into the study of fluorescent receptors for anion [24] and ion-pair recognition. Thus, dihomooxacalix[4]arene-based fluorescent sensors bearing (thio)urea groups as a binding site and naphthalene moieties as a fluorophore unit were obtained for the first time. This paper describes the synthesis of three disubstituted dihomooxacalix[4]arenes containing naphthylurea (compounds **5a** and **5b**) or naphthylthiourea (compound **5c**) residues at the 1,3- or 3,4-positions of the lower rim, via a butyl spacer. These derivatives were obtained in the cone conformation in solution, confirmed by NMR. The cone conformation was also observed in the solid state (for **5a** and **5b**) by X-ray diffraction. Their binding properties towards several relevant anions were assessed by proton NMR, UV-Vis absorption and fluorescence spectroscopy. The urea compounds (**5a** and **5b**) were also tested as heteroditopic receptors for biologically relevant alkylammonium salts, such as the amino acid γ-aminobutyric acid (GABA·HCl) and the betaine deoxycarnitine·HCl. GABA is an important neurotransmitter with inhibitory activity in mammal central nervous system. Deoxycarnitine results from enzymatic methylation of GABA, and as other betaines is used as an osmotic regulator in plants. Chiral recognition towards the chiral guest *sec*-butylamine·HCl was also investigated taking advantage of the intrinsic chirality of urea **5a**. Computational studies were also performed to add further insight to the binding process.

2. Results and Discussion

2.1. Synthesis and Structural Analysis

A few years ago we reported the reaction of parent compound **1** with bromobutyronitrile and K_2CO_3 to afford, after chromatographic separation, the asymmetric 1,3-dicyanodihydroxy derivative **2a** and the symmetric 3,4-dicyanodihydroxy derivative **2b** [19]. Following this synthetic route (Scheme 1), we undertook a three-step procedure from both the majority and the minority products **2a** and **2b**, respectively, obtaining in the last step naphthylurea **5a** and naphthylthiourea **5c** from the asymmetric diamine **4a** and naphthylurea **5b** from the symmetric **4b**. Comparing the ion affinity of receptors **5a** and **5b**, it is expected to obtain some insights about the role of the substitution pattern (distal vs. proximal) of the two ureido groups in a cooperative binding process.

The ^{1}H-NMR spectra of inherently chiral receptors **5a** and **5c** in CDCl$_3$ at room temperature show four singlets for the *tert*-butyl groups, five AB quartets for the CH_2 bridge protons, four pairs of doublets for the aromatic protons of the calixarene skeleton, and two triplets and two singlets for the NHa and NHb protons, respectively. Beside these peaks, the spectra display also two triplets and several multiplets for the methyl and methylene protons of the *n*-butyl groups and butyl spacers, as well as for the aromatic protons of the naphthyl groups. The proton assignments were confirmed by COSY spectra. Receptors **5a** and **5c** were obtained in the cone conformation, as indicated by the three ArCH$_2$Ar resonances in the range 29.5–30.8 ppm of the ^{13}C-NMR spectra [25].

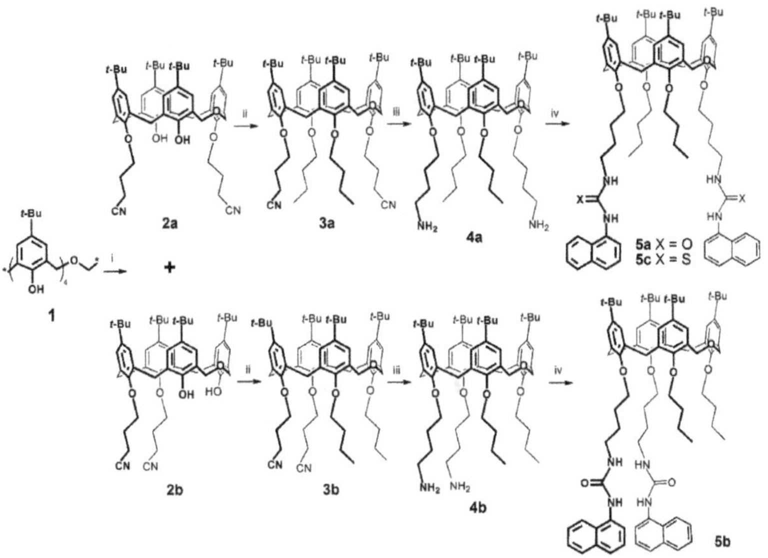

Scheme 1. Synthesis of naphthyl(thio)ureas **5a–5c**. *Reaction conditions*: (i) Br(CH$_2$)$_3$CN, K$_2$CO$_3$, MeCN, Δ; (ii) n-BuI, NaH, THF/DMF, Δ; (iii) NaBH$_4$/CoCl$_2$, MeOH, rt; (iv) Naph-NCX, CHCl$_3$, rt.

In contrast, receptor **5b** presents symmetric NMR spectra. The ^{1}H-NMR spectrum displays two singlets for the *tert*-butyl groups, three AB quartets (in a 2:2:1 ratio) for the CH$_2$ bridge protons, two pairs of doublets for the aromatic protons of the calixarene platform and one triplet and one singlet for the NHa and NHb protons, respectively, besides one triplet and several multiplets for the CH$_3$ and CH$_2$ protons of the butyl groups and spacers, and also for the aromatic protons of the naphthyl groups. The ^{13}C-NMR spectrum exhibits two ArCH$_2$Ar resonances at 30.3 ppm (two carbon atoms) and at 30.5 ppm (one carbon atom), indicating a cone conformation also for **5b**.

Small single crystals of naphthylureas **5a** and **5b** were analyzed using synchrotron radiation at 100 K. The X-ray structures confirm that both **5a** and **5b** adopt the expected cone conformation, also in the solid state. The structural model of **5a** clearly show that it is inherently chiral due to the 1,3-substitution pattern on the lower rim, which is asymmetric with respect to the dihomooxa bridge (Figure 1). As the space group is centrosymmetric, a racemic mixture of the two inherently chiral enantiomers is present in the crystals. With regard to the cone conformation, the planes of the two ureido-substituted phenyl rings **A** (connected to the dihomooxa bridge) and **C** make large dihedral angles of 124° and 143°, respectively, with respect to the mean plane of the methylene bridging groups. Angles greater than 90° indicate that the *tert*-butyl groups on the upper rims lean outwards from the centre of the cone (Figure 1). With respect to the butoxy-substituted phenyl rings, the plane of the one adjacent to the dihomooxa bridge (**B**) makes at a dihedral angle of 69° with the mean plane of the methylene bridging groups, with the upper rim inclined inwards. The last phenyl ring (**D**) is tilted slightly outwards, with a dihedral angle of 99°.

Consistently with what we have previously observed for analogous calixarenes containing ureido or thioureido units on the lower rim, the two ureido groups form a bifurcated intramolecular N–H···O hydrogen bond with N···O distances of 2.866 Å and 2.895 Å. In the present case, the hydrogen bond is quite symmetric, indicative of a strong interaction. The mean planes formed by the NCON atoms of the urea moieties show a dihedral angle of 24° (38° for the second orientation of the disordered naphthyl group of ring **A**), while the terminal naphthyl groups form dihedral angles of about 69° (Ring **A**, 86° for the second orientation) and 42° (Ring **C**) with respect to their corresponding planar NCON groups.

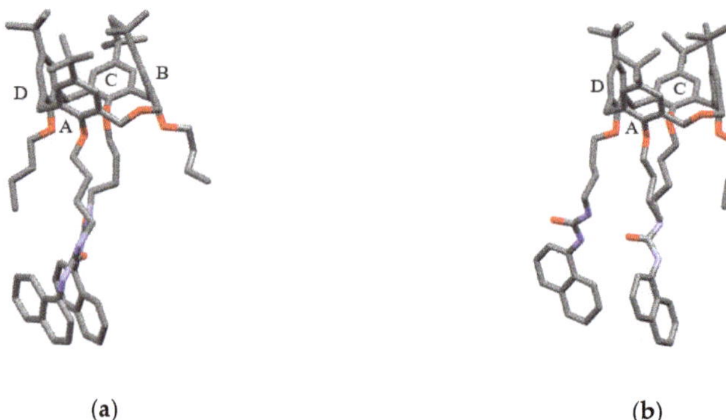

(a)   (b)

Figure 1. Solid state structures of (a) **5a** and (b) **5b**. The structures, with very similar cone conformations, show significant differences in conformation of the lower rim substituents. The atomic species are represented in CPK colours. Hydrogen atoms are omitted for clarity.

The overall result is that the two naphthyl groups are almost parallel, forming a dihedral angle of about 4° between their mean planes (7° for the second orientation) (Figure 1). The disordered naphthyl group (Ring **A**) is oriented head-to-tail (head-to-head in the second orientation) with respect to the other naphthyl group (Ring **C**). **5b** adopts a similar cone conformation (Figure 1). In contrast to **5a**, the 3,4-substitution pattern on the lower rim maintains the C_s point symmetry of the macrocycle. Comparison of the dihedral angles of the four phenyl rings A, B, C and D with the mean plane of the methylene bridging groups, indicate an analogous cone conformation for **5a** and **5b** (Table 1). Thus, only slight differences are evident for the B (14°) and D (−11°) rings. For comparison, we have previously reported two analogous dihomooxacalix[4]arenes with a 1,3-substitution pattern on the lower rim and which differ from **5a** for the presence of two p-CF$_3$-phenylurea moieties [23] or unsubstituted phenylurea moieties [20] in place of the naphthyl urea groups. All structures with the 1,3-substitution pattern on the lower rim exhibit conformations which are comparable (Table 1). Thus, the small difference in the cone conformation observed for **5b** can be attributed to the different substitution pattern. More significant difference between the two structures **5a** and **5b** are apparent in the relative orientations of the ureido substituents. The planes of the two naphthyl rings are almost perpendicular, with a dihedral angle of 88°, as opposed to the almost parallel situation found for **5a** (Table S1). There are significant differences in the hydrogen bonds formed by the various molecules discussed here (Table S2). **5a** and **5b** both form strong bifurcated intramolecular and intermolecular hydrogen bond, while for the p-CF$_3$-Phurea moieties the intramolecular bonds are quite asymmetric, and in the case of the Phurea moieties a solvent molecule is involved and breaks the intermolecular H-bond pattern.

Table 1. Comparison of cone conformations: Dihedral angles between corresponding aryl planes of the calixarene cones (A, B, C and D) and the mean planes of the bridging methylene carbon atoms for various dihomooxacalix[4]arenes.

	A (°)	B (°)	C (°)	D (°)
5a	124 [a]	69	143 [a]	99
5b	125	83	140 [a]	88 [a]
p-CF$_3$-Phurea [b]	125 [a]	66	131 [a]	101
Phurea [c] (I)	123 [a]	64	133 [a]	100
Phure a[c] (II)	121 [a]	74	137 [a]	97

[a] Ureido substituent on lower ring; [b] data taken from ref. 23; [c] data taken from ref. 20 for two independent molecules in the asymmetric unit. See Figure **2** and text for labelling of rings A, B, C and D.

With regard to the crystal packing, in **5a** each molecule acts as both an *N*-donor (on one ureido group) and an *O*-acceptor (on the other ureido group) in the formation of two symmetry equivalent bifurcated intermolecular N–H⋯O hydrogen bonds (2.839 Å and 2.974 Å) with two other molecules generated by the 2_1 screw axis symmetry operation, thereby forming alternating intramolecular /intermolecular H-bond chains parallel to the crystallographic *b*-axis. Two inversion-related anti-parallel chains are formed (Figure 2a). Each chain is composed of molecules with the same inherent chirality. Like **5a**, the urea group of **5b** forms an intramolecular bifurcated H-bond with N⋯O distances of 2.891 Å and 2.975 Å, and two intermolecular bifurcated H-bonds with N⋯O distances of 2.909 Å and 2.952 Å (Figure 2b). The intermolecular H-bonds are formed with molecules generated by the diagonal glide plane, forming antiparallel chains of H-bonds, parallel to the *a*-c cell diagonal.

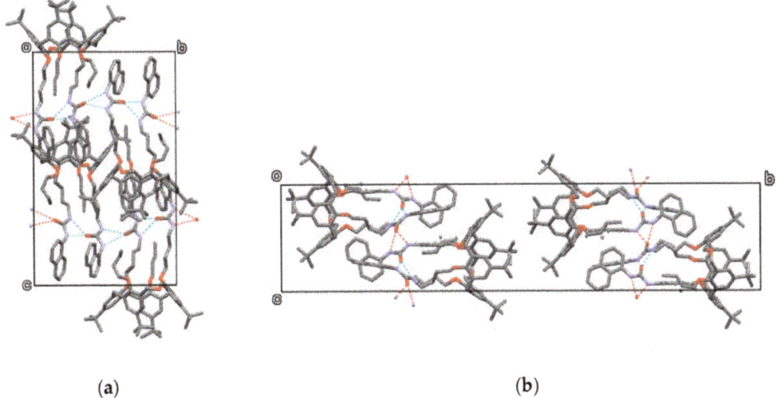

(a) (b)

Figure 2. Crystal packing in the unit cells of (**a**) **5a** and (**b**) **5b**. In both cases the crystal packing is characterized by inversion-related antiparallel chains of bifurcated H-bonds. In **5a**, each molecule in the chain is generated by a 2_1 screw operation and the chains are parallel to the *b*-axis. Each chain is composed of molecules with the same chirality and the antiparallel chain is composed of opposite chirality. In **5b**, the molecules in each chain are generated by the glide planes and the chains are parallel to the *a*-c cell diagonal. The atomic species are represented in CPK colours. Hydrogen atoms are omitted for clarity.

2.2. Anion Complexation

2.2.1. Proton NMR Studies

Complexation abilities of naphthyl(thio)ureas **5a–5c** toward relevant anions of different geometries (spherical, trigonal planar and tetrahedral) were investigated in $CDCl_3$ by proton NMR titrations with tetrabutylammonium (TBA) salts. The association constants (as log K_{ass}) were determined following the urea NH chemical shifts through the WinEQNMR2 program [26] and are reported in Table 2.

Table 2. Association constants (log K_{ass}) [a] of dihomooxa naphthyl(thio)ureas **5a-5c** in $CDCl_3$ at 25 °C.

	Spherical				Trigonal Planar			Tetrahedral	
	F^-	Cl^-	Br^-	I^-	NO_3^-	AcO^-	BzO^-	HSO_4^-	$H_2PO_4^-$
I. Radius/Å [b]	1.33	1.81	1.96	2.20	1.79	2.32	—	1.90	2.00
5a	2.80	2.60	2.12	1.78	2.12	2.51	2.76	2.54	2.25
5b	3.12	2.91	2.46	1.94	2.38	3.17	3.07	2.52	2.67
5c	2.67	1.75	1.06	1.09	1.06	2.17	2.01	1.89	2.02

[a] Estimated error < 10%; [b] Data quoted in Marcus, I. *Ion Properties*; Marcel Dekker: New York, pp. 50–51,1997.

Significant downfield shifts of the NH protons were observed upon addition of TBA salts to the receptors, clearly indicating hydrogen bonding interactions between the (thio)urea groups and the anions, as illustrated in Figure 3 and Figure S1. Only one set of signals was observed during the titrations, showing fast exchange rate between the free and the complexed receptor on the NMR time scale at room temperature. The titration curves obtained (Figure S2) evidence the 1:1 complexes, this stoichiometry being also confirmed by Job plots (Figures S3 and S4).

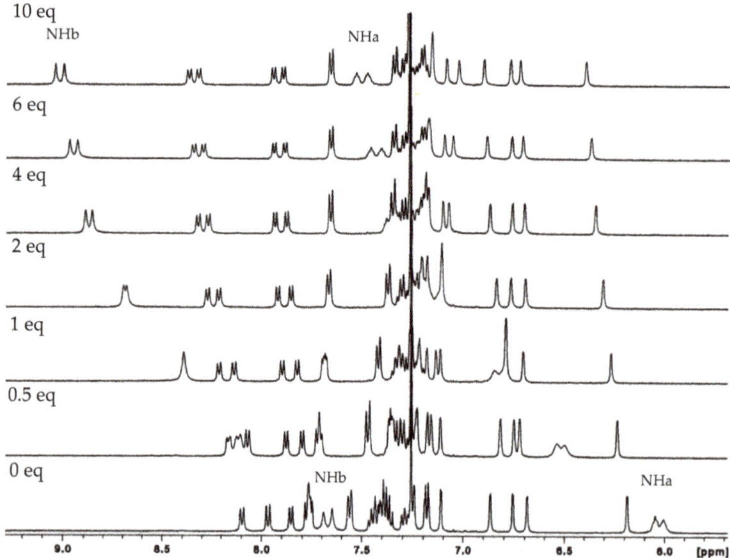

Figure 3. ^{1}H-NMR partial spectra (500 MHz, CDCl$_3$, 25 °C) of Naph-urea **5a** with several equiv of TBA chloride.

The comparison between naphthylureas **5a** and **5b** allowed us to make some considerations about the cooperative action of the two ureido moieties on alternate vs. adjacent positions of the calixarene framework. The results displayed in Table 1 indicate that proximal naphthylurea **5b** is a more efficient receptor for all the anions (except HSO$_4^-$). The association constants were, in average, 0.32 log units higher than those obtained for distal naphthylurea **5a** for the majority of the anions. In the case of H$_2$PO$_4^-$ and AcO$^-$ this enhancement was even higher (0.42 and 0.66 log units, respectively). A similar behaviour was previously observed with distal and proximal calix[4]arene diphenylurea analogues [27]. Concerning the spherical halides, the data reveal that both ureas **5a** and **5b** form the strongest complexes with F$^-$ (log K_{ass} = 2.80 and 3.12, respectively), and the association constants increase with the anion basicity. With regard to the trigonal planar and tetrahedral anions, urea **5b** displays the same behaviour, showing the highest affinity for the carboxylate AcO$^-$ and the inorganic oxoanion H$_2$PO$_4^-$, respectively (log K_{ass} = 3.17 and 2.67). In the case of urea **5a**, and as observed before with other dihomooxa bidentate [19,23] and tetraurea receptors [21], there is a slight inversion of the basicity order (AcO$^-$/BzO$^-$ and H$_2$PO$_4^-$/HSO$_4^-$). π stacking interactions may contribute to the increased binding of BzO$^-$ over that of AcO$^-$.

The anion binding results reported in Table 2 also show that naphthylthiourea **5c** is a weaker receptor than naphthylurea **5a**, despite the increased acidity of its NH groups. Thiourea **5c** exhibits however a similar trend to **5a**, with the anions bound according to their basicity. The association constants were, in average, 0.84 log units lower than those obtained for **5a**, except for the best bound anions F$^-$, AcO$^-$ and H$_2$PO$_4^-$, whose differences were smaller (Δ log K_{ass} = 0.13, 0.34 and 0.23, respectively). Similar results were reported before for different homooxacalixarene thiourea receptors [21,28,29], as well as thioureido-calix[4] and [6]arene analogues [30,31]. This fact may be related to the larger size of sulfur atom, that destabilizes the cis-cis geometry required for anion binding, causing a lower preorganization and consequently a high energy demand of the thiourea groups compared to the urea ones [32].

The calixarene skeleton of symmetric urea **5b** seems to undergo no conformational changes upon the addition of 8 equiv of the salts, as the *tert*-butyl and the aromatic protons show very small downfield or upfield chemical shift variations ($\Delta\delta \leq 0.06$ and 0.03 ppm, respectively). By contrast, asymmetric urea **5a** undergoes deeper conformational changes upon complexation. One of the four *t*-Bu groups experiences downfield variations from 0.05 to 0.20 ppm, while the other three display smaller upfield variations (from 0.01 to 0.14 ppm). The maximum chemical shifts were observed for BzO^- anion. Concerning the aromatic protons, some of them are overlapped by other peaks and difficult to follow during all the titration. However, it is possible to observe that two of them show significant downfield variations, from 0.10 to 0.29 ppm, the highest chemical shifts being observed for BzO^- (0.29 and 0.25 ppm).

2.2.2. UV-Vis Absorption and Fluorescence Studies

The interactions between naphthyl(thio)ureas **5a–5c** and the previous anions as TBA salts have also been studied in dichloromethane by UV-Vis absorption and fluorescence titrations. Naphthylureas **5a** and **5b** showed identical behaviours with respect to anion complexation. Both ureas display absorption bands centred at approximately 283 nm in the absence of anions. These bands decrease in intensity upon addition of increasing amounts of F^-, while a new one gradually appears at longer wavelength, reaching a maximum at approximately 315 nm (red shift of 32 nm). Isosbestic points can also be observed, as for example in Figure 4a for receptor **5a**. Concerning Cl^-, AcO^-, BzO^- and $H_2PO_4^-$ anions similar absorption spectral changes were obtained (Figure S5), leading to red shifts of 20 nm, but with no isosbestic points. Finally, additions of Br^-, NO_3^- and HSO_4^- anions induced progressive increases of the absorption, but no shifts in their maxima were recorded (Figure S6). Naphthylthiourea **5c** behaved differently for all the anions (Figure S7). In this case, the absorption band centred at 283 nm decreases as the anion concentration increases, presenting isosbestic points, but no significant shifts of its maximum.

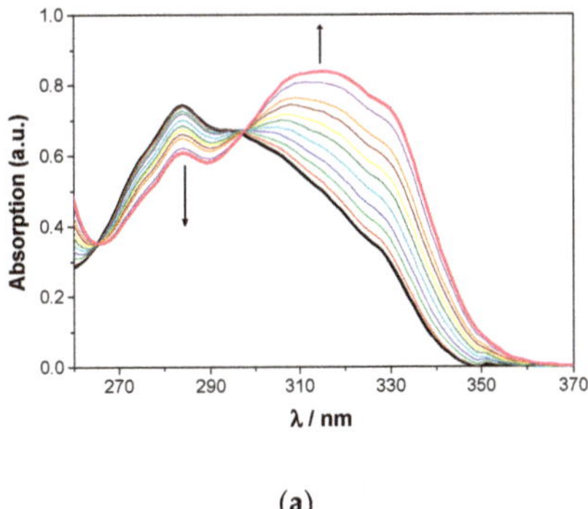

(a)

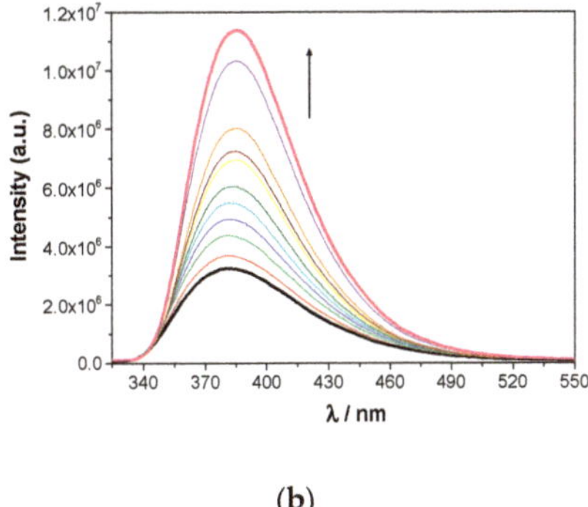

(b)

Figure 4. (**a**) Changes in the UV (**a**) and emission (**b**) spectra of Naph-urea **5a** (5.0×10^{-5} M) upon addition of TBA F (up to 10 equiv.) in CH_2Cl_2. The arrows indicate the decreasing or increasing amounts of salt.

With regard to steady-state fluorescence studies, receptors **5a** and **5b** exhibit emission bands centred at approximately 380 nm, characteristic of the naphthylurea groups [33]. Similar absorption and fluorescence spectra were reported in the literature for a ureido-calix[5]arene analogue [11]. No intramolecular excimer is observed in these fluorescence spectra, indicating the absence of π-π^* stacking between the naphthyl moieties [33]. Both **5a** and **5b** display significant fluorescence lifetimes and quantum yields (Table 3). Successive additions of F^-, AcO^- and $H_2PO_4^-$ anions caused an increase of the emission intensity, as shown in Figure 4b. For Br^-, NO_3^- and HSO_4^- anions this increase was less pronounced (Figure S8), and in the case of Cl^- and BzO^- (Figure S9) a quenching of the fluorescence intensity, with a concomitant decrease of the fluorescence quantum yield (Table 3), was observed. This decrease is stronger for asymmetric urea **5a**. In the case of Cl^- and **5a**, the fluorescence lifetime of the 1:1 complex shows a moderate decrease (about 2/3) with respect to the pristine receptor (Table 3), whereas the quantum yield drops by a factor of almost 4. This implies the existence of a marked static quenching in the complex, with only a few configurations being able to emit. The same applies to **5b**, although to a lesser extent. In the case of BzO^-, the quenching is similar with respect to both intensity and lifetimes (Table 3). It is seen that the quenching arises mainly from an increase in the nonradiative decay constant in the complex, where aromatic moieties of receptor and anion appear to interact. The fluorescence of thiourea **5c** could not be studied in detail as this receptor is unstable upon irradiation.

Table 3. Photophysical properties of Naphureas **5a** and **5b** in CH_2Cl_2 at 25 °C.

	$\lambda_{max,abs}$ (nm)	ε $(M^{-1}\,cm^{-1})$	$\lambda_{max,em}$ (nm)	Stokes Shift [a] (nm)	τ_f (ns)	Φ_F	k_r (ns^{-1})	k_{nr} (ns^{-1})
5a	283	1.5×10^4	382	99	8.93	0.31 [b]	0.035	0.077
5a + Cl^-	303	2.4×10^4	382	79	6.26	0.085 [b]	—	—
5a + BzO^-	303	2.0×10^4	382	79	1.98	0.048 [b]	0.024	0.48
5b	282	1.8×10^4	379	97	7.57	0.26 [b]	0.034	0.098
5b + Cl^-	302	2.3×10^4	379	77	6.63	0.16 [b]	—	—
5b + BzO^-	302	1.7×10^4	379	77	3.27	0.078 [b]	0.024	0.28

[a] Computed as $\lambda_{max,em} - \lambda_{max,abs}$; [b] Against quinine sulfate $\Phi_F = 0.546$ in H_2SO_4 0.5 M.

Important spectral variations were observed for the three receptors in the presence of all the anions, allowing the calculation of the corresponding binding constants by absorption and emission (for the ureas) data (Table 4). The association constants are higher than those obtained by NMR (different concentration range), but follow the same trend. The more diluted solutions used in the UV/fluorescence titrations favour the dissociation of the salts, producing a higher concentration of the anions available for complexation and resulting in higher association constants [11]. Similar results were obtained by absorption and emission, in the same concentration range, showing that fluorescence can also be a useful method for the determination of the association constants. Proximal naphthylurea **5b** is a slightly better receptor for all the anions (except HSO_4^-) than distal urea **5a**, and F^-, AcO^- and BzO^- are the best bound anions. Naphthylthiourea **5c** displayed the same trend as its urea analogue, with the anions bound according to their basicity (Table 4). However, **5c** is a weaker receptor, except in the cases of F^-, AcO^- and $H_2PO_4^-$ anions, for which it showed similar log K_{ass} values.

Table 4. Association constants (log K_{ass}) [a] of dihomooxa naphthyl(thio)ureas **5a–5c** in C H_2Cl_2 at 25 °C.

		Spherical			Trigonal Planar			Tetrahedral	
		F^-	Cl^-	Br^-	NO_3^-	AcO^-	BzO^-	HSO_4^-	$H_2PO_4^-$
5a	Abs	4.21	3.59	3.20	3.19	3.79	3.94	3.13	3.06
	Emi	4.05	3.48	3.14	3.12	3.66	4.00	3.01	2.90
5b	Abs	4.36	3.69	3.37	3.31	4.16	4.08	2.90	3.16
	Emi	4.34	3.67	3.23	3.23	4.21	3.84	2.97	3.14
5c	Abs	4.18	3.01	2.71	3.07	3.66	3.71	2.87	3.14

[a] Estimated error < 10%.

2.3. Organic Ion Pair Recognition

Naphthylureas **5a** and **5b** have also been tested as ditopic receptors for *n*-propyl and *n*-butylammonium chlorides (Figure 5) in an exploratory study to estimate their ion pair binding efficiency.

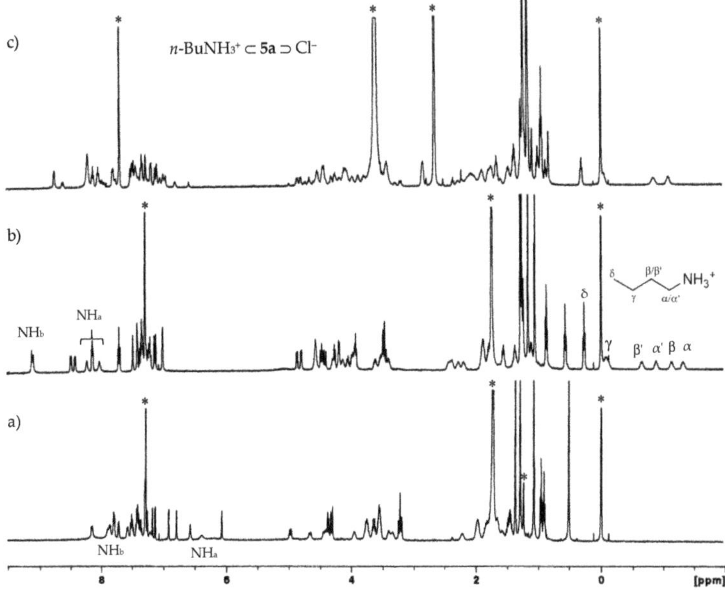

Figure 5. Structures of the ammonium hydrochloride guests studied.

Proton NMR titrations were performed by adding increasing amounts (up to two equiv.) of the salts to $CDCl_3$ solutions of **5a** and **5b** at room temperature. The addition of the first salt aliquot produced doubling of the receptor peaks, and also a new set of signals corresponding to the guest bound to the host. Alkylammonium cation inclusion inside the dihomoxa cavity is shown by the appearance of high field resonances for the alkyl groups (from −1.32 to −0.33 ppm for n-PrNH$_3^+$ and from −1.32 to 0.26 ppm for n-BuNH$_3^+$). In the case of receptor **5a**, due to its intrinsic chirality, the pairs of enantiotopic hydrogen atoms of the α- and β-CH$_2$ groups of the included guest display chemically non-equivalent signals, as shown in Figure 6 for n-BuNH$_3^+$. On the other hand, chloride binding to the urea groups is demonstrated by the downfield shifts observed for all the NH protons ($\Delta\delta \geq 1$ ppm), indicating anion complexation through hydrogen-bond interactions. These observations are compatible with a slow binding process on the NMR time scale. The percentages of complex formation and the corresponding association constants could thus be determined by direct integration of the peaks. The temperature was lowered to 263 K/253 K to get a more sound integration of the signals, as they were slightly broad at room temperature. All host-guest pairs studied displayed percentages of complexation higher than 95% (corresponding to $K_{ass} > 10^9$ M^{-2}), preventing a more accurate calculation of the association constants in chloroform.

Figure 6. ^{1}H-NMR spectra (500 MHz) of: (**a**) [**5a**] = 1.0 mM at 253 K in $CDCl_3$; (**b**) [**5a**] = [n-BuNH$_2$·HCl] = 1 mM at 253 K in $CDCl_3$; (**c**) [**5a**] = [n-BuNH$_2$·HCl] = 1 mM at 243 K in $CDCl_3$/DMSO-d$_6$, 5:1, v/v. * Denotes residual solvent signals.

These titration experiments were repeated in a more competitive solvent (CDCl$_3$/DMSO-d$_6$, 5:1) for the anion binding site and also that increases the ammonium solvation. It was necessary to lower the temperature until 273 K to observe the appearance of the peaks at the negative region of the ^{1}H NMR spectra, and to lower until 243 K to a better integration of the signals. Thus, the results obtained (K_{ass} = 18,000 and 29,000 M^{-2}, corresponding to 79% and 83% of complex formation for *n*-PrNH$_3$$^+$·Cl$^-$ with **5a** and **5b**, respectively, and K_{ass} = 10,000 and 18,000 M^{-2}, corresponding to 73% and 79% of complex formation for *n*-BuNH$_3$$^+$·Cl$^-$ with **5a** and **5b**, respectively) show that receptor **5b** is more efficient than **5a** for both ion pairs, and the former guest is better bound than the latter by both receptors. This trend was also observed by the theoretical calculations (see below, Section 2.4). It is worth noting that in this solvent mixture the guests are less fixed within the aromatic cavity of the host, as indicated by the chemically equivalent signals for the α- and β-CH$_2$ groups of the included guests (Figure 6c).

The binding affinities of ureas **5a** and **5b** were also extended to the aminoacid GABA·HCl and to the betaine deoxycarnitine·HCl (Figure 5) in a CDCl$_3$/CD$_3$OD (5:1, *v/v*) solvent mixture. The former guest was first tested in its zwitterionic form at room temperature and at 233 K. The NMR spectra remained almost unchanged after the addition of two equiv. of the guest, indicating no host-guest interaction. However, resonances at the negative region of the spectrum appeared when GABA·HCl was used, revealing the ammonium cation inclusion inside the aromatic cavity of the host (Figure 7).

For the inherently chiral receptor **5a**, four high field signals for the β- and γ-CH$_2$ protons of the guest were observed (Figure S10), in analogy with the alkylammonium chloride cases seen previously. To obtain a more reliable integration of the signals, the NMR spectra were registered at 233 K. The data (81 and 88% of complex formation, corresponding to K_{ass} = 22,000 and 60,000 M^{-2} for **5a** and **5b**, respectively) indicate that receptor **5b** is more efficient than **5a**, being in line with the previous anion and alkylammonium chloride binding results. Concerning the latter guest, no interaction at all was detected with both receptors, suggesting that the more bulky groups (CH$_3$ vs. H) of the betaine guest prevent the inclusion inside the macrocycle cavity.

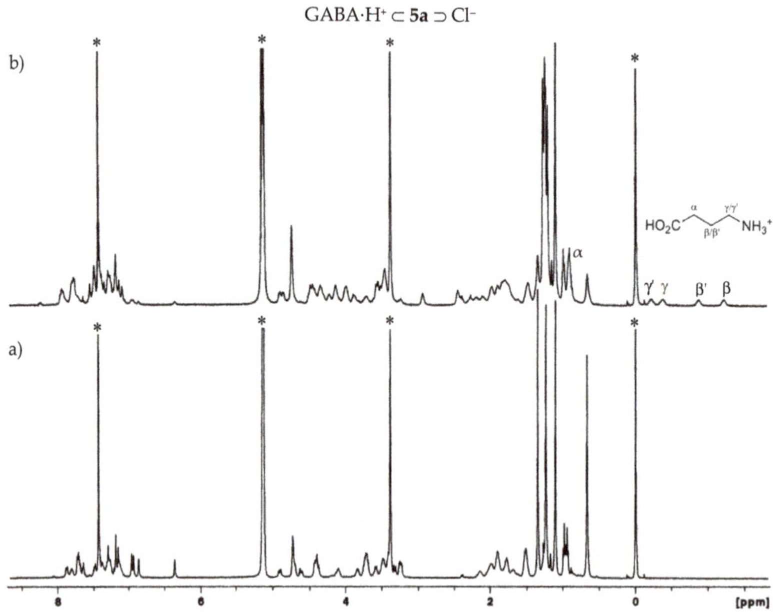

Figure 7. ^{1}H-NMR spectra (500 MHz, 233 K, CDCl$_3$/CD$_3$OD, 5:1, *v/v*) of: (a) [**5a**] = 1.0 mM; (b) [**5a**] = [GABA·HCl] = 1 mM. * Denotes residual solvent signals.

Chiral recognition towards racemic *sec*-butylamine·HCl guest (Figure 5) was also investigated with both urea receptors. The NMR binding studies (CDCl$_3$/CD$_3$OD, 5:1, *v/v*) performed at room temperature showed only a slight broadening of the host peaks. However, by lowering the temperature (until 223 K) it was possible to see the appearance of upfield resonances belonging to the guest inside the aromatic cavity of the hosts. In the case of the achiral urea **5b**, the addition of one equiv. of the guest gives rise to the appearance of an asymmetric structure. For example, six singlets for the *t*-Bu groups, corresponding to the free (two peaks) and to the complexed receptor (four peaks) can be observed. Five high field resonances for the *sec*-BuNH$_3^+$ guest, including two multiplets for the diastereotopic β methylene protons are also shown in the proton spectrum (Figure S11). The splitting of the *t*-Bu signals is not complete, preventing a very precise integration. However, a complex formation of approximately 70% could be determined. As reported before for the binding of *sec*-butylammonium ion with another achiral dihomooxacalixarene [34], the inclusion of the branched *sec*-BuNH$_3^+$ ion into the dihomooxa cavity should restrict its free motion, producing this chiral complex. Concerning racemic urea **5a** and in the same conditions as before, the proton NMR spectrum displays at least ten singlets for the *t*-Bu groups, four corresponding to the free host and the remaining ones corresponding to the two diastereotopic complexes formed [host(*P*)/guest(*R*) ≡ host(*M*)/guest(*S*) + host(*P*)/guest(*S*) ≡ host(*M*)/guest(*R*)] [35] (Figure 8). The percentage of complex formation is approximately of 65%. The same situation [host(*P*)/guest(*R*) + host(*M*)/guest(*R*)] was obtained when we used an enantiomerically pure guest [(*R*)-(−)-*sec*-butylamine·HCl]. Two sets of shielded resonances (8 signals instead of the expected 10 due to overlapping) for the *sec*-Bu group of the guest included into the cavity for the two diastereotopic complexes were seen in the high field region of the spectrum. A complete assignment of these peaks was obtained by a COSY spectrum (Figure S12). Their integration indicated a diastereomeric ratio of about 5:2.

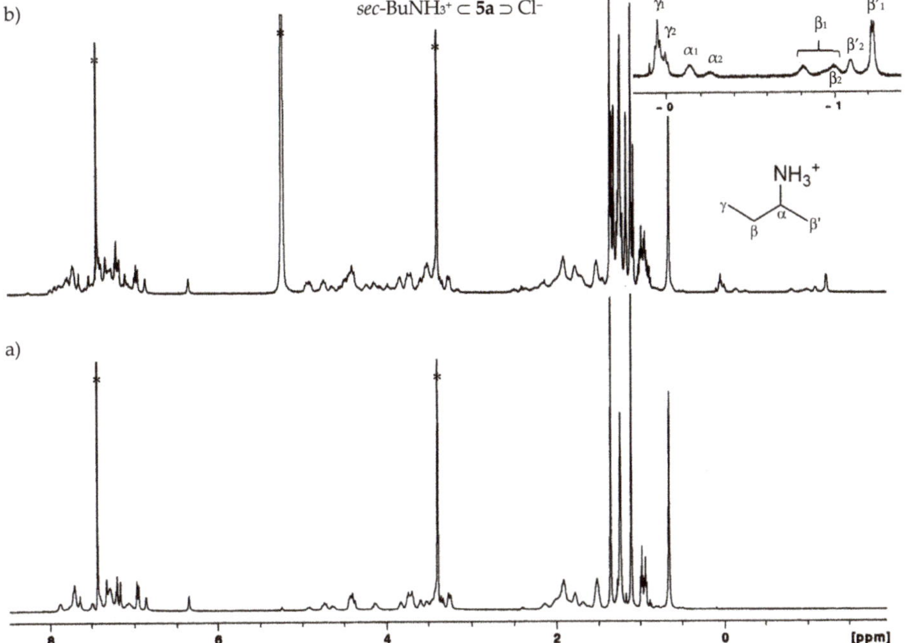

Figure 8. ^{1}H-NMR spectra (500 MHz, 223 K, CDCl$_3$/CD$_3$OD, 5:1, *v/v*) of: **(a)** [**5a**] = 1.0 mM; **(b)** [**5a**] = [*sec*-BuNH$_2$·HCl] = 1 mM. Inset: 1 and 2 mean the two sets of signals for the *s*-Bu group of the guest inside the cavity for the two diastereotopic complexes formed. * Denotes residual solvent signals.

By NMR is not possible to determine which complex [host(P)/guest(R) or host(M)/guest(R)] is more stable. Thus, DFT calculations were performed (see below) and showed an higher energy for the [host(M)/guest(S)] complex. On this basis, the more intense NMR signals were assigned to the more stable complex [(S)-sec-BuNH$_3^+$·Cl$^-$/(M)-**5a**], that can be directly transferred to its enantiomeric pair [host(P)/guest(R)], and a selectivity of about 5:2 could be deduced.

2.4. Theoretical Studies

In order to get further insights into the anion binding ability of **5a** and **5b**, we performed quantum mechanical calculations on the complexed receptors with an extensive range of anions, including spherical halides (F$^-$ vs. Cl$^-$), trigonal (AcO$^-$ vs. BzO$^-$) and tetrahedral (HSO$_4^-$ vs. H$_2$PO$_4^-$). Heteroditopic complexation properties of **5a** and **5b** were also studied with alkylammonium salts, comparing *n*-PrNH$_3^+$·Cl$^-$ and *n*-BuNH$_3^+$·Cl$^-$, and the affinity of the asymmetric host **5a** for the chiral guest *sec*-BuNH$_3^+$·Cl$^-$ was also investigated.

Each anion, whatever is its geometry, is bonded to the urea groups of **5a** and **5b** via four hydrogen bonds, as illustrated by snapshots of the optimized structures for fluoride, acetate and hydrogenophosphate anions in Figure 9. Similar structures were found for the other anions, as shown in Figure S13. The urea moieties always interact with the coordinated anions that sit in a hole formed by the four hydrogens of the NH groups. AcO$^-$ and BzO$^-$ are recognized via their carboxylate groups, while the methyl and benzyl groups point away from the binding cavity. The tetrahedral anions interact via their non protonated oxygens.

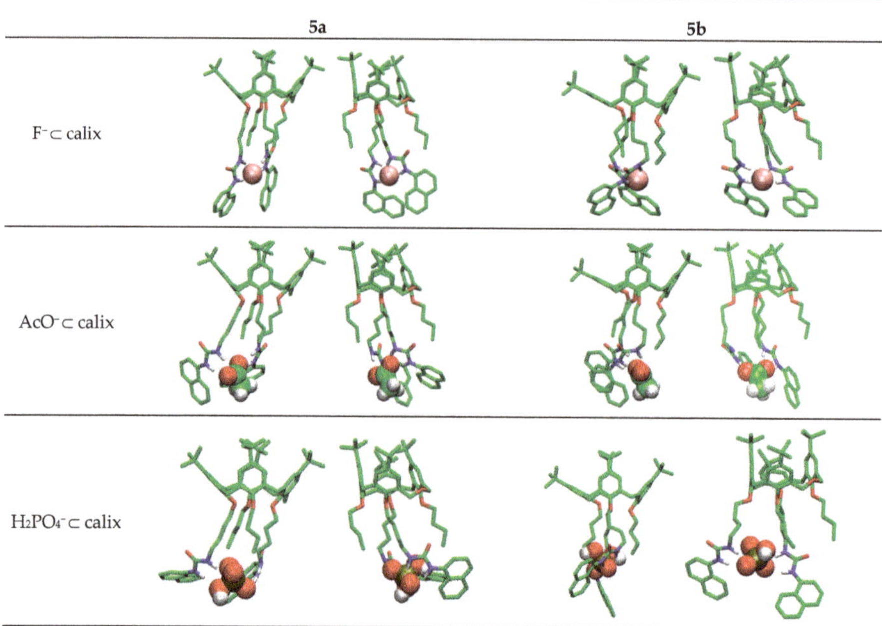

Figure 9. Structures of the X$^-$ ⊂ Naph-urea complexes after QM optimization (orthogonal views). The atomic species are represented in CPK colours, except carbons that are in green.

The ΔE calculated complexation energies (ΔE = E(complex) − E(free ligand) − E(ions), Table 5 and Table S3) nicely follow the association constants from Table 2. Naph-urea **5b** is always a better receptor than Naph-urea **5a**, but the ΔE differences depend on the nature of the anions and go from less than 15 kJ.mol^{-1} for HSO$_4^-$ to almost 60 kJ.mol^{-1} for the F$^-$ ion. For the latter anion and for the trigonal planar the energy discrimination between **5a** and **5b** is quite high (more than 50 kJ.mol^{-1} for F$^-$ and BzO$^-$ and 30 kJ.mol^{-1} for AcO$^-$), while it is smaller (less than 17 kJ.mol^{-1}) for Cl$^-$ and the tetrahedral anions. To analyse these differences the H-bond distances between the receptor and the anions were measured (Table S4). As expected, comparing the anions within the same geometry group, the averaged H-bond distances are correlated to the interaction energies: for the spherical F$^-$ vs. Cl$^-$ anions the mean values are 1.747 Å vs. 2.300 Å and 1.746 Å vs. 2.338 Å for **5a** and **5b**, respectively. This trend is also observed for the trigonal planar and tetrahedral ions. What is more surprising is the fact that the H-bond distances are equal or shorter for **5a**, although **5a** is always a weaker receptor than **5b**, for the same anion. Regarding the deformation energies of the calixarenes (i.e., the energy loss of the ligand) upon complexation (Table S5), it can be seen that the deformation is higher for **5a** than for **5b** with the spherical and trigonal planar anions ($\Delta_{def}E$ = 49 and 18 kJ.mol^{-1} for F$^-$ and Cl$^-$, respectively, and around 40 kJ.mol^{-1} for AcO$^-$ and BzO$^-$). These differences in the destabilisation of **5a** may explain its lower complexation energies for these anions, although its H-bond network is stronger. This tendency is inverted for the tetrahedral anions, the deformation being higher for **5b** than for **5a**, although this difference is smaller ($\Delta_{def}E \leq 13$ kJ.mol^{-1}). In this case, **5b** is much more destabilized (Table S5) than for the other groups of anions, presumably because the tetrahedral geometry of the anions is less adapted to the complexation by the two urea moieties.

Table 5. B3LYP/6-31G(d.p) + BG3BJ complexation energies ΔE (in kJ.mol^{-1}) for the guest-host complexes.

	ΔE (kJ.mol^{-1})	
Host	5a	5b
F$^-$	−498.9	−557.1
Cl$^-$	−202.5	−219.0
AcO$^-$	−261.5	−291.4
BzO$^-$	−232.1	−283.3
HSO$_4$$^-$	−192.3	−206.2
H$_2$PO$_4$$^-$	−220.6	−236.6
n-PrNH$_3$$^+$·Cl$^-$	−736.3	−811.6
n-BuNH$_3$$^+$·Cl$^-$	−656.5	−709.5
(*R*)-*sec*-BuNH$_3$$^+$·Cl$^-$/(*M*)-**5a**	−704.1	—
(*S*)-*sec*-BuNH$_3$$^+$·Cl$^-$/(*M*)-**5a**	−712.4	—
(*R*)-*sec*-BuNH$_3$$^+$·Cl$^-$/(*P*)-**5a**	−691.1	—
(*S*)-*sec*-BuNH$_3$$^+$·Cl$^-$/(*P*)-**5a**	−695.4	—

Concerning the heteroditopic complexation with the alkylammonium salts, the position of the chloride anion is the same than for the single anion complexes (see Figure 10). The H-bonding network is similar: Cl$^-$ interacts with the four hydrogen atoms of the urea groups with bond distances of about 2.3 Å each. The alkylammonium cations are positioned in the centre of the upper rim of the calixarenes. The ammonium group is asymmetrically H-bonded to the phenoxy oxygen atoms, as illustrated by the H-bond length given in Table S4. As NH$_3$$^+$ does not perfectly suit with the topology of the macrocycle cavity, it always displays two short H-bonds (less than 1.9 Å) and a longer one (up to 2.8 Å). The bridging ether oxygen atom is never involved in these interactions. The interaction energies obtained also indicate that **5b** is a better host than **5a**, and the *n*-PrNH$_3$$^+$·Cl$^-$ salt is better bound than the *n*-BuNH$_3$$^+$·Cl$^-$ one. Calculations with both enantiomers (*R*) and (*S*) of the chiral guest *sec*-BuNH$_3$$^+$·Cl$^-$ show no clear differences in the coordination mode of receptor **5a**, displaying however higher energy for the latter enantiomer (ΔE = 8.3 and 4.3 kJ.mol^{-1} for (*M*) and (*P*) enantiomers of **5a**, respectively). The (*M*)-**5a** enantiomer leads to higher coordination energy than the (*P*)-**5a** one (ΔE = 17 and 13 kJ·mol^{-1} for (*S*) and (*R*) guests, respectively), indicating the [(*S*)-*sec*-BuNH$_3$$^+$·Cl$^-$/(*M*)-**5a**] complex as the most stable.

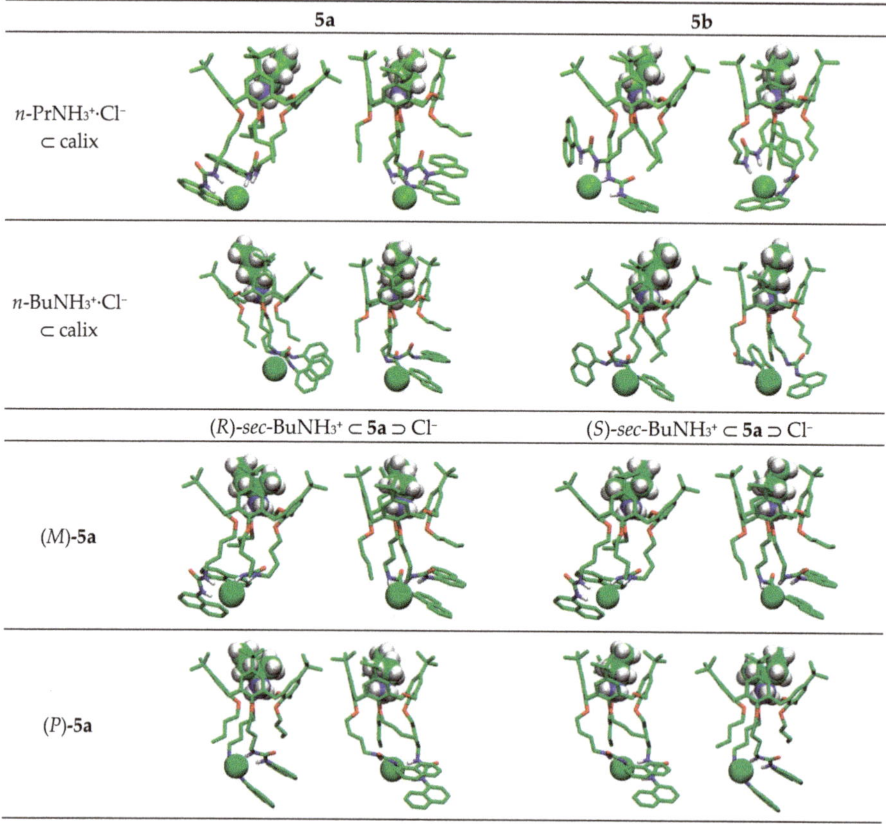

Figure 10. Structures of the R-NH$_3^+$ ⊂ **calix** ⊃ ∩ Cl$^-$ complexes after QM optimization (orthogonal views). The atomic species are represented in CPK colours, except carbons that are in green.

3. Materials and Methods

3.1. Synthesis

3.1.1. Procedure for the Synthesis of (thio)ureas **5a** and **5c**

To a solution of **4a** [19] (0.77 g, 0.83 mmol) in CHCl$_3$ (30 mL) was added 1.65 mmol of naphthyl isocyanate or naphthyl isothiocyanate, respectively. The mixture was stirred at room temperature under N$_2$ for 4 h. Evaporation of the solvent yielded the crude products which were purified as described below.

7,13,19,25-Tetra-tert-butyl-27,29-bis[[(N'-1-naphthylureido)butyl]oxy]-28,30-dibutoxy-2,3-dihomo-3-oxacalix[4]arene (**5a**). Flash chromatography (SiO$_2$, eluent CH$_2$Cl$_2$/MeOH, from 99.5:0.5 to 95:5) followed by recrystallization from CH$_2$Cl$_2$/n-hexane: it was obtained in 30% yield (0.31 g); m.p. 258–259 °C; IR (KBr) 3314 cm^{-1} (NH), 1638 cm^{-1} (CO); ^{1}H-NMR (CDCl$_3$, 500 MHz) δ 0.58, 1.04, 1.28,1.36 [4s, 36H, C(CH$_3$)$_3$], 0.89, 0.94 (2t, 6H, J = 7.45 Hz, CH$_3$), 1.45 (m, 4H, OCH$_2$CH$_2$**C**H$_2$CH$_3$), 1.64, 1.71, 1.82, 1.95, 2.13 (5m, 12H, OCH$_2$**C**H$_2$**C**H$_2$CH$_2$NH$_a$ and OCH$_2$**C**H$_2$**C**H$_2$CH$_3$), 3.18, 4.33 (ABq, 2H, J = 13.9 Hz, ArCH$_2$Ar), 3.20, 4.38 (ABq, 2H, J = 12.7 Hz, ArCH$_2$Ar), 3.21, 4.35 (ABq, 2H, J = 12.9 Hz, ArCH$_2$Ar), 3.35–3.62, 3.68, 3.76, 3.94 (several m, 12H, OCH$_2$CH$_2$CH$_2$**C**H$_2$NH$_a$ and OCH$_2$CH$_2$CH$_2$CH$_3$), 4.45, 4.54 (ABq, 2H, J = 13.3 Hz, CH$_2$OCH$_2$), 4.48, 4.87 (ABq, 2H, J = 12.8 Hz, CH$_2$OCH$_2$), 6.01, 6.05 (2t, 2H, NH$_a$), 6.19, 6.69, 6.76, 6.86, 7.11, 7.18, 7.19, 7.24 (8d, 8H, ArH), 7.29 (t, 1H, Napht), 7.35–7.47 (m, 5H,

Napht), 7.56, 7.85, 7.97, 8.09 (4d, 5H, Napht), 7.65, 7.69 (2s, 2H, NH$_b$), 7.76 (m, 3H, Napht); ^{13}C-NMR (CDCl$_3$, 125.8 MHz) δ 13.9, 14.1 (OCH$_2$CH$_2$CH$_2$*C*H$_3$), 19.35, 19.41 (OCH$_2$CH$_2$*C*H$_2$CH$_3$), 26.4, 26.7, 27.0, 28.5 (OCH$_2$*C*H$_2$CH$_2$CH$_2$NH$_a$), 29.5, 30.7, 30.8 (ArCH$_2$Ar), 31.2, 31.3, 31.6, 31.7 [C(*C*H$_3$)$_3$], 32.3, 32.6 (OCH$_2$*C*H$_2$CH$_2$CH$_3$), 33.7, 33.9, 34.15, 34.22 [*C*(CH$_3$)$_3$], 40.3, 40.6 (OCH$_2$CH$_2$CH$_2$*C*H$_2$NH$_a$), 69.2 (2C) (*C*H$_2$O*C*H$_2$), 73.1, 74.0, 74.6, 75.4 (O*C*H$_2$CH$_2$CH$_2$CH$_2$NH$_a$ and O*C*H$_2$CH$_2$CH$_2$CH$_3$), 119.7, 121.3, 121.5, 121.6, 123.6, 124.2 (2C), 124.4, 125.0, 125.2, 125.5, 125.7 (2C), 125.84, 125.87, 125.94, 126.0, 126.3, 126.8, 127.2, 128.3, 128.48 (ArH), 127.6, 128.52, 129.6, 131.8, 132.3, 132.6 (2C), 133.8, 134.1, 134.2 (3C), 134.3, 135.7, 144.4, 145.0, 145.1, 145.2, 152.4, 152.5, 153.0, 153.9 (Ar), 157.1, 157.6 (CO). Anal. Calcd for C$_{83}$H$_{106}$N$_4$O$_7$: C, 78.39; H, 8.40; N, 4.41. Found: C, 77.97; H, 8.68; N, 4.28.

7,13,19,25-Tetra-tert-butyl-27,29-bis[[(N′-1-naphthylthioureido)butyl]oxy]-28,30-dibutoxy-2,3- dihomo-3-oxa calix[4]arene (**5c**). Flash chromatography (SiO$_2$, eluent CH$_2$Cl$_2$/MeOH, from 99.5:0.5 to 97:3); the product obtained was chromatographed again (SiO$_2$, eluent CH$_2$Cl$_2$/MeOH, from 99.7:0.3 to 97:3), yielding **5c** in 29% (0.31 g); m.p. 111–113 °C; ^{1}H-NMR (CDCl$_3$, 500 MHz) δ 0.75, 1.08, 1.11, 1.25 [4s, 36H, C(CH$_3$)$_3$], 0.87 (t, 6H, *J* = 7.45 Hz, CH$_3$), 1.23–1.42, 1.54–1.87 (several m, 16H, OCH$_2$*C*H$_2$*C*H$_2$CH$_3$ and OCH$_2$*C*H$_2$*C*H$_2$CH$_2$NH$_a$), 3.08, 4.26 (ABq, 2H, ArCH$_2$Ar), 3.08, 4.28 (ABq, 2H, ArCH$_2$Ar), 3.17, 4.24 (ABq, 2H, ArCH$_2$Ar), 3.43–3.82 (several m, 12H, O*C*H$_2$CH$_2$CH$_2$CH$_3$ and O*C*H$_2$CH$_2$CH$_2$*C*H$_2$NH$_a$), 4.30, 4.43 (ABq, 2H, CH$_2$OCH$_2$), 4.50, 4.53 (ABq, 2H, CH$_2$OCH$_2$), 6.02, 6,07 (2t br, 2H, NH$_a$), 6.41, 6.87, 6.89, 6.99, 7.06, 7.07 (6d, 8H, ArH), 7.39–7.53 (several m, 8H, Napht), 7.57, 7.62 (2s, 2H, NH$_b$), 7.82–7.96 (several m, 6H, Napht); ^{13}C-NMR (CDCl$_3$, 125.8 MHz) δ 14.10, 14.12 (OCH$_2$CH$_2$CH$_2$*C*H$_3$), 19.26, 19.34 (OCH$_2$CH$_2$*C*H$_2$CH$_3$), 25.7, 25.8, 27.3, 27.8 (OCH$_2$*C*H$_2$CH$_2$CH$_2$NH$_a$), 29.7, 30.1, 30.7 (ArCH$_2$Ar), 31.3, 31.4, 31.51, 31.54 [C(*C*H$_3$)$_3$], 32.2, 32.4 (OCH$_2$*C*H$_2$CH$_2$CH$_3$), 33.8, 33.9, 34.0, 34.1 [*C*(CH$_3$)$_3$], 45.7 (2C) (OCH$_2$CH$_2$CH$_2$*C*H$_2$NH$_a$), 68.4, 68.5 (*C*H$_2$O*C*H$_2$), 73.4, 73.8, 74.1, 74.9 (O*C*H$_2$CH$_2$CH$_2$CH$_2$NH$_a$ and O*C*H$_2$CH$_2$CH$_2$CH$_3$), 122.5 (2C), 123.4, 123.6, 124.9, 125.2, 125.3, 125.4, 125.5, 125.7, 125.8, 125.9, 126.2, 126.3, 127.0, 127.1, 127.4, 127.5, 128.5 (2C), 128.9, 129.0 (ArH), 129.96, 130.02, 130.5, 131.5, 131.8, 132.8, 132.9, 133.4, 133.5, 133.7, 134.6, 134.7 (2C), 144.5, 144.9 (2C), 145.1, 152.0, 152.1, 152.5, 153.2 (Ar), 181.6 (CS). Anal. Calcd for C$_{83}$H$_{106}$N$_4$O$_5$S$_2$: C, 74.46; H, 8.19; N, 4.30; S, 4.92. Found: C 75.18; H, 8.72; N, 4.05; S, 4.25.

3.1.2. Procedure for the Synthesis of Symmetric Urea **5b**. Precursor **2b** has Already been Obtained and **3b** and **4b** were Synthesised as Described for **3a** and **4a**

7,13,19,25-Tetra-tert-butyl-28,29-bis[(cyanopropyl)oxy]-27,30-dibutoxy-2,3-dihomo-3-oxacalix[4]arene (**3b**): Flash chromatography (SiO$_2$, eluent gradient from *n*-hexane/ethyl acetate 95:5 to 90:10), 61% yield; RMN ^{1}H (CDCl$_3$, 500 MHz) δ 0.93, 1.19 [2s, 36H, C(CH$_3$)$_3$], 1.00 (t, 6H, *J* = 7.4 Hz, CH$_3$), 1.45 (m, 4H, OCH$_2$CH$_2$*C*H$_2$CH$_3$), 1.76 (m, 4H, OCH$_2$*C*H$_2$CH$_2$CH$_3$), 2.28 (m, 4H, OCH$_2$*C*H$_2$CH$_2$CN), 2.65 (m, 4H, OCH$_2$CH$_2$*C*H$_2$CN), 3.24, 4.31 (ABq, 4H, *J* = 13.5 Hz, ArCH$_2$Ar), 3.28, 4.26 (ABq, 2H, *J* = 13.0 Hz, ArCH$_2$Ar), 3.62, 3.75 (2m, 4H, O*C*H$_2$CH$_2$CH$_2$CH$_3$), 3.84, 3.91 (2m, 4H, O*C*H$_2$CH$_2$CH$_2$CN), 4.49, 4.68 (ABq, 4H, *J* = 13.5 Hz, CH$_2$OCH$_2$), 6.66, 7.00, 7.07 (3d, 8H, ArH); RMN ^{13}C (CDCl$_3$, 125,8 MHz) δ 14.1 (OCH$_2$CH$_2$CH$_2$*C*H$_3$), 14.4 (OCH$_2$CH$_2$CH$_2$CN), 19.3 (OCH$_2$CH$_2$*C*H$_2$CH$_3$), 26.1 (OCH$_2$CH$_2$*C*H$_2$CN), 29.7, 30.3 (ArCH$_2$Ar), 31.4 [C(*C*H$_3$)$_3$], 32.4 (OCH$_2$*C*H$_2$CH$_2$CH$_3$), 34.0, 34.1 [*C*(CH$_3$)$_3$], 67.1 (*C*H$_2$O*C*H$_2$), 72.1, 74.9 (O*C*H$_2$CH$_2$CH$_2$CN and O*C*H$_2$CH$_2$CH$_2$CH$_3$), 119.9 (CN), 123.2, 125.6, 125.8, 126.0 (ArH), 130.9, 132.8, 133.3, 133.7, 145.3, 145.6, 151.9, 152.2 (Ar).

7,13,19,25-Tetra-tert-butyl-28,29-bis[(aminobutyl)oxy]-27,30-dibutoxy-2,3-dihomo-3-oxacalix[4] arene (**4b**): 0.83 g (87% yield) of product pure enough to be immediately used in the next step; RMN ^{1}H (CDCl$_3$, 500 MHz) δ 0.94, 1.18 [2s, 36H, C(CH$_3$)$_3$], 1.01 (t, 6H, *J* = 7.38 Hz, CH$_3$), 1.49–1.60 (m, 8H, OCH$_2$CH$_2$*C*H$_2$CH$_3$ and OCH$_2$CH$_2$*C*H$_2$CH$_2$NH$_a$), 1.79 (m, 4H, OCH$_2$*C*H$_2$CH$_2$CH$_3$), 1.98 (m, 4H, OCH$_2$*C*H$_2$CH$_2$CH$_2$NH$_a$), 2.78 (t, 4H, OCH$_2$CH$_2$CH$_2$*C*H$_2$NH$_a$), 3.17, 4.40 (ABq, 4H, *J* = 13.5 Hz, ArCH$_2$Ar), 3.22, 4.39 (ABq, 2H, *J* = 13.0 Hz, ArCH$_2$Ar), 3.60, 3.69 (2m, 4H, O*C*H$_2$CH$_2$CH$_2$CH$_3$), 3.78,

3.84 (2m, 4H, OCH_2CH$_2$CH$_2$CH$_2$NH$_a$), 4.60, 4.66 (ABq, 4H, J = 13.5 Hz, CH$_2$OCH$_2$), 6.69, 6.96, 6.97, 7.05 (4d, 8H, ArH).

7,13,19,25-Tetra-tert-butyl-28,29-bis[[(N'-1-naphthylureido)butyl]oxy]-27,30-dibutoxy-2,3-dihomo -3-oxacalix[4]arene (**5b**). To a solution of **4b** (0.83 g, 0.89 mmol) in CHCl$_3$ (35 mL) was added 0.26 mL (1.77 mmol) of naphthyl isocyanate. The mixture was stirred at room temperature under N$_2$ for 4 h. Evaporation of the solvent yielded the crude product which was purified by flash chromatography (SiO$_2$, eluent CH$_2$Cl$_2$/MeOH, from 99.7:0.3 to 98:2), followed by recrystallization from CH$_2$Cl$_2$/*n*-hexane: it was obtained in 40% yield (0.45 g); mp 211–213 °C; IR (KBr) 3329 cm^{-1} (NH), 1647 cm^{-1} (CO); ^{1}H-NMR (CDCl$_3$, 500 MHz) δ 0.87 (t, 6H, J = 7.35 Hz, CH$_3$), 0.95, 1.17 [2s, 36H, C(CH$_3$)], 1.42 (m, 4H, OCH$_2$CH$_2$CH_2CH$_3$), 1.70 (m, 8H, OCH$_2$CH_2CH$_2$CH$_3$ and OCH$_2$CH$_2$CH_2CH$_2$NH$_a$), 1.98, 2.06 (2m, 4H, OCH$_2$CH_2CH$_2$CH$_2$NH$_a$), 3.17, 4.33 (ABq, 4H, J = 13.1 Hz, ArCH$_2$Ar), 3.20, 4.39 (ABq, 2H, J = 12.9 Hz, ArCH$_2$Ar), 3.34, 3.43 (2m, 4H, OCH$_2$CH$_2$CH$_2$CH_2NH$_a$), 3.56, 3.72 (2m, 4H, OCH_2CH$_2$CH$_2$CH$_2$NH$_a$), 3.74, 3.77 (2m, 4H, OCH_2CH$_2$CH$_2$CH$_3$), 4.52, 4.61 (ABq, 4H, J = 13.4 Hz, CH$_2$OCH$_2$), 5.83 (t, 2H, NH$_a$), 6.70, 6.95, 7.03 (3d, 8H, ArH), 7.37, 7.38, 7.43 (3t, 6H, Napht), 7.48 (s, 2H, NH$_b$), 7.61, 7.74, 7.81, 8.01 (4d, 8H, Napht); ^{13}C-NMR (CDCl$_3$, 125.8 MHz) δ 14.1 (OCH$_2$CH$_2$CH$_2$CH_3), 19.3 (OCH$_2$CH$_2$CH_2CH$_3$), 27.3, 27.9 (OCH$_2$CH_2CH$_2$CH$_2$NH$_a$), 30.3, 30.5 (ArCH$_2$Ar), 31.5 [C(CH$_3$)$_3$], 32.6 (OCH$_2$CH_2CH$_2$CH$_3$), 34.0 [C(CH$_3$)$_3$], 40.7 (OCH$_2$CH$_2$CH$_2$CH_2NH$_a$), 67.4 (CH$_2$OCH$_2$), 74.21, 74.24 (OCH$_2$CH$_2$CH$_2$CH$_2$NH$_a$ and OCH$_2$CH$_2$CH$_2$CH$_3$), 121.1, 121.6, 123.2, 125.2, 125.56, 125.59, 125.61, 125.97, 126.04, 126.1, 128.5 (ArH), 128.2, 131.0, 133.2, 133.3, 133.7, 134.1, 134.4, 144.8, 145.0, 152.3, 153.0 (Ar), 157.3 (CO). Anal. Calcd for C$_{83}$H$_{106}$N$_4$O$_7$: C, 78.39; H, 8.40; N, 4.41. Found: C,78.51: H, 8.38; N, 4.40).

3.2. Determination of the Crystallographic Structures of 5a and 5b

Single crystals suitable for an X-ray investigation were obtained by slow evaporation of solutions containing compound **5a** and **5b** using dichloromethane/ethanol solvent mixtures. Data collection was carried out at the XRD1 beamline of the Elettra synchrotron (Trieste, Italy), employing the rotating-crystal method with a Dectris Pilatus 2M area detector (DECTRIS Ltd., Baden-Daettwil, Switzerland). Single crystals were dipped in paratone cryoprotectant, mounted on a loop and flash-frozen under a liquid nitrogen stream at a 100 K. Diffraction data were indexed and integrated using the XDS package [36], while scaling was carried out with XSCALE [37]. Structures was solved using the SHELXT program [38] and structure refinement was performed with SHELXL-14 [39], operating through the WinGX GUI [40], by full-matrix least-squares (FMLS) methods on F^2. Non-hydrogen atoms with occupancy of more than 50% were anisotropically refined, while non-hydrogen atoms with a lower occupancy were refined isotropically. Hydrogen atoms were added at the calculated positions and refined using the riding model. Crystallographic data and refinement details are reported in Table S6.

3.3. H-NMR Titrations

The anion association constants (as log K_{ass}) were determined in CDCl$_3$ by ^{1}H-NMR titration experiments. Several aliquots (up to 10 equiv.) of the anion solutions (as TBA salts) were added to 0.5 mL solution of the receptors (2.5 × 10^{-3} –5 × 10^{-3} M) directly in the NMR tube. The spectra were recorded after each addition of the salts, and the temperature of the NMR probe was kept constant at 25 °C. The association constants were evaluated using the WinEQNMR2 program [26] and following the urea NH chemical shifts. The Job methods were performed keeping the total concentration in the same range as before. In the case of ion-pair recognition studies, the percentage of complexation was determined by direct ^{1}H-NMR integration of the free and complexed peaks of the hosts and/or the guests, present at equilibrium. The samples were prepared by mixing aliquots of stock solutions of the host (600 μL) and guests (60 μL) to obtain a final equimolar host-guest solution of 1.0 × 10^{-3} M. For each host-guest system titrations were repeated at least two times. Details related to these experiments have already been described [20].

3.4. UV-Vis Absorption and Fluorescence Studies

Absorption and fluorescence studies were done using an UV-3101PC UV-Vis-NIR spectrophotometer (Shimadzu, Kyoto, Japan) and a Fluorolog F112A fluorimeter (Spex Industries, Edison, NJ, USA) in right-angle configuration, respectively. The association constants were determined in CH_2Cl_2 by UV-Vis absorption spectrophotometry and by steady-state fluorescence at 25 °C. The absorption spectra were recorded between 260 and 370 nm and the emission ones between 325 and 550 nm, and using quartz cells with an optical path length of 1 cm. Several aliquots (up to 10 equiv) of the anion solutions (as TBA salts) were added to a 2 mL solution of the receptors (3.0×10^{-5}–5.0×10^{-5} M) directly in the cell. The spectral changes were interpreted using the HypSpec 2014 program [41]. Details concerning the photophysical properties determination has already been described [23].

3.5. Quantum Chemistry Calculations

Stationary points were optimized with the Gaussian 09 program [42] with the B3LYP [43] density functional with the 6-31G(d,p) basis set. A D3-Grimme correction [44] was also used. Experimental X-ray diffraction structure determinations were employed as the starting structures for the calixarene hosts and different starting positions for the ions were used for the geometry optimization. All reported structures were confirmed as energy minima, with no negative eigenvalue in the Hessian matrix. The structures and energies given are the most stable conformations obtained after optimization. The interaction energy ΔE between the calixarenes and the ions ($\Delta E = E$(complex) − E(free calix) − E(ion)) was calculated with respect to the optimized geometries of all species.

4. Conclusions

New fluorescent dihomooxacalix[4]arene receptors containing two (thio)urea moieties in distal and proximal positions (1,3-dinaphthylurea **5a**, 3,4-dinaphthylurea **5b** and 1,3-dinaphthylthiourea **5c**) at the lower rim linked by a butyl spacer were obtained in the cone conformation in solution. The X-ray structures of **5a** and **5b** were reported and revealed only small differences in the cone conformation. The main difference in the structures is ascribable to the mutual orientation of the naphthyl rings of the ureido substituents, which are almost parallel in **5a** and almost perpendicular in **5b**. Both crystal structures are characterized by intra- and inter-molecular bifurcated H-bonds involving the ureido groups. The anion binding affinity of these derivatives was established by ^{1}H NMR, UV-Vis, fluorescence and DFT studies. 1:1 complexes between anions of different geometries and the receptors through hydrogen bonding were obtained. The results revealed that for all receptors the association constants increase with the anion basicity, and the strongest complexes were obtained with F^-, followed by the carboxylates AcO^- and BzO^-. Symmetric urea **5b** is a better anion receptor compared to the asymmetric urea **5a**, as shown by all the spectroscopic techniques used and corroborated by the quantum mechanical calculations. Both ureido compounds are more efficient than thiourea **5c**. As ditopic receptors, ureas **5a** and **5b** showed a very high affinity for the guests n-$PrNH_3^+ \cdot Cl^-$ and n-$BuNH_3^+ \cdot Cl^-$ ($K_{ass} = 1.0 \times 10^4$ − 2.9×10^4 M^{-2} in CDCl$_3$/DMSO-d_6, 5:1), as well as for the neurotransmitter GABA·HCl ($K_{ass} = 2.2 \times 10^4$ and 6.0×10^4 M^{-2} respectively, in CDCl$_3$/CD$_3$OD, 5:1). The bulkier CH_3 groups of the betaine guest prevented its inclusion inside the macrocycle cavity. Concerning chiral recognition, the enantiopure (R)-sec-$BuNH_3^+ \cdot Cl^-$ guest displayed a 5:2 selectivity towards (P) and (M) enantiomers of the inherently chiral host **5a**. Based on DFT calculations, the [(S)-sec-$BuNH_3^+ \cdot Cl^-/(M)$-**5a**] complex was deduced as the more stable.

Supplementary Materials: The following are available online: Crystallographic data and refinement details; titration curves with TBA salts in CDCl$_3$; Job's plots; absorption and emission spectra with TBA salts; COSY spectra of **5a** + GABA·HCl and **5a** + sec-BuNH$_2$·HCl; ^{1}H, ^{13}C and COSY spectra of compounds **3b**, **5a**, **5b** and **5c**.

Author Contributions: A.S.M.: investigation, acquisition and analysis of data, editing. P.M.M.: conceptualization, supervision, acquisition, analysis and interpretation of data, writing, review and editing. J.R.A.: analysis and interpretation of NMR data, writing and review. M.N.B.-S.: analysis, interpretation and writing of photophysics

Molecules **2020**, *25*, 4708

data. R.S.: performance of DFT calculations, analysis and writing. N.H. and S.G.: analysis of structural data, writing and review. All authors have read and agreed to the published version of the manuscript.

Funding: Authors thank Fundação para a Ciência e a Tecnologia, Projects ref. UID/QUI/00100/2013, UIDB/00100/2020 and UIDB/04565/2020; A. S. Miranda thanks a PhD Grant ref. SFRH/BD/129323/2017.

Conflicts of Interest: The authors declare no conflict of interest.

References and Notes

1. Gutsche, C.D. *Calixarenes: An Introduction: Edition 2*, 2nd ed.; The Royal Society of Chemistry: Cambridge, UK, 2008; Monographs in Supramolecular Chemistry.
2. Neri, P.; Sessler, J.L.; Wang, M.-X. *Calixarenes and Beyond*; Springer International Publishing: Cham, Switzerland, 2016.
3. Kumar, R.; Jung, Y.; Kim, J.S. Fluorescent calixarene hosts. In *Calixarenes and Beyond*; Neri, P., Sessler, J.L., Wang, M.-X., Eds.; Springer International Publishing: Cham, Switzerland, 2016; pp. 743–760.
4. Kumar, R.; Sharma, A.; Singh, H.; Suating, P.; Kim, H.S.; Sunwoo, K.; Shim, I.; Gibb, B.C.; Kim, J.S. Revisiting fluorescent calixarenes: From molecular sensors to smart materials. *Chem. Rev.* **2019**, *119*, 9657–9721. [CrossRef]
5. Kim, S.K.; Bok, J.H.; Bartsch, R.A.; Lee, J.Y.; Kim, J.S. A fluoride-selective PCT chemosensor based on formation of a static pyrene excimer. *Org. Lett.* **2005**, *7*, 4839–4842. [CrossRef] [PubMed]
6. Jeon, N.J.; Ryu, B.J.; Lee, B.H.; Nam, K.C. Fluorescent sensing of tetrahedral anions with a pyrene urea derivative of calix[4]arene chemosensor. *Bull. Kor. Chem. Soc.* **2009**, *30*, 1675–1677.
7. Ryu, B.J.; Jeon, N.J.; Nam, K.C. 1,3-Alternate calix[4]arene bifuntional fluorescent receptor containing urea and crown ether moieties. *Bull. Korean Chem. Soc.* **2010**, *31*, 3445–3447. [CrossRef]
8. Patra, S.; Gunupuru, R.; Lo, R.; Suresh, E.; Ganguly, B.; Paul, P. Cation-induced fluorescent excimer emission in calix[4]arene-chemosensors bearing quinoline as a fluorogenic unit: Experimental, molecular modeling and crystallographic studies. *New J. Chem.* **2012**, *36*, 988–1002. [CrossRef]
9. Sutariya, P.G.; Pandya, A.; Lodha, A.; Menon, S.K. A pyrenyl linked calix[4]arene fluorescence probe for recognition of ferric and phosphate ions. *RSC Advances* **2014**, *4*, 34922–34926. [CrossRef]
10. Uttam, B.; Kandi, R.; Hussain, M.A.; Rao, C.P. Fluorescent lower rim 1,3-dibenzooxadiazole conjugate of calix[4]arene in selective sensing of fluoride in solution and in biological cells using confocal microscopy. *J. Org. Chem.* **2018**, *83*, 11850–11859. [CrossRef]
11. Capici, C.; De Zorzi, R.; Gargiulli, C.; Gattuso, G.; Geremia, S.; Notti, A.; Pappalardo, S.; Parisi, M.F.; Puntoriero, F. Calix[5]crown-3-based heteroditopic receptors for n-butylammonium halides. *Tetrahedron* **2010**, *66*, 4987–4993. [CrossRef]
12. Jeon, N.J.; Ryu, B.J.; Park, K.D.; Lee, Y.J.; Nam, K.C. Tetrahedral anions selective fluorescent calix[6]arene receptor containing urea and pyrene moieties. *Bull. Korean Chem. Soc.* **2010**, *31*, 3809–3811. [CrossRef]
13. Brunetti, E.; Picron, J.-F.; Flidrova, K.; Bruylants, G.; Bartik, K.; Jabin, I. Fluorescent chemosensors for anions and contact ion pairs with a cavity-based selectivity. *J. Org. Chem.* **2014**, *79*, 6179–6188.
14. Brunetti, E.; Moerkerke, S.; Wouters, J.; Bartik, K.; Jabin, I. A selective calix[6]arene-based fluorescent chemosensor for phosphatidylcholine type lipids. *Org. Biomol. Chem.* **2016**, *14*, 10201–10207. [CrossRef] [PubMed]
15. Busschaert, N.; Caltagirone, C.; Van Rossom, W.; Gale, P.A. Applications of supramolecular anion recognition. *Chem. Rev.* **2015**, *115*, 8038–8155. [CrossRef] [PubMed]
16. Gale, P.A.; Howe, E.N.W.; Wu, X. Anion receptor chemistry. *Chem* **2016**, *1*, 351–422. [CrossRef]
17. Kim, S.K.; Sessler, J.L. Ion pair receptors. *Chem. Soc. Rev.* **2010**, *39*, 3784–3809. [CrossRef] [PubMed]
18. McConnell, A.J.; Beer, P.D. Heteroditopic receptors for ion-pair recognition. *Angew. Chem. Int. Ed.* **2012**, *51*, 5052–5061. [CrossRef] [PubMed]
19. Marcos, P.M.; Teixeira, F.A.; Segurado, M.A.P.; Ascenso, J.R.; Bernardino, R.J.; Michel, S.; Hubscher-Bruder, V. Bidentate urea derivatives of *p-tert*-butyldihomooxacalix[4]arene: Neutral receptors for anion complexation. *J. Org. Chem.* **2014**, *79*, 742–751. [CrossRef]
20. Gattuso, G.; Notti, A.; Parisi, M.F.; Pisagatti, I.; Marcos, P.M.; Ascenso, J.R.; Brancatelli, G.; Geremia, S. Selective recognition of biogenic amine hydrochlorides by heteroditopic dihomooxacalix[4]arenes. *New J. Chem.* **2015**, *39*, 817–821. [CrossRef]

21. Teixeira, F.A.; Marcos, P.M.; Ascenso, J.R.; Brancatelli, G.; Hickey, N.; Geremia, S. Selective binding of spherical and linear anions by tetraphenyl(thio)urea-based dihomooxacalix[4]arene receptors. *J. Org. Chem.* **2017**, *82*, 11383–11390. [CrossRef]

22. Augusto, A.S.; Miranda, A.S.; Ascenso, J.R.; Miranda, M.Q.; Félix, V.; Brancatelli, G.; Hickey, N.; Geremia, S.; Marcos, P.M. Anion recognition by partial cone dihomooxacalix[4]arene-based receptors bearing urea groups: Remarkable affinity for benzoate ion. *Eur. J. Org. Chem.* **2018**, *2018*, 5657–5667. [CrossRef]

23. Miranda, A.S.; Serbetci, D.; Marcos, P.M.; Ascenso, J.R.; Berberan-Santos, M.N.; Hickey, N.; Geremia, S. Ditopic receptors based on dihomooxacalix[4]arenes bearing phenylurea moieties with electron-withdrawing groups for anions and organic ion pairs. *Front. Chem.* **2019**, *7*, 758. [CrossRef]

24. Miranda, A.S.; Martelo, L.M.; Fedorov, A.A.; Berberan-Santos, M.N.; Marcos, P.M. Fluorescence properties of p-tert-butyldihomooxacalix[4]arene derivatives and the effect of anion complexation. *New J. Chem.* **2017**, *41*, 5967–5973. [CrossRef]

25. Jaime, C.; de Mendoza, J.; Prados, P.; Nieto, P.; Sanchez, C. Carbon-13 NMR chemical shifts. A single rule to determine the conformation of calix[4]arenes. *J. Org. Chem.* **1991**, *56*, 3372–3376. [CrossRef]

26. Hynes, M.J. EQNMR: A computer program for the calculation of stability constants from nuclear magnetic resonance chemical shift data. *J.Chem. Soc., Dalton Trans.* **1993**, 311–312. [CrossRef]

27. Stibor, I.; Budka, J.; Michlova, V.; Tkadlecova, M.; Pojarova, M.; Curinova, P.; Lhotak, P. Systematic approach to new ligands for anion recognition based on ureido-calix[4]arenes. *New J. Chem.* **2008**, *32*, 1597–1607. [CrossRef]

28. Marcos, P.M.; Teixeira, F.A.; Segurado, M.A.P.; Ascenso, J.R.; Bernardino, R.J.; Brancatelli, G.; Geremia, S. Synthesis and anion binding properties of new dihomooxacalix[4]arene diurea and dithiourea receptors. *Tetrahedron* **2014**, *70*, 6497–6505. [CrossRef]

29. Teixeira, F.A.; Ascenso, J.R.; Cragg, P.J.; Hickey, N.; Geremia, S.; Marcos, P.M. Recognition of anions, monoamine neurotransmitter and trace amine hydrochlorides by ureido-hexahomotrioxacalix[3]arene ditopic receptors. *Eur. J. Org. Chem.* **2020**, 1930–1940. [CrossRef]

30. Scheerder, J.; Fochi, M.; Engbersen, F.J.; Reinhoudt, D.N. Urea-derivatized p-tert-butylcalix[4]arenes: Neutral ligands for selective anion complexation. *J. Org. Chem.* **1994**, *59*, 7815–7820. [CrossRef]

31. Hamon, M.; Ménand, M.; Le Gac, S.; Luhmer, M.; Dalla, V.; Jabin, I. *Calix[6]tris(thio)ureas: Heteroditopic receptors for the cooperative binding of organic ion pairs. J. Org. Chem.* **2008**, *73*, 7067–7071.

32. Bryantsev, V.S.; Hay, B.P. Conformational preferences and internal rotation in alkyl- and phenyl-substituted thiourea derivatives. *J. Phys. Chem. A* **2006**, *110*, 4678–4688. [CrossRef]

33. Mei, M.; Wu, S. Fluorescent sensor for α,ω-dicarboxylate anions. *New J Chem.* **2001**, *25*, 471–475. [CrossRef]

34. Gaeta, C.; Talotta, C.; Farina, F.; Teixeira, F.A.; Marcos, P.M.; Ascenso, J.R.; Neri, P. Alkylammonium cation complexation into the narrow cavity of dihomooxacalix[4]arene macrocycle. *J. Org. Chem.* **2012**, *77*, 10285–10293. [CrossRef] [PubMed]

35. Due to the inherent chirality of dihomooxacalix[4]arenes, (*P*) and (*M*) descriptors were used as an alternative to the usual (*R*) and (*S*) notation. For this subject see: (a) Szumna, A. Inherently chiral concave molecules – from synthesis to applications. *Chem. Soc. Rev.* **2010**, *39*, 4274–4285. (b) ref 22.

36. Kabsch, W. XDS. *Acta Crystallogr. Sect. D Biol. Crystallogr.* **2010**, *66*, 125–132. [CrossRef] [PubMed]

37. Kabsch, W. Integration, scaling, space-group assignment and post-refinement. *Acta Crystallogr. Sect. D Biol. Crystallogr.* **2010**, *66*, 133–144. [CrossRef] [PubMed]

38. Sheldrick, G.M. SHELXT–Integrated space-group and crystal-structure determination. *Acta Crystallog. Sect. A Found. Crystallogr.* **2015**, *71*, 3–8. [CrossRef] [PubMed]

39. Sheldrick, G.M. A short history of SHELX. *Acta Crystallog. Sect. A Found. Crystallogr.* **2008**, *64*, 112–122. [CrossRef] [PubMed]

40. Farrugia, L.J. WinGX and ORTEP for Windows: An update. *J. Appl. Crystallog.* **2012**, *45*, 849–854. [CrossRef]

41. Gans, P.; Sabatini, A.; Vacca, A. Investigation of equilibria in solution. Determination of equilibrium constants with the HYPERQUAD suite of programs. *Talanta* **1996**, *43*, 1739–1753. [CrossRef]

42. Frisch, M.J.; Trucks, G.W.; Schlegel, H.B.; Scuseria, G.E.; Robb, M.A.; Cheeseman, J.R.; Scalmani, G.; Barone, V.; Mennucci, B.; Petersson, G.A.; et al. *Gaussian 09 Rev. D.01*; Gaussian Inc.: Wallingford, CT, UAS, 2016.

43. Becke, A.D. Density-functional thermochemistry. III. The role of exact exchange. *J. Chem. Phys.* **1993**, *98*, 5648–5652. [CrossRef]

Molecules 2020, 25, 4708

44. Grimme, S.; Antony, J.; Ehrlich, S.; Krieg, H. A consistent and accurate ab initio parametrization of density functional dispersion correction (DFT-D) for the 94 elements H-Pu. *J. Chem. Phys.* **2010**, *132*, 154104. [CrossRef]

Sample Availability: Samples of the compounds are not available from the authors.

Publisher's Note: MDPI stays neutral with regard to jurisdictional claims in published maps and institutional affiliations.

 *molecules*

Article

Highly Sensitive and Selective Fluorescent Probes for Cu(II) Detection Based on Calix[4]arene-Oxacyclophane Architectures

Alexandra I. Costa [1,2], Patrícia D. Barata [1,2], Carina B. Fialho [1,†] and José V. Prata [1,2,*]

[1] Departamento de Engenharia Química, Instituto Superior de Engenharia de Lisboa,
     Instituto Politécnico de Lisboa, R. Conselheiro Emídio Navarro, 1, 1959-007 Lisboa, Portugal;
     acosta@deq.isel.ipl.pt (A.I.C.); pbarata@deq.isel.ipl.pt (P.D.B.); carina.fialho@tecnico.ulisboa.pt (C.B.F.)

[2] Centro de Química-Vila Real, Universidade de Trás-os-Montes e Alto Douro, 5001-801 Vila Real, Portugal

*   Correspondence: jvprata@deq.isel.ipl.pt; Tel.: +351-218317172

†   Present address: Centro de Química Estrutural, Instituto Superior Técnico, Universidade de Lisboa,
     Av. Rovisco Pais, 1049-001 Lisboa, Portugal.

Academic Editor: Paula M. Marcos
Received: 4 May 2020; Accepted: 22 May 2020; Published: 25 May 2020

Abstract: A new topological design of fluorescent probes for sensing copper ion is disclosed. The calix[4]arene-oxacyclophane (Calix-OCP) receptor, either wired-in-series in arylene-*alt*-ethynylene conjugated polymers or standing alone as a sole molecular probe, display a remarkable affinity and selectivity for Cu(II). The unique recognition properties of Calix-OCP system toward copper cation stem from its pre-organised cyclic array of *O*-ligands at the calixarene narrow rim, which is kept in a conformational rigid arrangement by a tethered oxacyclophane sub-unit. The magnitude of the binding constants ($K_a = 5.30 - 8.52 \times 10^4 \ M^{-1}$) and the free energy changes for the inclusion complexation ($-\Delta G = 27.0 - 28.1 \ kJmol^{-1}$), retrieved from fluorimetric titration experiments, revealed a high sensitivity of Calix-OCP architectures for Cu(II) species. Formation of supramolecular inclusion complexes was evidenced from UV-Vis spectroscopy. The new Calix-OCP-conjugated polymers (polymers **4** and **5**), synthesized in good yields by Sonogashira–Hagihara methodologies, exhibit high fluorescence quantum yields ($\Phi_F = 0.59 - 0.65$). Density functional theory (DFT) calculations were used to support the experimental findings. The fluorescence on–off behaviour of the sensing systems is tentatively explained by a photoinduced electron transfer mechanism.

Keywords: calix[4]arene; copper; supramolecular; inclusion complex; fluorescence; sensor; density functional theory

---

1. Introduction

Molecular fluorescent systems for signalling supramolecular interactions have been in use for more than forty years [1,2]. When aimed for ion detection, the system usually comprises either a fluorogenic unit covalently linked through some sort of a spacer to an ion receptor or an integrated fluorophore-ion recognition site assembly [1–3]. The photophysical changes that occur upon cation binding, which are the basis for signalling the binding event, may have different origins. The main underlying mechanisms are based on photoinduced electron transfer (PET), photoinduced charge transfer (PCT), excimer-formation/disappearance and Förster resonance energy transfer (FRET) processes [4]. The electronic and stereochemical characteristics of the recognition unit and the fluorophore, as well as the type of metal ion, dictate the actual mechanism. With transition metal ions, examples covering all the above-mentioned processes are known [4].

Calixarenes are cyclic oligomers widely investigated due to their ability to recognize and differentiate neutral and ionic guests through the formation of inclusion complexes [5,6].

Various topologies of fluorescent sensors for metal cation recognition based on calixarene-derived molecular receptors have been developed [7]. Of these, those targeting transition metal ions are of particular relevance for the present study. In general, sensors are built by attaching the fluorogenic unit(s) to the narrow rim of the calixarene receptor. The most used signalling elements belong to naphthyl, anthracenyl, pyrenyl, dansyl, quinolinyl, and benzimidazolyl groups [4,7]. On the receptor side, calixarenes having cone or 1,3-alternate conformations are usually employed [4,7].

Copper(II), lead(II) and mercury(II) ions are ubiquitous in nature, either associated to important physiological processes [8] or representing environmental hazards and health issues [9–13], making their sensitive and selective detection highly desirable. The literature regarding the use of calixarene-based fluorescence sensors for recognition of the above transition metal cations have been recently reviewed and will not be further detailed here [14].

Both off–on (fluorescence enhancement) [15–17] and on–off (fluorescence quenching) [18–20] responses have been used for signalling the binding events with copper ion, using the above described assemblies of calixarene receptors and fluorophore units, usually with high sensitivities.

The above reports stimulated our interest to screen the ability of a series of fluorescent calix[4]arene platforms we have synthesized in the past, mostly designed as probes for nitroaromatic explosives [21,22], nitroaliphatic explosives and taggants [23,24], nitroanilines [25,26], and proteins [27], as sensors of transition metal ions, specifically Cu(II), Pb(II) and Hg(II). A particular family of calixarene derivatives having an oxacyclophane (OCP) unit tethered to the narrow rim of a calixarene unit [25], either as single-molecules (**Calix-OCP-2-CBZ**; Chart 1a) [28] or integrated in a polymer (**Calix-OCP-PPE**; Chart 1b) [26] was envisaged as having potential for transition metal ions recognition. This stems from their well-defined cone conformations, which may offer suitable pre-organised ion binding sites at the Calix-OCP skeleton. The enhanced rigidity of the calix[4]arene structure provided by the OCP sub-unit, which concomitantly defines the size of the recognition centre, may potentiate the selectivity of binding to metal ions as a function of their size and charge. Therefore, they were selected for the current study. In addition, two new conjugated polymers differing in the substitution at the *para* position of aryl rings (with and without *tert*-butyl groups) were constructed from the above calixarene scaffolds, having in their chains phenylene-*alt*-ethynylene-*alt*-2,7-carbazolylenes as repeating units (***p*-H-Calix-OCP-PAE-2,7-CBZ** (4) and ***p*-*tert*-butyl-Calix-OCP-PAE-2,7-CBZ** (5); Scheme 1). It is expected that the good donor ability of the carbazole moieties could enhance the sensing capabilities of the whole assembly in comparison to similar calixarene architectures possessing phenylene-*alt*-ethynylene conjugated units along the polymer chain [*cf.* **Calix-OCP-PPE**].

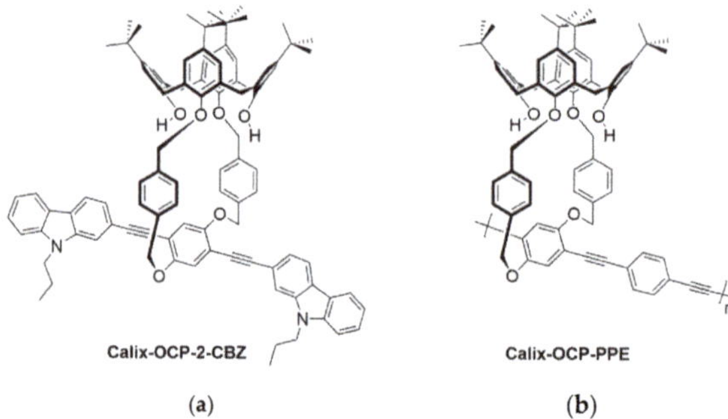

Chart 1. Chemical structures of **Calix-OCP-2-CBZ** (a) [28] and **Calix-OCP-PPE** (b) [26].

Screening of this pool of potential supramolecular probes will possibly allow a series of interesting observations and useful conclusions to be reached. One will be what entity is responsible for recognition of the metal ion (site specificity). The other will focus on any sensitive enhancement of sensing behaviour promoted by conjugated polymer chains (two types in appreciation) in comparison to the corresponding sole molecular probe. A third will be to what extent the electronic/stereochemical nature of the conjugated polymer backbone influences the signal transduction of the recognition event. The last will result from the evaluation of the impact of the substitution pattern at the upper rim of the calixarene unit on the sensitivity and selectivity of the sensor system.

2. Results and Discussion

2.1. Synthesis of Polymeric Fluorescent Probes

The new fluorescent polymers (**4** and **5**) were synthesized by a Pd-catalysed Sonogashira–Hagihara methodology. Cross-coupling of oxacyclophane tethered calix[4]arenes (Calix-OCPs) **1** [29] and **2** [25] (Scheme 1) with 2,7-diethynyl-9-propyl-9H-carbazole (2,7-CBZ) monomer **3** [30] furnished the desired polymers. Polymerisations were carried out in toluene/NEt$_3$ at 35 °C for several hours (24–48 h), using PdCl$_2$(PPh$_3$)$_2$/CuI as the catalytic system (Scheme 1). The polymers were isolated as yellow solids in good yields (58–66%). The polymers are moderately soluble in CH$_2$Cl$_2$, CHCl$_3$, and THF, and only sparingly soluble in CH$_3$CN.

Scheme 1. Cross-coupling of calix[4]arene diiodo derivatives **1** and **2** with 2,7-diethynyl-9-propyl-9H-carbazole (**3**) using PdCl$_2$(PPh$_3$)$_2$/CuI catalytic system in toluene/NEt$_3$ at 35 °C for 24–48 h.

Gel permeation chromatography (GPC) was used to estimate the polymers' molecular weights (see GPC traces in Figure S1; Supplementary Materials). GPC analysis of the isolated polymers showed the presence of less than 1% of monomers in their composition. Data for polymers **4** and **5** are summarised in Table 1.

Table 1. Data from the cross-coupling polymerisation of Calix-OCPs (**1** and **2**) with 2,7-CBZ (**3**) [1].

Polymer	Yield (%) [2]	M_w (g/mol) [3]	M_n (g/mol) [3]	M_w/M_n [3]
4	58	14842	5276	2.81
5	66	66468	15587	4.26

[1] Typical conditions: PdCl$_2$(Ph$_3$P)$_2$ (7 mol%), CuI (7 mol%), toluene, NEt$_3$, and 2,7-CBZ units (1.1 equivalent), 35 °C, 24–48 h; [2] Refer to yields of isolated soluble fractions with less than 1% of **1** or **2** (GPC evaluation); [3] Number-average (M_n) and weight-average (M_w) molecular weights and polydispersity (M_w/M_n) were retrieved from by GPC data, using THF as eluent at 35 °C and monodisperse polystyrene standards.

Formation of a reddish gel-like insoluble material was noted during the isolation of polymers, being this fraction higher for polymer **4**. This may explain the lower yield of its isolated soluble fraction, and also the lower DP attained in its isolated fraction (DP = 5).

2.2. Structural Characterization of Polymers

FTIR and ^{1}H NMR analyses were used to structurally characterise the new materials. One characteristic FTIR feature of the synthesized polymers is the disappearance of terminal ethynylic C≡C-H stretching vibrations characteristic of 2,7-CBZ monomer unit [30] and the concomitant presence of internal alkyne frequencies at 2202 cm^{-1} (**4**) and 2204 cm^{-1} (**5**). This indicates that the majority of the polymer chains are end-terminated by iodine. Other relevant vibrational frequencies may be found in the Experimental section. Analysis of FTIR spectra of the insoluble fractions resulting from the polymerisations leading to polymers **4** and **5** showed they are similar to those obtained from the corresponding soluble fractions. The insoluble fractions should then represent polymers with higher degrees of polymerisation, which render them insoluble.

The somewhat limited solubility of polymers **4** and **5** in $CDCl_3$, CD_2Cl_2, or DMSO, in proper concentrations for NMR experiments, allied to the usual broadening of signals in complex polymer matrices, prevent the acquisition of good quality ^1H NMR spectra. Nonetheless, the spectral assignments that were tentatively made (see Experimental section) fully confirm the proposed structures. In particular, the cone conformation of the calixarene units in both polymers was substantiated by the existence of a set of characteristic resonances for the diastereotopic protons of bridged methylene groups in the calixarene skeleton. For polymer **5** the equatorial protons present signals at 3.22 ppm (2H, d, J = 13.6 Hz), 3.33 ppm (2H, d, J = 13.2 Hz), and at 4.25–4.50 ppm (4H, two unresolved doublets) for the axial protons. The corresponding resonances in **4** appear at 3.25-3.45 ppm (4H, two types of unresolved doublets) and 4.32–4.42 ppm (4H, two unresolved doublets), respectively. The two types of signals for equatorial protons present in **4** and the pattern of resonances for the diastereotopic benzylic protons (4.90–5.62 ppm) is an indication of a much less conformationally fixed calixarene structure. NMR spectra of **4** and **5** are displayed in Figures S2 and S3.

2.3. Photophysical Properties

The ground–state absorption, excitation and emission spectra of the polymers **4** and **5** are depicted in Figure 1.

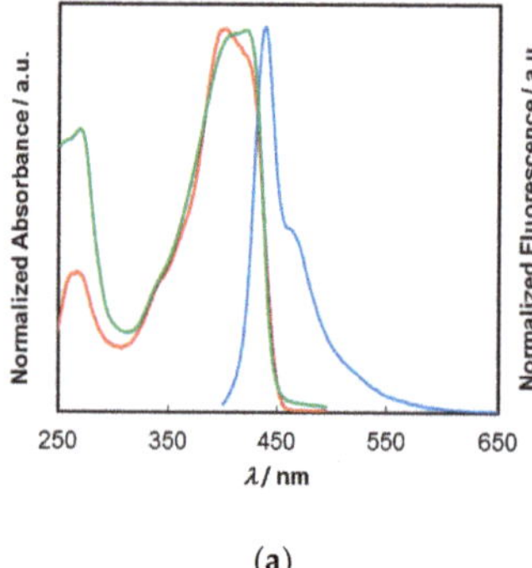
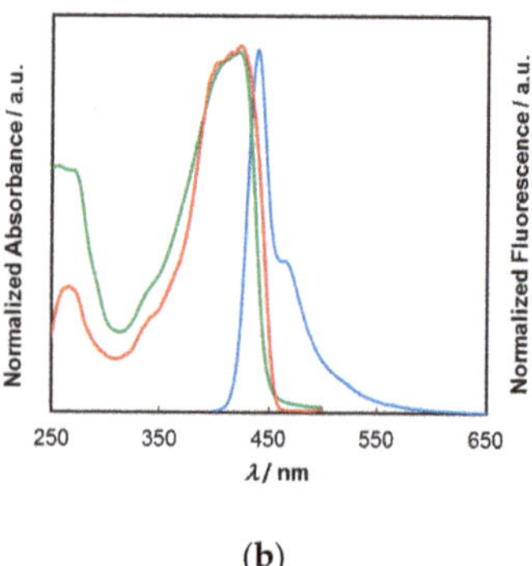

(a)   (b)

Figure 1. Absorption (green line), excitation (red line, monitored at emission maxima) and emission (blue line) spectra (λ_{exc} = 380 nm, $CHCl_3$) of polymer **4** (a) and polymer **5** (b).

Polymers **4** and **5** have similar absorption profiles, both peaking around 420 nm at their absorption maxima. A large molar absorptivity is observed for **5**. Note the similarity of the absorption maxima of polymers despite their different degrees of polymerisation; this means that the mean conjugation length of the polymer chain reaches a plateau within at least around five repeating units. The optical HOMO-LUMO energy gaps (E_g), calculated from the low energy onset of the absorption bands of the polymers ($E_{g(4)}$ = 2.87 eV; $E_{g(5)}$ = 2.86 eV) are in line with the above observations.

Polymers **4** (Φ_F = 0.65) and **5** (Φ_F = 0.59) are highly fluorescent in $CHCl_3$ solution. Their emission profiles are again very similar, presenting 0–0 transitions at 431 nm for **4** and a slightly redshift emission for polymer **5** at 439 nm, followed by discernible 0-1 vibronic progressions at 458 and 467 nm, respectively. In each case, relatively short Stokes shifts were found. The main photophysical properties of **4** and **5** are gathered in Table 2.

Table 2. Photophysical properties of polymers **4** and **5** in $CHCl_3$.

Polymer	$\lambda_{abs\,max}$/nm ($\log\varepsilon_{max}$)	E_g/eV [1]	$\lambda_{em\,max}$/nm	Stokes Shift/nm [2]	Φ_F [3]
4	420 (4.621)	2.87	431 [4] 458 [5]	11	0.65
5	422 (4.682)	2.86	439 [4] 467 [5]	17	0.59

[1] Optical energy gaps (E_g) were calculated from the low energy onset of the absorption bands; [2] Stokes shifts were calculated from $\lambda_{F,\,max}$ (0-0) − $\lambda_{A,\,max}$; [3] Fluorescence quantum yields were determined in $CHCl_3$ using 9,10-diphenylanthracene as reference in air-equilibrated conditions; [4] At the maximum of the first vibronic (0-0) band; [5] At the maximum of the second vibronic (0-1) band.

2.4. Inclusion Complexes and Sensing of Metal Cations

The sensing ability of our integrated receptor–fluorophore systems toward metal ions are expected to closely follow their aptness for supramolecular complex formation. Their sensorial activity was assessed through fluorimetric titrations. The experiments were carried out by titrating diluted CH_3CN solutions of fluorescent polymers (mostly at 5.0×10^{-6} M) with metal cations as perchlorates salts. The photostability of polymers **4** and **5** in CH_3CN was first evaluated under the conditions used in the titration experiments (λ_{exc} = 380 nm, right angle illumination). No photodegradation occurs in any of the synthesized polymers up to 1 h of continuous irradiation.

The Stern–Volmer expression

$$F_0/F = 1 + K_{SV}\,[Q] \tag{1}$$

derived for static quenching (with $K_{SV} = K_a$) [31], or any of its modified forms, such as the reciprocal fitting (Benesi–Hildebrand or Lineweaver–Burk) or double logarithmic plots (Scatchard), are widely used to calculate the association constant in binding events. However, they may furnish erroneous results. In Equation (1), K_{SV} is the static Stern–Volmer constant (also denoted as K_S) and [Q] is the free quencher concentration. As several authors have noticed [32–34], there are two main reasons why these formalisms may fail to produce accurate results. One is to consider that the host–guest complex once formed is completely non-fluorescent and the other is the assumption that the concentration of added guest during a titration experiment (and later used in Stern–Volmer plots from which K_{SV} is extracted) is identical to that of the free guest (the one featuring in Equation (1). Deviations from the true K_a value are more significant in systems with large binding affinities, in which the concentration of free guest is considerably lower than that of added guest. Following the above reasoning, all the association constants presented in this work were calculated by a non-linear fitting approach (see below) because it is a more accurate fitting method, and no a priori assumptions need to be made.

The extent of complexation in fluorimetric assays was quantified as follows. The fluorescence intensity of the unbound host (H, calixarene species) in diluted solutions is expressed through the equation

$$F = \varepsilon_F[H] = (\ln 10)\Phi_F I_0 \varepsilon\, l\, [H] \tag{2}$$

where ε_F is the molar fluorescence intensity [35]. Changes in fluorescence quantum yield (Φ_F) and/or in the molar absorptivity (ε) after exposure to a given analyte, while keeping the exciting intensity (I_0) and path length (l) constant, may lead to observable changes in the fluorescence intensity, which constitutes the basis of fluorescence titration. The association constant (K_a) for the formation of a supramolecular complex (HG) between the host and the metal guest (G), from the 1:1 equilibria H + G $\rightleftarrows$ HG, is described by

$$K_a = [HG]/[H][G] \tag{3}$$

By combining Equations (2) and (3), the quadratic equation

$$\Delta F^2 - \Delta\varepsilon_F([H]_0 + [G]_0 + 1/K_a)\,\Delta F + \Delta\varepsilon_F^{\,2}[H]_0[G]_0 = 0 \tag{4}$$

may be derived [36]. The association constant of the binding event may then be calculated by solving Equation (4) for ΔF (Equation (5)):

$$\Delta F = 1/2\{\Delta\varepsilon_F([H]_0 + [G]_0 + 1/K_a) - [\Delta\varepsilon_F^2([H]_0 + [G]_0 + 1/K_a)^2 - 4\Delta\varepsilon_F^2[H]_0[G]_0]^{1/2}\} \qquad (5)$$

where ΔF and $\Delta\varepsilon_F$ are the changes in fluorescence intensity and molar fluorescence intensity of the host upon complexation with metal ions, K_a is the association constant, and $[H]_0$ and $[G]_0$ denote the initial concentrations of the host and the guest, respectively. Using a non-linear regression analysis, the association constant of the complex can be retrieved using ΔF, $[G]_0$ and $[H]_0$ as input parameters. This equation holds for a 1:1 host-to-guest equilibrium and considers a neglectable dynamic quenching component for the system under study.

Before applying this model, it was verified that the reduction of emission intensity is only due to host-guest interactions and, therefore, artefacts such as hetero inner-filter effects (h-IFEs) by metal salts were absent. Indeed, the metal salts used in this work do not exhibit absorption of radiation at the excitation (380 nm) or emission wavelengths (400–470 nm) of the fluorophores; thus, no h-IFEs could occur in these systems.

Polymer **5** was assayed in first place. As depicted in Figure 2a, the emission of **5** is strongly quenched upon contact with increasing amounts of Cu(II), reaching a 44% reduction of the initial fluorescence intensity with the addition of 5 equivalent of copper ion.

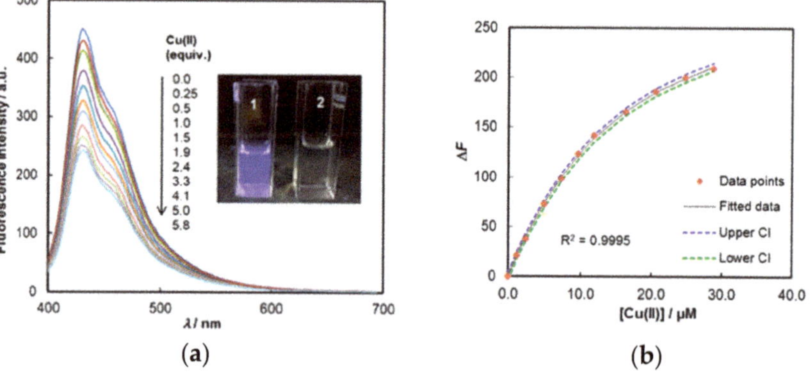

Figure 2. (a) Emission spectra of polymer **5** (5.0×10^{-6} M in CH_3CN) upon addition of increasing amounts (0.25 – 5.8 equivalent) of $Cu(ClO_4)_2$ (λ_{exc} = 380 nm). Inset: photo of polymer **5** fluorescence under UV irradiation (366 nm) before (**1**) and after (**2**) $Cu(ClO_4)_2$ addition (5 equivalent); (b) Binding isotherm for the fluorimetric titration of **5** with Cu(II) with fitted curve and confidence intervals (see text).

Titration data and curve-fitting plots for the supramolecular titration of polymer **5** with Cu(II) are shown in Figure 2b. Regression analysis was performed by the Solver function in Microsoft Excel [37] using the non-linear generalised reduced gradient (GRG) algorithm, from which a $K_a = (8.52 \pm 0.82) \times 10^4$ M^{-1} was derived. The excellent goodness of fit (R^2 = 0.9995) confirms the 1:1 stoichiometry of the supramolecular complex. The degree of uncertainty was estimated from the upper and lower confidence intervals (CI) for fitted data (Figure 2b), calculated from the standard error of residuals and the critical t-value [37,38]. The stoichiometry of the complex was also assessed by the continuous variation method (Job's method). By plotting the concentration of the host–guest complex ([HG]) against the mole fraction (f) of the guest, the stoichiometry of the complex can be found for f corresponding to the maximum of [HG]. The relative complex concentration was calculated from [HG] = $(F_0 - F)/F_0 \times$ [H], where $(F_0 - F)/F_0$ is the relative fluorescence intensities and [H] the concentration of pure **5**. The Job plot (Figure 3) exhibits a maximum at a mole fraction of 0.5 indicating the formation of a complex with n:n stoichiometry (1:1, 2:2, or higher) between **5** and Cu(II). Full evidence for a 1:1 complex was obtained after application of a described method (Table S1) [39].

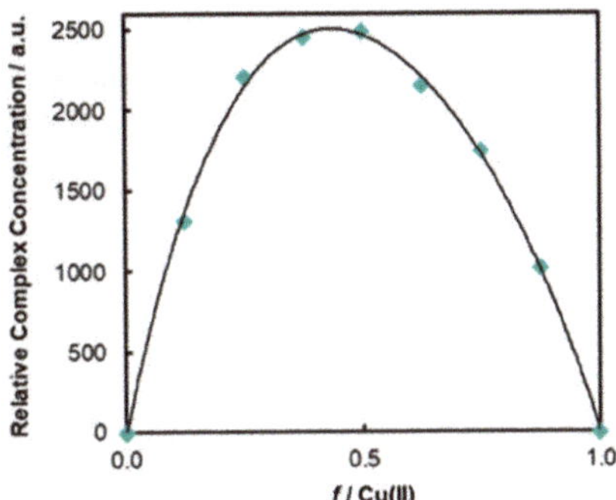

Figure 3. Job plot for complex formation between host **5** and Cu(II) in CH$_3$CN (at constant 1.0×10^{-5} M total concentration) as obtained from changes in fluorescence (λ_{exc} = 380 nm).

A previously designed conjugated polymer (**Calix-OCP-PPE (6)**; DP = 35) [26], possessing the same recognition unit (*tert*-butyl-calix[4]arene-oxacyclophane) as **5** but holding instead phenyleneethynylene entities along the polymer chain (Chart 1b), was next evaluated for its recognition ability. The titration data (Figure S4a) were fitted by the non-linear procedure yielding an association constant of $(5.30 \pm 1.27) \times 10^4$ M^{-1} (Figure S4b). As with polymer **5**, a very high sensitivity was observed.

In order to get further insights about the role of conjugated polymer backbones in polymers **5** and **6** on the hypothetical amplification of the binding event, likely occurring at the narrow rim of the calixarene sub-unit (see below), the molecular receptor **Calix-OCP-2-CBZ** [28] (Chart 1a), having in common with the former polymers the calixarene-OCP framework, was investigated. Following similar fitting procedures, it was found that this sole molecule also displayed a high binding affinity for copper ion ($K_a = (6.62 \pm 1.92) \times 10^4$ M^{-1}), albeit with a higher statistical uncertainty when compared to **5** (Figure S5a,b). On the other hand, a bis-calix[4]arene derivative, **bis-Calix-TriPr-2-CBZ** [23] (Chart 2), having the same fluorophore element (bis-2-CBZ-phenyleneethynylene) but lacking the calix-oxacyclophane architecture did not show any responsiveness to Cu(II) (Figure S6). It results obvious that the fluorophore unit is not able by itself to give any direct response to the presence of the analyte.

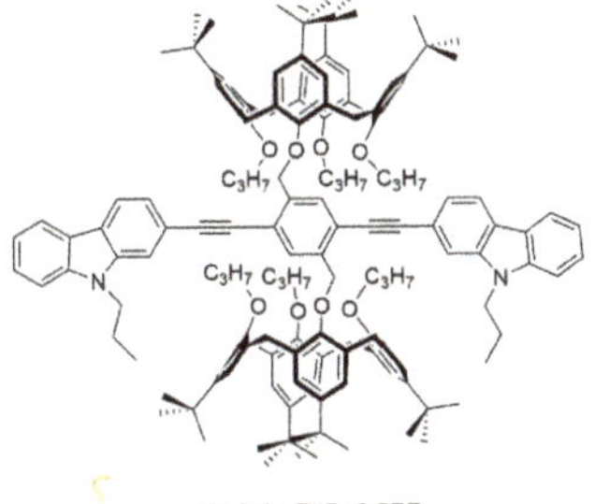

Chart 2. Chemical structure of **bis-Calix-TriPr-2-CBZ** [23].

From the above findings, one may formulate that for a strong binding event to occur, and subsequently a high sensitivity in detection could be observed, it is absolutely essential the existence of a conformationally rigid cyclic array of *O*-ligands at the narrow rim of calixarene derivative functioning as the cation receptor. Calixarene-OCP structures provide such a binding site in which the four oxygen ligands are very well positioned, and possess the appropriate inter-atomic distances between them that will incidentally allow an effective interaction with incoming cations. Of course, the proper match between the size of the ligand pocket and the size of the cation will dictate the relative success of the binding, and also any observable selectivity. This is what apparently occurs in the case of polymers **5** and **6**, and **Calix-OCP-2-CBZ**, where the copper ion fits well into the host cavity.

It has been shown by single-crystal X-ray diffraction that in polynuclear clusters formed from *p-tert*-butyl-calix[4]arenetetraol and Fe(III) [40] or Cu(II) [41], the four narrow rim oxygen atoms bind to either of the metal ions in a square-planar geometry. Density functional theory (DFT) calculations on *p*-H-calix[4]arenes containing appended coumarins [42] also predicted binding through the phenolic oxygens to Cu(II) and Fe(III) cations.

Support for the occurrence of a ground-state complex between polymer **5** and Cu(II) was brought about by UV-Vis spectroscopic titrations. As shown in Figure 4a, a stepwise increment of the amount of copper ion (up to 10 equivalent) causes appreciable changes in the absorption profile. In the range 275–350 nm, an increase in absorbance is assignable to a ligand-to-metal charge transfer (LMCT) absorption band, whereas an intense hypochromic effect appears within the π-π^* transition region of the conjugated polymer system. Two clear isosbestic points appear at 370 and 439 nm denoting the establishment of the binding equilibrium. Similar features appear in the titration of **Calix-OCP-2-CBZ** with Cu(II), also showing the LMCT band and a less pronounced chromic effect on the region of phenyleneethynylene-2-CBZ chromophore (Figure 4b). The two isosbestic points are located at 340 and 398 nm.

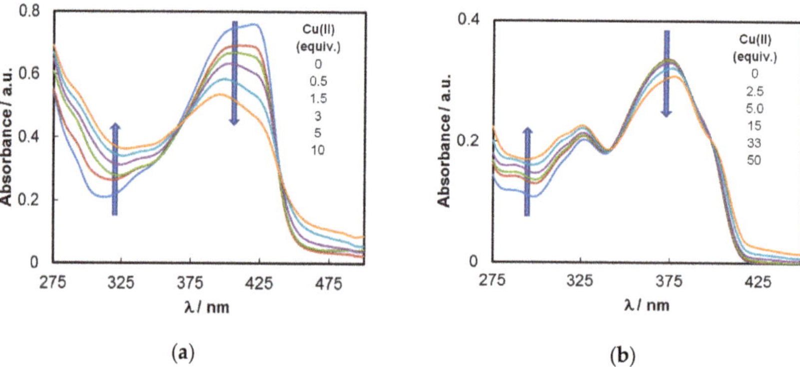

Figure 4. Absorption spectra of (a) polymer **5** (8.2×10^{-5} M in CH$_3$CN:CHCl$_3$, 1:1) and (b) **Calix-OCP-2-CBZ** (5.0×10^{-6} M in CH$_3$CN) upon addition of increasing amounts of Cu(ClO$_4$)$_2$.

To further elucidate the above interactions, a Calix-OCP-Cu(II) model was constructed. It is known that Cu(II) forms coordination compounds with different geometries (tetra- to hexa-coordinate) [43]. In the present case, and for the analysis of any conformational changes occurring in the calixarene structure upon binding, a square planar tetra-coordinate geometry for Cu(II) was considered for modelling purposes. Even if solvated forms of the complex truly exist, for example with CH$_3$CN bound at one or two apical positions of Cu(II) leading to putative square pyramidal or octahedral geometries, respectively, this should not interfere with the main conclusions to be drawn from the model.

The energy and geometry of the complex were optimised by DFT calculations at B3LYP/6-31G(d) level of theory [44] (see details in Materials and Methods section). On complexation with Cu(II) ion, the structure of the host, having in its unbound state a pinched cone conformation (C_{2v} symmetry) in the calixarene unit (Figure 5a), adopts an almost perfect cone arrangement (C_{4v} symmetry) (Figure 5b). Besides, the attached oxacyclophane ring changes its pristine conformation dramatically in order to better accommodate the copper ion.

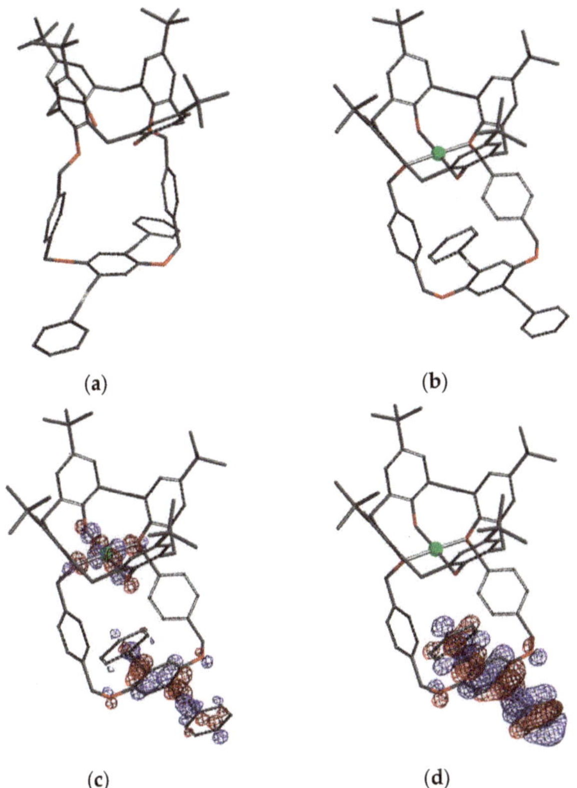

Figure 5. Energy and geometry optimised structures of Calix-OCP (**a**) and Calix-OCP-Cu(II) (**b**) models (side views). β-HOMO (**c**) and α-HOMO (**d**) molecular orbitals mapped on the structure of the complex (side views). DFT calculations run at the B3LYP/6-31G (d) level of theory in vacuum [44]. Hydrogens omitted for clarity. Colour codes for elements: red = oxygen, grey = carbon, green = copper.

The average distance between the distal oxygen ligands (d_{O1-O3} = 3.222 Å and d_{O2-O4} = 4.953 Å) at the narrow rim of the free Calix-OCP-model in the pinched cone conformation is 4.088 Å, meaning a mean distance of 2.044 Å to their centroid. On the other hand, the sum of the cation ionic radius of Cu(II) (r_{ion} = 0.57 Å) [45] and the van der Waals oxygen atom radius (r_{vdW} = 1.52 Å) [44] is 2.09 Å. This shows that the binding pocket of Calix-OCP is very well suited for receiving the copper ion, after proper conformational rearrangement. Inspection of the model's structure showed that the copper complex adopts a square planar configuration with a very low degree of distortion (dihedral angle (θ) between the planes formed by O-Cu-O is 177°), having O-Cu bond lengths of O(H)-Cu = 2.001 Å, O(H)-Cu = 2.019 Å, O(CH$_2$Ar)-Cu = 2.002 Å and O(CH$_2$Ar)-Cu = 1.990 Å, yielding a mean distance between the *O*-ligands and the central Cu(II) in the complex equal to 2.003 Å.

The strong binding of Cu(II) with Calix-OCP systems that has been experimentally observed is supported by the high binding energy (ΔE = − 1616.6 kJ/mol) obtained by DFT calculations for the model compound with Cu(II) in gas-phase. The coordinative bonding of *O*-ligands to Cu(II) is reflected on the resulting ground-state Mulliken charge of the metal ion after complexation, which is reduced from +2.000, in its unbound state, to +0.812.

Upon complexation, the copper guest is held in close proximity (<6 Å) to the transduction centres located either at the polymers backbone (polymers **5** and **6**) or at the bis-2-CBZ-phenyleneethynylene units (**Calix-OCP-2-CBZ**). Given the electron acceptor characteristic of the complexed Cu(II), where the β-LUMO contains an appreciable orbital coefficient in the copper ion ($d_{x^2-y^2}$ orbital) (Figure 5c), and the electron donor nature of the conjugated phenyleneethynylene-aryl systems (where the α-HOMO is located; Figure 5d), it is conceivable that, after complexation and light excitation, a rapid photoinduced electron transfer (PET) ensues from the singlet excited state (^{1}SOMO*) of the donor to the β-LUMO, followed by a non-radiative back electron transfer to the fluorophore, leading to the observed fluorescence quenching (for a schematic illustration, see Figure S7).

The absence of any shift in the maxima of excitation or emission spectra of polymer **5**-Cu(II) complex (Figure S8), or in its ground-state absorption spectra (Figure 4), reinforces the idea that the observed fluorescence quenching occurs by a PET mechanism and not by a photoinduced charge transfer (PCT) mechanism [4].

From the above sensing results a few conclusions may be drawn. First, the structures containing phenyleneethynylene-CBZ elements (polymer **5** and **Calix-OCP-2-CBZ**) are better reporters for signalizing the formation of the complex, than polymer **6**. This is because the systems bearing carbazole units, which possess electron rich N atoms, raise the HOMO and LUMO energies [24]. The HOMO, on excitation, lead to a higher-energy ^{1}SOMO*, which, in turn, result in more exergonic electron transfers. Secondly, the higher response of polymer **5** in comparison to **Calix-OCP-2-CBZ** (*ca.* 30% of K_a enhancement) can be attributed to the ability of **5** to amplify the transduction signal through energy migration along the conjugated polymer backbone [46,47].

The sensorial ability of polymers **5** and **6** and **Calix-OCP-2-CBZ** toward Cu(II) is quite remarkable particularly when compared to other already reported excellent sensors. For non-polymeric probes, a calix[4]arene containing two pyrenyl amine groups showed a K_{sv} of 1.49×10^4 M^{-1} [48], while a series of calixarene receptors with appended dansyl groups display association constants ranging from $1.45 \times 10^3 - 6.31 \times 10^4$ M^{-1} [49]. Pyrenyl-isoxazolo-tethered calix[4]arenes also display significant responses to Cu(II) with K_a in the range $6.00 \times 10^3 - 4.71 \times 10^4$ M^{-1} [50]. When considering conjugated polymers, either fluorene-*alt*-phenylene and phenyleneethynylene-*alt*-thienyleneethynylene type polymers, containing attached, respectively, *N*-cyclohexyl carbamates ($K_{sv} = 1.10 \times 10^4$ M^{-1}) [51] and tolylterpyridyl ($K_{sv} = 2.01 \times 10^5$ M^{-1}) [52] groups as recognition elements, have been used for Cu(II) detection.

Having established the noteworthy sensitivity of polymer **5** toward Cu(II), the next step dealt with the selectivity of the recognition. For this, its fluorescence response toward two other transition metal ions, Pb(II) and Hg(II), was evaluated. Notably, titration experiments with these ions do not reveal any significant or measurable response (Figures S9 and S10). A similar behaviour was followed by **Calix-OCP-2-CBZ**, whereas no affinity for the lead ion was displayed (Figure S11). These high selectivities, in addition to their outstanding sensitivity already revealed, turns polymer **5** and **Calix-OCP-2-CBZ** very efficient platforms for Cu(II) detection. It is worth mentioning that in several previous works using calix[4]arene-based ionophores, particularly those relying in dansyl groups [49] sensitivities for Cu(II) and Pb(II) are of the same order of magnitude, whereas for Hg(II) the affinity was in general even higher, overall making the attained selectivity for Cu(II) in these systems rather poor. In other works [50], a good selectivity was achieved for Cu(II) in comparison, for instance, with Pb(II) or Hg(II), although in a lower sensitivity when compared to polymer **5**.

The reason behind the lack of affinity of polymer **5** toward Pb(II) should be due to a mismatch between the cation size and the available space in the ligand, leading to a weak binding interaction. DFT calculations were used to shed some light on this issue. Pb(II) is a much larger ion (ionic radius = 98 pm; four-coordinate) [45] than Cu(II) (ionic radius = 57 pm; tetra-coordinate) [45]. Figure S12 presents the molecular structure of a putative Calix-OCP-Pb(II) complex (model). As is clear from the model, in order to accommodate the lead ion in a tetra-coordinate complex, a severe distortion of the square planar configuration (θ = 149°) has to occur. This has profound consequences in the binding strength. A measure of this is provided by the predicted Mulliken charge (+1.146) of Pb(II) in the complex. The lower charge transfer from the oxygen ligands to lead cation, in comparison to Cu(II), results in a much less stable complex.

The same observation applies to the absence of measurable binding between polymer **5** and Hg(II), which has a similar ionic radius (ionic radius = 96 pm; four-coordinate) [45] to that of Pb(II).

2.5. Effect of Calixarene Structure on Sensing

Polymer **4** is the de-*tert*-butylated analogue of **5**. Owing to the absence of *tert*-butyl groups in the calixarene upper rim, the conformational mobility of the calixarene moiety is much higher. This structural lability may have an impact on the binding affinity to metal ions. It is, thus, of interest to compare the behaviour of polymer **4** with polymer **5** in terms of its sensitivity and selectivity. The spectra of **4** upon successive additions of Cu(II) is shown in Figure 6a. After the addition of 5 equivalents of copper ion, the initial emission only drops by 14%. This represents a huge loss of affinity for Cu(II). The curve-fitting plots are depicted in Figure 6b. A $K_a = (5.29 \pm 0.78) \times 10^3$ M^{-1} was obtained from the regression analysis, which is 16 times lower than that of polymer **5**.

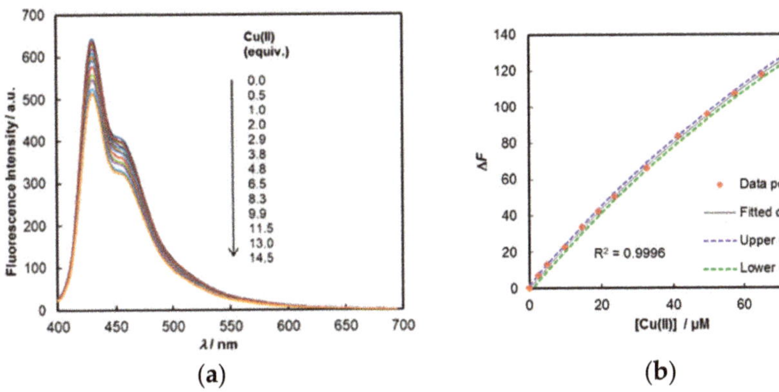

Figure 6. (a) Emission spectra of polymer **4** (5.0×10^{-6} M in CH_3CN) upon addition of increasing amounts (up to 14.5 eq.) of $Cu(ClO_4)_2$ (λ_{exc} = 380 nm); (b) Binding isotherm for the fluorimetric titration of **4** with Cu(II) along with the fitted curve and confidence intervals.

The stoichiometry of the complex is also 1:1, as judged from the goodness of fit ($R^2 = 0.9971$) on the non-linear model. The Job plot (Figure S13) exhibits a maximum for the concentration of the complex at a mole fraction of 0.5, again pointing to the formation of a 1:1 supramolecular complex (Table S2).

The association constants and the free energy changes (ΔG) for complexation of the various screened hosts with Cu(II) are gathered in Table 3.

Table 3. Association constants (K_a) and free energy changes (ΔG) for the inclusion complexation with Cu(II).

Host	K_a/M^{-1}	$\Delta G/kJ/mol$ [1]
Polymer **4**	5.29×10^3	−21.2
Polymer **5**	8.52×10^4	−28.1
Polymer **6**	5.30×10^4	−27.0
Calix-OCP-2-CBZ	6.62×10^4	−27.5

[1] ΔG values were calculated at 25 °C (298 K).

From the thermodynamic point of view, and assuming similar entropic losses (e.g., decrease of motion freedom of calixarene-OCP units and rearrangement of solvent molecules around the host and guest upon complexation) for the inclusion complexation of Cu(II) by polymers **4** and **5**, the higher free energy change (ΔG) associated to polymer **5** should be driven by the more favourable enthalpy changes, which are a direct result of stronger binding interactions developed in this complex. The same rational applies to polymer **6** and **Calix-OCP-2-CBZ**.

In titrations experiments of polymer **4** with Pb(II), and contrary to the results from the assays with polymer **5** and **Calix-OCP-2-CBZ**, a measurable affinity for this ion was found, with a $K_a = (4.47 \pm 1.61) \times 10^3$ M^{-1} (Figure S14a,b). The drastic reduction of binding ability of polymer **4** toward Cu(II) and its lower selectivity, for instance in comparison to polymer **5**, can only be traced to the different structure of the calixarene recognition unit. Polymer **4**, being more conformationally mobile than **5**, do not bind so tightly to the metal ion. Indeed, the lack of the *tert*-butyl substituents at the larger rim, allows that the phenyl rings in the cyclic oligomer can adopt a more open geometry at the narrow rim, due to the removal of steric hindrance at the wider rim. This was in part corroborated by modelling studies (same theory as above), from which a lower binding energy ($\Delta E = -1581.9$ kJ/mol) was obtained for the complexation of *p*-H-Calix-OCP-Cu(II) model, turning the complex of polymer **5** with Cu(II) more stable by 34.7 kJ/mol.

The same line of thought helps to explain why **4** is also responsive to Pb(II) whereas polymer **5** is almost silent. The conformational flexibility of **4** is apparently sufficient to hold Pb(II) in its cavity, in spite of the larger ionic radius of lead ion.

The above effects are not completely unprecedented for calix[4]arenes in the cone conformation. A survey of the literature showed that an increase of the number of *tert*-butyl groups at the upper rim of thiacalix[4]arenes containing dansyl derivatives decreases the sensing ability of the hosts toward Co(II), Ni(II) and Cd(II) cations [53]. In another report, using dansyl-modified calix[4]arenes, a slight increase in selectivity for Pb(II) over Hg(II) and Fe(III) was registered when the hosts were devoid of *tert*-butyl groups [54].

3. Materials and Methods

3.1. Instruments and Methods

FTIR were measured on a Bruker Vertex 70 as KBr pellets (transmission mode). ^{1}H NMR spectra were collected on Brüker AVANCE II$^+$ spectrometers (400 MHz); reported chemical shifts (δ/ppm) are internally referenced to CDCl$_3$ (^{1}H, 7.26 ppm). The splitting parameters for ^{1}H NMR are denoted as follows: s (singlet), d (doublet), t (triplet), and m (multiplet).

GPC was performed on a Jasco Liquid Chromatograph system equipped with a Jasco UV Absorption Detector 1575 (selected to 275 nm), using Polymer Standards Service (PSS) styrene-divinylbenzene copolymer (SDV) (PSS SDV) columns (10^3 and 10^5 Å) and THF as eluent at 35 °C. Calibration was done with monodisperse polystyrene standards. Number-average (M_n) and weight-average (M_w) molecular weights and polydispersity (M_w/M_n) were calculated from GPC data based on UV response at 270 nm. Polymer concentrations are expressed as per the molecular weight of the repeating unit.

UV-vis spectra were recorded on a Nicolet Evolution 300 or on a Jasco J-815 spectrophotometer using 1-cm quartz cells at 25 °C.

Steady-state fluorescence spectra were acquired on a Perkin Elmer LS45 fluorimeter using a 1-cm quartz cuvette in right angle geometry (RA) at 25 °C in air-equilibrated conditions. Fluorescence quantum yields were measured using 9,10-diphenylanthracene as fluorescence reference standard ($\Phi = 0.72$, EtOH, air equilibrated conditions, RA) [55]. The quantum yields were determined by the slope method [56], keeping the optical densities of the sample and reference below 0.05 at the excitation wavelength to prevent inner filter effects.

Titration experiments (UV-Vis and fluorescence) were carried out at a constant concentration of the hosts (5.0×10^{-6} M in CH$_3$CN, except when noticed otherwise) by adding known amounts of metal perchlorates in CH$_3$CN solutions at 25 °C, being the spectra repeatedly acquired after each addition.

3.2. Materials

Calixarenes **1** [29] and **2** [25], 2,7-diethynyl-9-propyl-9*H*-carbazole (**3**) [30], **Calix-OCP-PPE (6)** [26], **Calix-OCP-2-CBZ** [28], and **bis-Calix-TriPr-2-CBZ** [23] were synthesized according to our reported methods and fully characterized by FTIR, UV-Vis and NMR spectroscopies.

Dichlorobis(triphenylphosphine)palladium (II) (98%, Aldrich), copper(I) iodide (98%, Aldrich), $Cu(ClO_4)_2$ (Acros Organics, 98%), $Hg(ClO_4)_2.3H_2O$ (Alfa Aesar, reagent grade), $Pb(ClO_4)_2$ (Acros Organics, 99%), and 9,10-diphenylanthracene (scintillation grade, Nuclear Enterprises Ltd.) were used as received. Triethylamine (99%, Riedel-de-Haën) was previously dried from CaH_2 and distilled under N_2 prior to use. Toluene was previous dried from Na, distilled under N_2 and stored over Na. All other reagents and solvents were reagent grade and were purified and dried by standard methods. Organic extracts were dried over anhydrous magnesium sulphate.

3.3. Polymers' Synthesis

***p*-H-Calix-OCP-PAE-2,7-CBZ (4)**: To an argon degassed solution containing 40 mg (40.4 µmol) of **1** in dry toluene (1.6 mL) and freshly distilled NEt_3 (1.6 mL) were added $PdCl_2(PPh_3)_2$ (1.98 mg, 2.83 µmol), CuI (0.54 mg, 2.83 µmol) and 11.4 mg (44.4 µmol) of carbazole monomer **3** under argon. The mixture was degassed, the flask sealed, and the contents stirred in a pre-heated bath at 35 °C. The mixture acquired a yellow colour that gradually turned to brown–orangish with turbidity. After 24 h, GPC showed an almost quantitative conversion of **1**. Solvents were removed by evaporation and the residue taken in CH_2Cl_2 and washed successively with aqueous solutions of 2% HCl, 0.1M $NaHSO_3$, 10% NH_4SCN and water. The organic extract was dried and evaporated to dryness. The residue was dissolved in a minimum amount of CH_2Cl_2 and the polymer precipitated by the addition of MeOH. The product was obtained as a yellow amorphous solid in 58% (23.2 mg). During the CH_2Cl_2 dissolution, an insoluble gel-type fraction was obtained. ν_{max}/cm^{-1} (KBr) 3435, 3060, 3020, 2960, 2925, 2874, 2202, 2137, 1622, 1598, 1559, 1465, 1404, 1371, 1324, 1250, 1214, 1192, 1088, 1016, 846, 805, 753; λ_{max}/nm ($\varepsilon_{max} \times 10^{-4}$ M^{-1} cm^{-1}) 420 (4.18), cutoff at 470 nm; δ_H/ppm ($CDCl_3$; 400.130 MHz; Figure S2) 0.92–1.05 (m, 3H, $N\text{-}CH_2CH_2C\underline{H}_3$), 1.85–1.99 (2H, m, $N\text{-}CH_2C\underline{H}_2CH_3$), 3.25–3.45 (4H, m, $ArC\underline{H}_{2eq}Ar$), 4.18–4.32 (2H, m, $N\text{-}C\underline{H}_2CH_2CH_3$), 4.32–4.42 (4H, m, $ArC\underline{H}_{2ax}Ar$), 4.90–5.62 (8H, several m, $Ar_{calix}OC\underline{H}_2Ar$), 6.61–6.72 (4H, m, $Ar_{calix}\underline{H}$), 6.73–6.83 (4H, m, $Ar_{calix}\underline{H}$), 7.01–7.08 (4H, m, $Ar_{calix}\underline{H}$ and 2H, s, $ArO\underline{H}$), 7.08–7.13 (2H, s, $Ar_{phenylene}\underline{H}$), 7.28–7.35 (4H, m, $Ar_{benzylene\text{-}metaArCalix}\underline{H}$, overlapped with $CDCl_3$), 7.40–7.48 (2H, m, $ArC_{(3,6)}\underline{H}\text{-}CBZ$), 7.56–7.71 (2H, m, $ArC_{(1,8)}\underline{H}\text{-}CBZ$, and 4H, m, $Ar_{benzylene\text{-}orthoArCalix}\underline{H}$), 7.94–8.05 (2H, m, $ArC_{(4,5)}\underline{H}\text{-}CBZ$).

GPC data (THF solution at 35 °C against polystyrene standards): M_n = 5276 $gmol^{-1}$; M_w/M_n = 2.81; degree of polymerisation (DP) = 5 based on $M_{n(GPC)}$ (Figure S1).

***p*-tert-Butyl-Calix-OCP-PAE-2,7-CBZ (5)**. The above procedure was used in the synthesis of **5**, using 40 mg (32.9 µmol) of **2**, 1.3 mL of dry toluene and freshly distilled NEt_3, 1.6 mg (2.3 µmol) of $PdCl_2(PPh_3)_2$, 0.43 mg (2.3 µmol) of CuI and 9.3 mg (36.2 µmol) of **3**. After 48 h, polymer **5** was isolated as a yellow amorphous solid in 66% (26.3 mg). During the CH_2Cl_2 dissolution, an insoluble gel-type fraction was obtained in vestigial amount. ν_{max}/cm^{-1} (KBr) 3435, 3050, 2961, 2926, 2903, 2867, 2204, 2138, 1599, 1556, 1483, 1461, 1406, 1363, 1323, 1208, 1124, 1097, 1019, 947, 872, 804; λ_{max}/nm ($\varepsilon_{max} \times 10^{-4}$ M^{-1} cm^{-1}) 422 (4.81), cutoff at 470 nm. δ_H/ppm ($CDCl_3$; 400.130 MHz; Figure S3) 0.87 (18H, s, $Ar_{calixOH}\text{-}C(C\underline{H}_3)_3$), 0.93–1.06 (3H, m, $N\text{-}CH_2CH_2C\underline{H}_3$), 1.29 (18H, s, $Ar_{calixOR}\text{-}C(C\underline{H}_3)_3$), 1.84–2.02 (2H, m, $N\text{-}CH_2C\underline{H}_2CH_3$), 3.22 (2H, d, $ArC\underline{H}_{2eq}Ar$, J = 13.6 Hz), 3.33 (2H, d, $ArC\underline{H}_{2eq}Ar$, J = 13.2), 4.10–4.35 (2H, m, $N\text{-}C\underline{H}_2CH_2CH_3$; partially overlapped with $ArC\underline{H}_{2ax}Ar$), 4.25–4.50 (4H, m, $ArC\underline{H}_{2ax}Ar$), 4.85–5.00 (2H, m, $Ar_{calix}OC\underline{H}_2Ar$), 5.00–5.15 (2H, m, $Ar_{calix}OC\underline{H}_2Ar$), 5.24–5.44 (2H, m, $ArOC\underline{H}_2Ar$), 5.46–5.67 (2H, m, $ArOC\underline{H}_2Ar$), 6.65 (2H, s, $ArO\underline{H}$), 6.69 (4H, s, $Ar_{calixOH}\underline{H}$), 7.04 (4H, s, $Ar_{calixOR}\underline{H}$), 7.08 (2H, s, $Ar_{phenylene}\underline{H}$), 7.22–7.39 (4H, m, $Ar_{benzylene\text{-}metaArCalix}\underline{H}$, overlapped with $CDCl_3$), 7.40–7.50 (2H, m, $ArC_{(3,6)}\underline{H}\text{-}CBZ$), 7.55–7.75 (2H, m, $ArC_{(1,8)}\underline{H}\text{-}CBZ$ and 4H, m, $Ar_{benzylene\text{-}orthoArCalix}\underline{H}$), 7.90–8.12 (2H, m, $ArC_{(4,5)}\underline{H}\text{-}CBZ$).

GPC data (THF solution at 35 °C against polystyrene standards): M_n = 15,587 $gmol^{-1}$; M_w/M_n = 4.26; DP = 13 based on $M_{n(GPC)}$ (Figure S1).

3.4. Computational Methods and Non-linear Regression Analysis

Lowest-energy conformers for Calix-OCP and *p*-H-Calix-OCP models were obtained from conformational searches (Monte Carlo method, MMFF94 force field). They were then subjected to full energy and geometry optimisation by density functional theory (DFT) calculations running on an hybrid model (B3LYP) using 6-31G(d) basis sets, as developed in Spartan'14 [44], using default grids and convergence criteria in vacuum. Calculations involving Pb(II) have used the LANL2DZ pseudopotentials as implemented by default in Spartan'14. The input geometries for the corresponding metal complexes (Calix-OCP-Metal(II) and *p*-H-Calix-OCP-Metal(II)) were prepared by placing the metal(II) at the centroid of the narrow rim of the calixarene structures in a tetra-coordinate geometry. No other ligands (CH_3CN) or counter-ions (perchlorate ions) were considered in the modelling studies. The lowest-energy conformers were found by the same molecular mechanics method, and then fully optimized by DFT/B3LYP/6-31G(d) level of theory.

The binding energy (ΔE) in gas-phase for the complexation has been calculated by the equation $\Delta E = E_{Complex} - (E_{Cu(II)} + E_{Calix-OCP})$ where $E_{Complex}$ is the energy of Calix-OCP-Cu(II) model, $E_{Cu(II)}$ the energy of the naked cation, and $E_{Calix-OCP}$ the energy of the most stable conformation of the unbound Calix-OCP model.

Non-linear regression analysis was performed by the Solver function in Microsoft Excel [37] using the non-linear generalised reduced gradient (GRG) algorithm, and a convergence criterion for $R^2 < 10^{-9}$. The critical *t*-value was calculated at a significance level of 95%.

4. Conclusions

New fluorescent probes for sensing copper ion in organic media were revealed. Calix[4]arene-oxacyclophane molecular receptors proved to be excellent platforms for the recognition of Cu(II) which, combined with highly responsive fluorophore units as transduction sites, led as a whole to an outstanding sensitivity and selectivity for copper(II). The best signalling entities comprise either a conjugated polymer integrating phenylene-*alt*-ethynylene-*alt*-2,7-carbazolylenes as repeating units (polymer **5**) or an individual molecular species (**Calix-OCP-2-CBZ**) containing a similar structural arrangement of the fluorescent element. The three-dimensionality of the host cavity, providing a conformationally rigid cyclic array of *O*-ligands at the narrow rim of the calixarene fostered by the tethered oxacyclophane sub-unit, was found to be essential for the development of a strong binding between the host and the copper cation. The existence of *tert*-butyl groups at the calixarene wider rim is essential for keeping the host in a more tight conformation, not only allowing much higher sensitive responses but also an enhancement on the selectivity of metal ions recognition based on their sizes.

Experimental evidence from UV-Vis have undoubtedly shown the formation of supramolecular inclusion complexes between the calixarene hosts and Cu(II). A photoinduced electron transfer mechanism between the excited fluorophore and the complexed copper ion was proposed to explain the fluorescence turn-off behaviour of the sensing systems used in this work.

The new sensors here described feature among the best ever reported for copper detection.

Studies are now underway to evaluate the behaviour of Calix-OCP-Cu(II)/Cu(I) inclusion complexes as redox-active centres in biomimetic chemistry.

Supplementary Materials: The following are available online at http://www.mdpi.com/1420-3049/25/10/2456/s1, Figures S1–S14, Tables S1 and S2.

Author Contributions: Conceptualization, J.V.P.; methodology, J.V.P.; formal analysis, J.V.P.; investigation, C.B.F., A.I.C., and P.D.B.; resources, J.V.P.; data curation, J.V.P.; supervision, A.I.C. and P.D.B.; validation, A.I.C and P.D.B.; writing—original draft preparation, J.V.P.; writing—review and editing, J.V.P., A.I.C., and P.D.B. All authors have read and agreed to the published version of the manuscript.

Funding: We are grateful to Fundação para a Ciência e a Tecnologia/Ministério da Ciência, Tecnologia e Ensino Superior (FCT/MCTES) for financial support (UIDB/00616/2020 and UIDP/00616/2020) and Instituto Politécnico de Lisboa (IPL) for funding this work (Project IPL/2016/NoSeMeTox/ISEL).

Conflicts of Interest: The authors declare no conflict of interest.

References

1. Bissell, R.A.; Prasanna de Silva, A.; Gunaratne, H.Q.N.; Lynch, P.L.M.; Maguire, G.E.M.; Sandanayake, K.R.A.S. Molecular Fluorescent Signalling with 'Fluor-Spacer-Receptor' Systems: Approaches to Sensing and Switching Devices via Supramolecular Photophysics. *Chem. Soc. Rev.* **1992**, *21*, 187–195. [CrossRef]
2. Fabbrizzi, L.; Poggi, A. Sensors and Switches from Supramolecular Chemistry. *Chem. Soc. Rev.* **1995**, *24*, 197–202. [CrossRef]
3. Prasanna de Silva, A.; Gunaratne, H.Q.N.; Gunnlaugsson, T.; Huxley, A.J.M.; McCoy, C.P.; Rademacher, J.T.; Rice, T.E. Signaling Recognition Events with Fluorescent Sensors and Switches. *Chem. Rev.* **1997**, *97*, 1515–1566. [CrossRef] [PubMed]
4. Valeur, B.; Berberan-Santos, M.N. *Molecular Fluorescence: Principles and Applications*, 2nd ed.; Wiley-VCH: Weinheim, Germany, 2012; pp. 420–436. [CrossRef]
5. Asfari, Z.; Böhmer, V.; Harrowfield, J.; Vicens, J. (Eds.) *Calixarenes 2001*; Kluwer Academic: Dordrecht, The Netherlands, 2001. [CrossRef]
6. Gutsche, C.D. Calixarenes—An Introduction. In *Monographs in Supramolecular Chemistry*; Stoddart, J.F., Ed.; The Royal Society of Chemistry: Cambridge, UK, 2008.
7. Valeur, B.; Leray, I. Design principles of fluorescent molecular sensors for cation recognition. *Coord. Chem. Rev.* **2000**, *205*, 3–40. [CrossRef]
8. Malmström, B.G.; Leckner, J. The chemical biology of copper. *Curr. Opin. Chem. Biol.* **1998**, *2*, 286–292. [CrossRef]
9. Uriu-Adams, J.Y.; Keen, C.L. Copper, oxidative stress, and human health. *Mol. Aspects Med.* **2005**, *26*, 268–298. [CrossRef]
10. Desai, V.; Kaler, S.G. Role of copper in human neurological disorders. *Am. J. Clin. Nutr.* **2008**, *88*, 855S–858S. [CrossRef]
11. Jaffe, E.S. The 2008 WHO classification of lymphomas: Implications for clinical practice and translational research. *Eur. J. Haematol.* **2008**, *80*, 523–531. [CrossRef]
12. Renzoni, A.; Zino, F.; Franchi, E. Mercury levels along the food chain and risk for exposed populations. *Environ. Res.* **1998**, *77*, 68–72. [CrossRef]
13. Toscano, C.D.; Guilarte, T.R. Lead neurotoxicity: From exposure to molecular effects. *Brain Res. Rev.* **2005**, *49*, 529–554. [CrossRef]
14. Kumar, R.; Sharma, A.; Singh, H.; Suating, P.; Kim, H.S.; Sunwoo, K.; Shim, I.; Gibb, B.C.; Kim, J.S. Revisiting Fluorescent Calixarenes: From Molecular Sensors to Smart Materials. *Chem. Rev.* **2019**, *119*, 9657–9721. [CrossRef] [PubMed]
15. Li, G.-K.; Xu, Z.-X.; Chen, C.-F.; Huang, Z.-T. A highly efficient and selective turn-on fluorescent sensor for $Cu2+$ ion based on calix[4]arene bearing four iminoquinoline subunits on the upper rim. *Chem. Commun* **2008**, 1774–1776. [CrossRef] [PubMed]
16. Kumar, M.; Babu, J.N.; Bhalla, V. Fluorescent chemosensor for Cu^{2+} ion based on iminoanthryl appended calix[4]arene. *J. Inclusion Phenom. Mol. Recognit. Chem.* **2010**, *66*, 139–145. [CrossRef]
17. Chawla, H.M.; Shukla, R.; Pandey, S. Novel fluorescein appended calix[4]arenes for preferential recognition of Cu^{2+} ions. *Tetrahedron Lett.* **2013**, *54*, 2063–2066. [CrossRef]
18. Chang, K.-C.; Luo, L.-Y.; Diau, E.W.-G.; Chung, W.-S. Highly selective fluorescent sensing of Cu^{2+} ion by an arylisoxazole modified calix[4]arene. *Tetrahedron Lett.* **2008**, *49*, 5013–5016. [CrossRef]
19. Fischer, C.; Stapf, M.; Seichter, W.; Weber, E. Fluorescent chemosensors based on a new type of lower rim-dansylated and bridge-substituted calix[4]arenes. *Supramol. Chem.* **2013**, *25*, 371–383. [CrossRef]
20. Yang, Q.; Zhu, X.; Yan, C.; Sun, J. Determination of trace copper(II) by Triton X-100 sensitized fluorescence quenching of a novel calix [4]arene Schiff base derivative. *Anal. Methods* **2014**, *6*, 575–580. [CrossRef]
21. Costa, A.I.; Pinto, H.D.; Ferreira, L.F.V.; Prata, J.V. Solid-state sensory properties of Calix-poly(phenylene ethynylene)s toward nitroaromatic explosives. *Sens. Actuators Chem. B* **2012**, *161*, 702–713. [CrossRef]
22. Costa, A.I.; Prata, J.V. Substituted p-phenylene ethynylene trimers as fluorescent sensors for nitroaromatic explosives. *Sens. Actuators Chem. B* **2012**, *161*, 251–260. [CrossRef]
23. Barata, P.D.; Prata, J.V. New Entities for Sensory Chemistry based on Calix[4]arene-Carbazole Conjugates: From Synthesis to Applications. *Supramol. Chem.* **2013**, *25*, 782–797. [CrossRef]

24. Barata, P.D.; Prata, J.V. Cooperative Effects in the Detection of a Nitroaliphatic Liquid Explosive and an Explosive Taggant in the Vapor Phase by Calix[4]arene-Based Carbazole-Containing Conjugated Polymers. *ChemPlusChem* **2014**, *79*, 83–89. [CrossRef]

25. Teixeira, C.M.; Costa, A.I.; Prata, J.V. A new fluorescent double-cavity calix[4]arene: Synthesis and complexation studies toward nitroanilines. *Tetrahedron Lett.* **2013**, *54*, 6602–6606. [CrossRef]

26. Prata, J.V.; Costa, A.I.; Teixeira, C.M. A Solid-State Fluorescence Sensor for Nitroaromatics and Nitroanilines Based on a Conjugated Calix[4]arene Polymer. *J. Fluoresc.* **2020**, *30*, 41–50. [CrossRef]

27. Prata, J.V.; Barata, P.D. Fostering protein-calixarene interactions: From molecular recognition to sensing. *RSC Adv.* **2016**, *6*, 1659–1669. [CrossRef]

28. Prata, J.V.; Barata, P.D.; Pescitelli, G. Inherently chiral calix[4]arenes with planar chirality: Two new entries to the family. *Pure Appl. Chem.* **2014**, *86*, 1819–1828. [CrossRef]

29. Teixeira, C.M. New Molecular Receptors Based on Calix[4]arenes—Application to Sensorial Chemistry. Master's Thesis, Instituto Politécnico de Lisboa, Instituto Superior de Engenharia de Lisboa, Lisboa, Portugal, 2013. Available online: http://hdl.handle.net/10400.21/3307 (accessed on 9 April 2020).

30. Barata, P.D.; Costa, A.I.; Prata, J.V. Calix[4]arene-carbazole-containing polymers: Synthesis and properties. *React. Funct. Polym.* **2012**, *72*, 627–634. [CrossRef]

31. Lakowicz, J.R. *Principles of Fluorescence Spectroscopy*, 3rd ed.; Springer: New York, NY, USA, 2006; p. 282. ISBN 978-0387-31278-1.

32. Weert, M. van de. Fluorescence Quenching to Study Protein-ligand Binding: Common Errors. *J. Fluor.* **2010**, *20*, 625–629. [CrossRef] [PubMed]

33. Thordarson, P. Determining association constants from titration experiments in supramolecular chemistry. *Chem. Soc. Rev.* **2011**, *40*, 1305–1323. [CrossRef]

34. van de Weert, M.; Stella, L. Fluorescence quenching and ligand binding: A critical discussion of a popular methodology. *J. Mol. Struct.* **2011**, *998*, 144–150. [CrossRef]

35. Turro, N.J.; Ramamurthy, V.; Scaiano, J.C. *Modern Molecular Photochemistry of Organic Molecules*; University Science Books: Sausalito, CA, USA, 2010; pp. 216–217. ISBN 978-1891389252.

36. Liu, Y.; Han, B.-H.; Chen, Y.-T. Molecular Recognition and Complexation Thermodynamics of Dye Guest Molecules by Modified Cyclodextrins and Calixarenesulfonates. *J. Phys. Chem. B* **2002**, *106*, 4678–4687. [CrossRef]

37. Brown, A.M. A step-by-step guide to non-linear regression analysis of experimental data using a Microsoft Excel spreadsheet. *Comput. Meth. Programs Biomed.* **2001**, *65*, 191–200. [CrossRef]

38. Motulsky, H.J.; Ransnas, L.A. Fitting curves to data using nonlinear regression: A practical and nonmathematical review. *FASEB J.* **1987**, *1*, 365–374. [CrossRef] [PubMed]

39. Olson, E.J.; Bühlmann, P. Getting More out of a Job Plot: Determination of Reactant to Product Stoichiometry in Cases of Displacement Reactions and n:n Complex Formation. *Org. Chem.* **2011**, *76*, 8406–8412. [CrossRef] [PubMed]

40. Sanz, S.; Ferreira, K.; McIntosh, R.D.; Dalgarno, S.J.; Brechin, E.K. Calix[4]arene-supported $Fe^{III}_2 Ln^{III}_2$ clusters. *Chem. Commun.* **2011**, *47*, 9042–9044. [CrossRef] [PubMed]

41. Karotsis, G.; Kennedy, S.; Dalgarno, S.J.; Brechin, E.K. Calixarene supported enneanuclear Cu(II) clusters. *Chem. Commun.* **2010**, *46*, 3884–3886. [CrossRef] [PubMed]

42. Patra, S.; Lo, R.; Chakraborty, A.; Gunupuru, R.; Maity, D.; Ganguly, B.; Paul, P. Calix[4]arene based fluorescent chemosensor bearing coumarin as fluorogenic unit: Synthesis, characterization, ion-binding property and molecular modelling. *Polyhedron* **2013**, *50*, 592–601. [CrossRef]

43. Garribba, E.; Micera, G. The Determination of the Geometry of Cu(II) Complexes An EPR Spectroscopy Experiment. *J. Chem. Educ.* **2006**, *83*, 1229–1232. [CrossRef]

44. *Spartan'14 Molecular Modeling Program*; Wavefunction, Inc.: Irvine, CA, USA, 2014.

45. Shannon, R.D. Revised effective ionic radii and systematic studies of interatomic distances in halides and chalcogenides. *Acta Cryst. A* **1976**, *32*, 751–767. [CrossRef]

46. Crawford, K.B.; Goldfinger, M.B.; Swager, T.M. Na^+ Specific Emission Changes in an Ionophoric Conjugated Polymer. *J. Am. Chem. Soc.* **1998**, *120*, 5187–5192. [CrossRef]

47. Fan, L.-J.; Zhanga, Y.; Murphya, C.B.; Angell, S.E.; Parker, M.F.L.; Flynn, B.R.; Jones, W.E., Jr. Fluorescent conjugated polymer molecular wire chemosensors for transition metal ion recognition and signalling. *Coord. Chem. Rev.* **2009**, *253*, 410–422. [CrossRef]

48. Sahin, O.; Yilmaz, M. Synthesis and fluorescence sensing properties of novel pyrene-armed calix[4]arene derivatives. *Tetrahedron* **2011**, *67*, 3501–3508. [CrossRef]

49. Talanova, G.G.; Talanov, V.S. Dansyl-containing fluorogenic calixarenes as optical chemosensors of hazardous metal ions: A mini-review. *Supramol. Chem.* **2010**, *22*, 838–852. [CrossRef]

50. O'Sullivan, J.; Colleran, J.; Twamley, B.; Heaney, F. Highly Selective Fluorimetric Turn-Off Detection of Copper(II) by Two Different Mechanisms in Calix[4]arene-Based Chemosensors and Chemodosimeters. *ChemPlusChem* **2019**, *84*, 1610–1622. [CrossRef] [PubMed]

51. Deng, W.; Sun, P.; Fan, Q.; Zhang, L.; Minami, T. Highly selective detection of copper(II) by a "ligand-free" conjugated copolymer in nucleophilic solvents. *Front. Chem. Sci. Eng.* **2020**, *14*, 105–111. [CrossRef]

52. Murphy, C.B.; Zhang, Y.; Troxler, T.; Ferry, V.; Martin, J.J.; Jones, W.E. Probing Förster and Dexter Energy-Transfer Mechanisms in Fluorescent Conjugated Polymer Chemosensors. *J. Phys. Chem. B* **2004**, *108*, 1537–1543. [CrossRef]

53. Higuchi, Y.; Narita, M.; Niimi, T.; Ogawa, N.; Hamada, F.; Kumagai, H.; NobuhikoIki, N.; Miyano, S.; Kabuto, C. Fluorescent Chemo-Sensor for Metal Cations Based on Thiacalix[4]arenes Modified with Dansyl Moieties at the Lower Rim. *Tetrahedron* **2000**, *56*, 4659–4666. [CrossRef]

54. Ocak, Ü.; Ocak, M.; Surowiec, K.; Bartsch, R.A.; Gorbunova, M.G.; Tu, C.; Surowiec, M.A. Metal ion complexation in acetonitrile by di-ionized calix[4]arenes bearing two dansyl fluorophores. *Inclusion Phenom. Macrocyclic Chem.* **2009**, *63*, 131–139. [CrossRef]

55. Eaton, D.F. Reference materials for fluorescence measurement. *Pure Appl. Chem.* **1988**, *60*, 1107–1114. [CrossRef]

56. A Guide to Recording Fluorescence Quantum Yields, Horiba Scientific. Available online: http://www.horiba.com/fileadmin/uploads/Scientific/Documents/Fluorescence/quantumyieldstrad.pdf (accessed on 2 March 2020).

Sample Availability: Samples of some compounds are available from the authors upon request.

Article

Synthesis, Crystal Structure, and Biological Activity of a Multidentate Calix[4]arene Ligand Doubly Functionalized by 2-Hydroxybenzeledene-Thiosemicarbazone

Ehsan Bahojb Noruzi [1], Behrouz Shaabani [1,*], Silvano Geremia [2,*], Neal Hickey [2], Patrizia Nitti [2] and Hossein Samadi Kafil [3]

[1] Department of Inorganic Chemistry, Faculty of Chemistry, University of Tabriz, 5166616471 Tabriz, Iran; ehsanbahojb@gmail.com
[2] Centre of Excellence in Biocrystallography, Department of Chemical and Pharmaceutical Sciences, University of Trieste, 34127 Trieste, Italy; nhickey@units.it (N.H.); pnitti@units.it (P.N.)
[3] Drug Applied Research Center, Tabriz University of Medical Sciences, 5154853431 Tabriz, Iran; Kafilhs@tbzmed.ac.ir
* Correspondence: shaabani.b@gmail.com (B.S.); sgeremia@units.it (S.G.); Tel.: +98-41-3339-3144 (B.S.); +39-04-0558-3936 (S.G.)

Academic Editors: Mario Berberan-Santos and Paula M. Marcos
Received: 11 November 2019; Accepted: 14 January 2020; Published: 16 January 2020

Abstract: The design and synthesis of a novel tert-butyl-calix[4]arene functionalized at 1, 3 positions of the lower rim with two terminal 2-hydroxybenzeledene-thiosemicarbazone moieties is reported. The new ligand with multi-dentate chelating properties was fully characterized by several techniques: ESI-Mass spectroscopy, FT-IR, 1H-NMR, and single crystal X-ray diffraction. The solid state structure confirms that the calix[4]arene macrocycle has the expected open cone conformation, with two opposite phenyl rings inclined outwards with large angles. The conformation of the two alkoxythiosemicarbazone arms produces a molecule with a C2 point group symmetry. An interesting chiral helicity is observed, with the two thiosemicarbazone groups oriented in opposite directions like a two-blade propeller. A water molecule is encapsulated in the center of the two-blade propeller through multiple H-bond coordinations. The antibacterial, antifungal, anticancer, and cytotoxic activities of the calix[4]arene-thiosemicarbazone ligand and its metal derivatives (Co2+, Ni2+, Cu2+, and Zn2+) were investigated. A considerable antibacterial activity (in particular against *E. coli*, MIC, and MBC = 31.25 µg/mL) was observed for the ligand and its metal derivatives. Significant antifungal activities against yeast (*C. albicans*) were also observed for the ligand (MIC = 31.25 µg/mL and MBC = 125 µg/mL) and for its Co^{2+} derivative (MIC = 62.5 µg/mL). All compounds show cytotoxicity against the tested cancerous cells. For the Saos-2 cell line, the promising anticancer activity of ligand L (IC_{50} < 25 µg/mL) is higher than its metal derivatives. The microscopic analysis of DAPI-stained cells shows that the treated cells change in morphology, with deformation and fragmentation of the nuclei. The hemo-compatibility study demonstrated that this class of compounds are suitable candidates for further in vivo investigations.

Keywords: thiosemicarbazone; calix[4]arene; metal complex; X-ray structure; antimicrobial; anticancer

---

1. Introduction

Calix[4]arenes are macrocyclic compounds prepared by cyclo-condensation of p-tert-butylphenols with formaldehyde under alkaline conditions [1]. The cone conformation of calix[4]arene enables them in principle to act as highly preorganized hosts, for different types of guests, leading to a vast host–guest chemistry, which can be exploited in technological applications in various fields, including the development of sensors [2], separation science [3], catalysis [4,5], and drug delivery [6,7]. The calixarene skeleton can be functionalized on both the upper and lower rims due to the presence of reactive hydroxyl groups on the lower rim and the easy substitution of tert-butyl groups on the upper rim, by reversed Friedel–Crafts reaction followed by insertion of new functional groups through electrophilic attack [8]. A wide variety of functional groups, such as carbonyl, amide, nitrile, thiourea, and Schiff-base, have been introduced onto the calixarene platform [9–14]. Functionalized calixarenes are thus able to act as hosts for neutral molecules [15], cations [16,17], and anions [18,19] and offer the possibility to act as multifunctional hosts [20,21]. For this reason, they have been widely investigated for applications based on molecular or ion recognition, including the development of highly preorganized chelating ligands for metal coordination [22–25] and hybrid materials [26,27].

The thiosemicarbazone group has been widely researched in coordination chemistry [28], analytical applications [29], and pharmacology [30]. Thiosemicarbazones usually act as flexible multi-dentate chelating ligands and can coordinate with different transition metal ions by bonding through the sulfur and hydrazinic nitrogen atoms [24,31–33]. Extensive investigations by different research groups have indicated that thiosemicarbazone derivatives and their transition metal complexes show remarkable biological properties, including antiviral, antimicrobial, and antitumoral activities [34–41]. Inspection of the literature shows that the functionalization of calixarenes with thiosemicarbazone has received little attention [42–44]. Due to the lack of calixarene-based thiosemicarbazone compounds, and considering their broad spectrum of biological applications, the design and synthesis of new thiosemicarbazone functionalized calix[4]arene compounds for use as multi-dentate chelating ligands is an area of interest. While the rigid structure of tert-butyl-calix[4]arenes means that up to four ligands can be introduced onto the framework through the hydroxy groups in a controlled manner, only two thiosemicarbazone groups would be necessary for each calixarene molecule to preorganize a coordination environment for metal ions. Therefore, in our strategy, the functionalization was limited to the 1, 3 positions of the lower rim of the tert-butyl-calix[4]arene in cone conformation, with the introduction of two terminal 2-hydroxybenzeledene-thiosemicarbazone functions, which increase the number of potential chelating atoms.

Herein, we report the synthesis, characterization, and crystal structure of a novel thiosemicarbazone calix[4]arene derivative. The biological properties, including the antimicrobial and anticancer activities of the synthesized compound and its metal derivatives (Co^{2+}, Ni^{2+}, Cu^{2+}, and Zn^{2+}), were evaluated using various microorganisms and human bone cancer cell lines (Saos-2 and MG-63). We also investigated the cytotoxic effects of the ligand and its metal derivatives on human red blood cells (HRBCs) by determination of their hemolysis rate and by application of the DAPI staining method.

2. Results and Discussion

2.1. Synthesis of the Ligand and Metal Derivatives

The multi-dentate thiosemicarbazone chelating ligand **L** was prepared by treatment of 5, 11, 17, 23-tetra-tert-butyl-26, 28-dihydroxy-25, and 27-bis(thiosemicarbazidoethoxy) calix[4]arene with salicylaldehyde in a 1:1 molar ratio, under reflux conditions (Scheme 1). The pale-yellow crystalline solid was obtained and further recrystallization in chloroform produced single crystals of **L** whose structure was determined by crystallography. Considering the coordinating potential of the ligand **L** toward the metal centers, several divalent transition metals (Co^{2+}, Ni^{2+}, Cu^{2+}, Zn^{2+}) were considered to form metal derivatives. To do this, we mixed one equivalent of the metal salt in MeOH with one equivalent of **L** in THF followed by reflux for 24 h. The products were finally purified by recrystallization from THF.

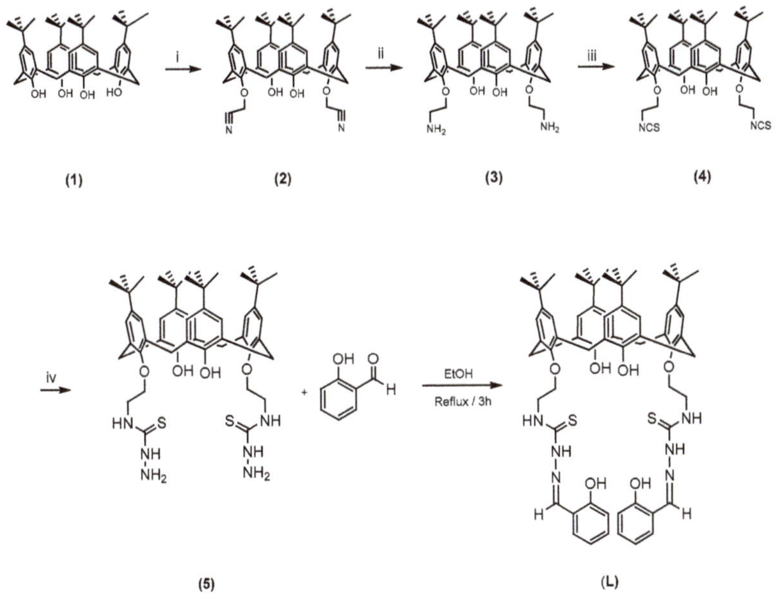

Scheme 1. The route for synthesis of **L**. Reaction conditions: (**i**) Chloroacetonitrile, K_2CO_3, NaI, reflux, 7h (**ii**) LiAlH4, (0 °C), 4h (**iii**) CSCl$_2$, BaCO$_3$, 24h (**iv**) Hydrazine hydrate, room temperature, 3 h.

2.2. Characterization

FTIR spectra of the Ligand **L** (Figure S1) show the absorption bands corresponding to the C=N (1633 cm^{-1}), N-C=S (1539 cm^{-1}), and C=S (898 cm^{-1}) groups [45,46]. Concerning the possible thione-thiol tautomerism in the ligand, the absence of the S-H vibration band at 2750 cm^{-1} and the presence of bands due to the C=S vibration at 898 cm^{-1} confirm that the CS group remains in the thione form [47]. The FT-IR spectra of the Co^{2+}, Ni^{2+}, Cu^{2+}, and Zn^{2+} complexes (Figures S2–S5) were analyzed in comparison with the free ligand **L** in the region 4000–500 cm^{-1}. The characteristic peaks related to the C=N and C=S groups shift to a lower wavenumber upon coordination of the metal centers to the ligand. On the other hand, the vibrations related to the N-C=S group shift to higher wavenumbers in all metal derivatives (Table S1). These changes in the C=N, N-C=S, and C=S vibrations support the suggestion of the metal coordination by **L** in all of the complexes.

The ^{1}H-NMR spectra of ligand **L** (Figure 1) show two singlet signals for two sets of 18 t-Bu protons (δ 0.92 and 1.22 ppm), two doublet signals (δ 3.16, 4 protons, and 4.13 ppm 4 protons) assigned to the calixarene -CH$_2$- bridging groups, a multiplet peak (δ 4.36–4.27 ppm) corresponding to the eight OCH$_2$CH$_2$NCS protons, two types of aromatic signals (δ 6.71, 4 protons, and 6.87 ppm, 4 protons) as two singlet signals assigned to the two types of calix[4]arene benzene ring protons, a multiplet peak (δ 6.83, 4 protons) and to two doublets of doublets (δ 6.99 ppm, 2 protons and 7.26 ppm, 2 protons) assigned to four types of aromatic protons of the phenol ring of the thiosemicarbazone arms, a singlet peak attributed to two OH protons of the calix[4]arene lower rim (δ 7.33 ppm), a singlet peak attributed to two thiosemicarbazone NH protons (δ 7.67 ppm), a singlet peak attributed to two NHCS protons (δ 8.68 ppm), a singlet peak attributed to two OH protons of the thiosemicarbazone phenol rings (δ 8.98 ppm), and a singlet peak related to two =CH imine protons (δ 9.74 ppm). The assignments of the ^{1}H-NMR spectrum are in agreement with literature data [48].

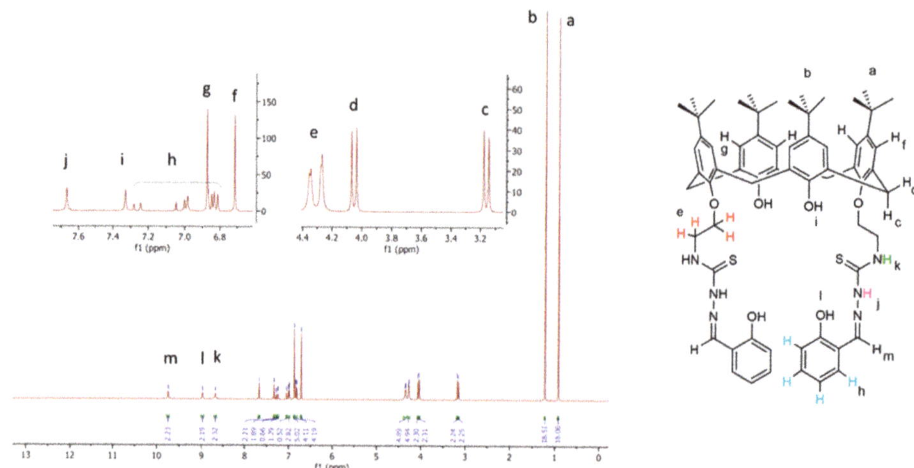

Figure 1. Selected portion of the ^{1}H-NMR spectrum of ligand **L**.

The high-resolution mass spectra (HRMS) of ligand **L** showed molecular ion peaks corresponding to the sodium adduct of the **L** (Figure S6). The most intense peak of 1113.5315 *m/z* perfectly agrees with a compound with formula $C_{64}H_{78}N_6O_6S_2$ (theoretical [L + Na]$^+$ equal to 1113.5316 *m/z*) and confirms the proposed structure for **L**. HRMS spectra of the metal derivatives show in all cases the formation of the metal adduct. The Co derivative spectrum (Figure S7) is in agreement with a di-cationic complex with formula LCo(II)Co(III)$^{2+}$, formed by the loss of 3 H$^+$ from L, $[C_{64}H_{75}N_6O_6S_2Co_2]^{2+}$. Analogous partial oxidation of the Co^{2+} species was also observed for the compound **5** derivative [49]. The HRMS spectrum of the Ni derivative (Figure S8) reveals the presence of multiple species, including a mixture of monomeric and dimeric complexes of a mono-deprotonated ligand, with formula LNi(II)$^+$ and L$_2$Ni(II)$_2{}^{2+}$, $[C_{64}H_{77}N_6O_6S_2Ni]^+$ and $[C_{128}H_{154}N_{12}O_{12}S_4Ni_2]^{2+}$ (Figure S9), and a di-cationic complex of L^{2-} with formula LNi(II)$_2{}^2$, $[C_{64}H_{76}N_6O_6S_2Ni_2]^{2+}$. The HRMS spectrum of the Cu derivative (Figure S10) shows the highest peak, which is consistent with an analogue di-cationic specie of the doubly deprotonated L ligand of formula LCu(II)$_2{}^{2+}$, $[C_{64}H_{76}N_6O_6S_2Cu_2]^{2+}$. Finally, the highest peak of the HRMS spectrum of the Zn derivative (Figure S11) is also attributable to a di-cationic species with the formula LZn(II)$_2{}^{2+}$, $[C_{64}H_{76}N_6O_6S_2Zn_2]^{2+}$.

The elemental analysis are consistent with nitrate salts of di-cationic species of the general formula [LM](NO$_3$)$_2$, with the exception of the Co derivative. In this case, the elemental analysis data can be rationalized as a di-hydrated nitrate salt [LCo](NO$_3$)$_2$. 2H$_2$O (Table S3).

2.3. Solid-State Structure of L

The X-ray structure of **L**, determined using synchrotron radiation with cryogenic techniques, shows that the asymmetric unit of the centrosymmetric triclinic crystals is composed of one molecule of **L**, a water molecule bound by multiple H-bond interactions, a co-crystallized chloroform solvent molecule disordered over at least two positions, and a partially occupied co-crystallized chloroform solvent molecule in close proximity to a crystallographic inversion center. Crystallographic data and refinement details are reported in Table S2.

The calix[4]arene macrocycle shows the expected cone conformation (Figure 2a). The mean planes of the two phenyl rings with alkoxythiosemicarbazone substituents (**A** and **C**) are almost orthogonal to the calix[4]arene mean plane, defined by the methylene bridging groups, with slight outward dihedral angles of 96° and 97°, respectively, (Figure 2b). With regard to the hydroxy-substituted phenyl rings (**B** and **D**), the mean planes of both phenyl rings are also both inclined outwards with respect to the calix[4]arene mean plane, with larger dihedral angles of 137° and 147°, respectively. Consequently, the cone exhibits a quite open overall conformation. The similar values of the above-discussed pairs of dihedral angles reflect the C2 symmetry of the calix[4]arene derivative symmetrically functionalized in the 1, 3 positions of the lower rim. This open structure is largely due to the formation of two intramolecular H-bonds between the hydroxy groups as donors and the adjacent alkoxy oxygen atoms as acceptors, with O⋯O distances of 2.871 (ring B-ring A) and 2.922 Å (ring D-ring C) (Table 1). Thus, the substitution pattern on the lower rim strongly influences the conformation of the calix[4]arene cone.

The conformations of the two alkoxythiosemicarbazone substituents on rings B and D are very similar, with comparable values of corresponding torsion angles along the chains. More specifically, the entire conformation of the arms is largely determined by the positive (or negative in the other enantiomeric conformer present in the centrosymmetric crystal structure) gauche conformation of both the ethyl spacer (61.4 and 57.6° for A and C, respectively). As a result, the thiosemicarbazone groups are oriented in opposite directions like a two-blade propeller, reflecting the C2 point group symmetry of the molecule, thereby forming an interesting chiral helicity (Figure 2c).

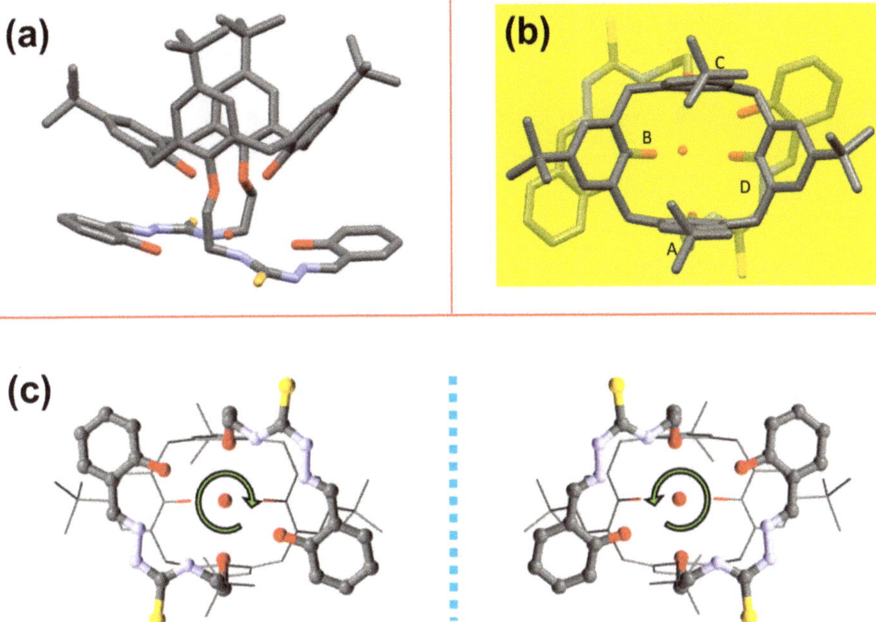

Figure 2. Solid-state structure of **L**. (**a**) The molecule shows a cone conformation. (**b**) Orthogonal view of **L** with respect to the calix[4]arene mean plane (yellow) defined by the methylene bridging groups. (**c**) The two enantiomeric conformations of **L** observed in the centrosymmetric crystal structure. Hydrogen atoms have been omitted for clarity.

Overall, the entire thiosemicarbazone substituents, including their terminal phenol groups, tend towards coplanarity: The mean planes of the two NCNN groups of atoms exhibit a dihedral angle of 19° while the dihedral angles between the mean plane of the NCNN atoms and the mean plane of its terminal phenyl ring are 19° and 13° for **A** and **C**, respectively. Similarly to the situation regarding the conformation of the cone, this coplanar conformation is influenced by the formation of two intramolecular H-bonds between the hydroxy groups of the terminal phenol groups as donors and the imine N atom of the thiosemicarbazone as an acceptor, with O⋯N distances of 2.710 (ring A substituent) and 2.742 Å (ring C substituent) (Table 1).

A significant amount of electron density approximately at the center of the oppositely oriented thiosemicarbazone groups was interpreted and well modelled as a water molecule. The water oxygen atom is within the H-bond distance of a number of the potential chelating atoms at various positions on the lower rim, including the N atoms of the thiosemicarbazone and hydroxy oxygens on both the terminal phenol groups and the calix[4]arene lower rim, as indicated in Table 1, in which details of the main short interactions observed in the crystal structure are summarized. This central water molecule, mimicking a possible coordinated metal ion, is surrounded by eight N,O atoms with the formation of multiple H-bonds, although some are very weak. It should be noted that water was not used as a solvent so its presence is presumably explained as an impurity of the methanol solvent or as adsorption from the atmospheric humidity.

Finally, the crystal packing shows one intermolecular H-bond between the hydroxy of one of the thiosemicabazone terminal phenol groups (ring **C**) and its inversion-related hydroxy group, with an O···O distance of 2.927 Å.

The overall conformation of L and the easy inclusion of the water molecules indicates the potential of **L** as a polydentate chelating ligand.

Table 1. Summary of the main short interactions found in the structure of **L**.

D •••A [a]	d (D•••A) (Å)
Intramolecular H-Bonds on Calix[4]arene Lower Rim	
O(1B)•••O(1A)	2.871
O(1D)•••O(1C)	2.922
Intramolecular H-Bonds on Thiosemicarbazone Substituents	
O(2A)•••N(3A)	2.710
O(2C)•••N(3C)	2.742
Interactions between Water Molecule and Thiosemicarbazone Substituents [b]	
N(1A)•••O(1W)	2.978
N(1C)•••O(1W)	3.185
O(2A)•••O(1W)	2.850 (phenol hydroxy)
O(2C)•••O(1W)	3.051 (phenol hydroxy)
Interactions between Water Molecule and Calix[4]arene Lower Rim Oxygen Atoms [b]	
O(1D)•••O(1W)	2.872 (hydroxy)
O(1B)•••O(1W)	2.986 (hydroxy)
O(1C)•••O(1W)	3.094 (alkoxy)
O(1A)•••O(1W)	3.258 (alkoxy)

[a] The atom labels in parenthesis are those used in the deposited structure. [b] For clarity, the type of non-water oxygen atom involved is identified in parenthesis.

2.4. Determination of Antimicrobial and Antifungal Activities

All of the synthesized compounds (**L** and Co, Ni, Cu, and Zn complexes) were evaluated for their quantitative antibacterial activity in the serial double dilution method against gram-positive bacteria (*S. aureus* and *B. subtilis*) and gram-negative bacteria (*E. coli* and *P. aeruginosa*). The anti-fungal activity was also investigated (*C. albicans* and *C. glabrata*). Gentamicin and nystatin were used as standard antibacterial and antifungal drugs, respectively. Considering literature data [50] and according to Table 2, all of the derivatives showed strong antibacterial activity against the tested bacteria except *S. aureus*, which only showed sensitivity against the copper complex. In many cases, the MIC of the synthesized compounds coincides with the MBC (31.25 μg/mL), which confirms the bactericidal antimicrobial activity of the ligand and its related coordination compounds. In general, the fungal strains were resistant against the synthesized compounds except for ligand L and its cobalt complex, which respectively showed strong and mild antifungal activity against *C. albicans*.

Comparing these results with our recently published paper [49] reveals that the thiosemicarbazone schiff-base ligand is a stronger antimicrobial agent than its thiosemicarbazide analogue. Its antibacterial activity (MIC value) increased 8-fold and 2-fold against *E. coli* and *P. aeruginosa*, respectively, in comparison with thiosemicarbazide. It also showed a significant antifungal activity against *C. albicans* while this fungal strain was resistant against the thiosemicarbaide ligand. In the case of metal derivatives, generally all of the thiosemicarbazone derivatives are stronger antibacterial agents than their corresponding thiosemicarbazide derivatives. The antibacterial activity of the copper derivative is eight times stronger than the thiosemicarbazide analogue against *S. aureus* and the cobalt derivative showed strong antimicrobial activity against *B. subtilis* strain while this bacteria was resistant against the corresponding thiosemicarbazide derivative. Ferthermore, the antimicrobial activity of the zinc derivative increased four-fold in comparison with its thiosemicarbazide derivative. In the case of gram-negative bacteria (*E. coli* and *P. aeruginosa*), the improvement of the antibacterial activities of the metal derivatives are more pronounced in comparison to the corresponding thiosemicarbazide derivatives. The MIC values of the Co, Ni, and Cu derivatives are modified 4 times, 2 times, and 4 times against *E. coli* bacteria, respectively, in comparison with their thiosemicarbazide equivalents. Considering *P. aeruginosa*, the MIC values of the Co and Cu derivatives are improved 4 times and 2 times, respectively.

Table 2. Antibacterial activity of synthesized compounds against different microorganisms in the micro-broth dilution method. The antibacterial activity is expressed as the MIC/MBC (µg/mL).

Microorganism	S. aureus	B. subtilis	E. coli	P. aeruginosa	C. albicans	C. glabrata
Ligand **L**	-	31.25	31.25/31.25	31.25/-	31.25/125	-
Complex **Co**	-	31.25/1000	31.25/31.25	31.25/250	62.5/-	-
Complex **Ni**	-	31.25	31.25/31.25	31.25/2000	-	-
Complex **Cu**	31.25	31.25	31.25/31.25	31.25/500	-	-
Complex **Zn**	-	31.25	31.25/31.25	31.25/31.25	-	-
Gentamicin *	0.12/0.25	2/2	0.5/0.5	2/2	-	-
Nystatin *	-	-	-	-	1.25	0.625

Staphylococcus aureus (ATCC® 29213™), *Bacillus subtilis* (ATCC® 6633™), *Escherichia coli* (ATCC® 25922™), *Pseudomonas aeruginosa* (ATCC® 27853™), *Candida albicans* (ATCC® 10231™), and *Candida glabrata* (ATCC® 2001™). * Gentamicin and nystatin used as a standard control for bacteria and fungi, respectively.

2.5. Cytotoxicity Assay

To determine the in vitro cytotoxicity, the IC_{50} values of the calix[4]arene-based ligand (**L**) and its complexes were evaluated against two different human bone cancer cell lines (Saos-2 and MG-63) by the MTT reduction assay. All of the compounds were dispersed in water (and 10% DMSO) and diluted with cell culture medium to reach four required concentrations of 25, 50, 100, and 200 µg/mL. The viability of cancer cells versus the compound concentrations is shown in Figure 3. In general, a significant dose-dependent antiproliferative activity was observed with the exclusion of the Zn derivative against Saos-2 cells. Considering the Saos-2 cell line, the anticancer activity of ligand **L** is higher than its metal derivatives at all concentrations, especially at a higher dosage (about 21%). With the exception of the Zn complex, all of the compounds showed increasing cytotoxicity against cancerous cells with an increasing concentration.

The copper derivate is nearly non-toxic at lower concentrations (around 96%), but its toxicity increased to a level comparable with the Ni and Co derivatives (nearly 39% at 200 µg/mL). In the case of the MG-63 cell line, all of the compounds showed very low toxicity at a lower dosage (80% on average), but their toxicity considerably changed with increasing concentration, especially the copper derivate, which showed 38% cell viability at the concentration of 200 µg/mL. In general, the anticancer activity of thiosemicarbazone compounds could be the consequence of the presence of the imine group in their structure [51] and to the intercalation between pairs of DNA bases, or the breaking of DNA strands [52,53].

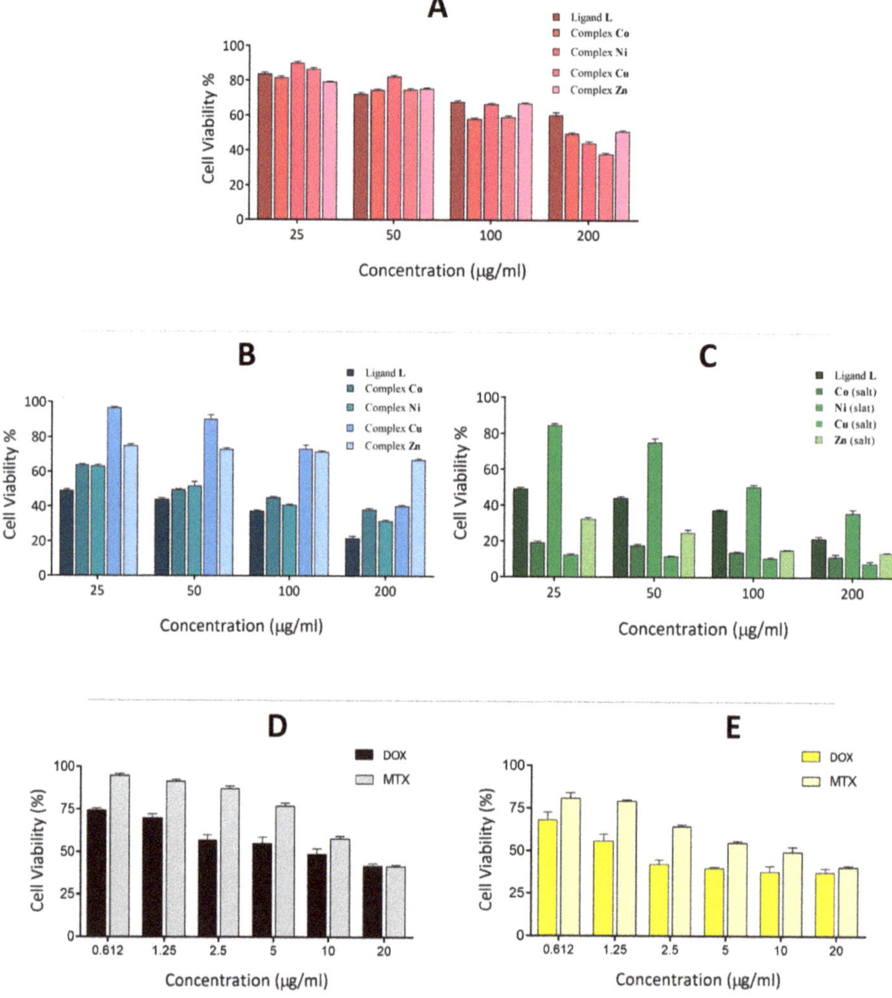

Figure 3. Cytotoxicity evaluation of synthesized compounds at various concentrations against cancer cell lines; (**A**) (MG-63 cell lines), (**B**), and (**C**) (Saos-2 cell lines). Cytotoxicity evaluation of MTX and DOX at various concentrations against MG-63 (**D**) and Saos-2 (**E**) cell lines.

Methotrexate (MTX) and doxorubicin (DOX) are well-known anti-cancer drugs, which are routinely used by research groups as effective anti-cancer drugs against human bone tissue cancer cell lines [54–57]. Herein, we used both of these anti-cancer drugs (with serial dilution from 20 to 0.612 µg/mL) as a well-studied criterion to compare our synthesized compound's cytotoxicity with their anti-cancer activities. As shown in Figure 3D,E, both of these drugs show strong anti-cancer effects against Saos-2 and MG-63 cell lines, especially at the concentration range 2.5–20 µg/mL. Although all of our compounds are not as strong anti-cancer agents as these drugs, the free ligand **L** and its Co, Ni, and Cu derivatives could be considered as somewhat effective anti-cancer agents against the Saos-2 cancer cell line with relatively low IC_{50} (Table 3).

The IC_{50} values were calculated from the curves (dose–response curves) constructed by plotting the cell viability (%) versus the concentration (µg/mL); therefore, the IC_{50} is equal to the concentration at which the cell viability equals 50%. The results of the cytotoxic activity (IC_{50} (µg/mL)) are summarized in Table 3. Ligand **L** is more toxic against Saos-2 cells with a lower IC_{50} value (<25 μg/mL) and the copper derivate is more toxic against the MG-63 cell with IC_{50} = 140 μg/mL.

Table 3. In vitro cytotoxic activity IC_{50} (μg/mL) of compounds against bone cancer cell lines.

Compound	Saos-2	MG-63
Ligand **L**	<25	>200
Complex **Co**	80	195
Complex **Ni**	62	173
Complex **Cu**	43	140
Complex **Zn**	>200	>200
MTX	7.9	14.5
DOX	2.1	10.1

Control assays performed with the corresponding inorganic salts show that the activity is mainly ascribable to the inorganic component rather than the organic calixarene component. Actually, the **L** ligand appears to protect the cells against the inherent cytotoxicity of the bivalent ions present in the inorganic salts of these metals.

Comparison of these results with those obtained from our previously published thiosemicarbazide ligand and its metal derivatives [49] indicates that the thiosemicarbazone Schiff-based ligand is a more effective anti-cancer agent against Saos-2 cancer cells versus thiosemicarbazide ligand (IC_{50} < 25 μg/mL versus IC_{50} > 200 μg/mL) even at lower concentrations. Considering the present Ni and Cu derivatives, it is apparent that the IC_{50} values (as a criterion of anticancer activity) decreased 3.22-fold (decrease from 200 μg/ml to 62 μg/ml) and 3.95-fold (decrease from 170 μg/mL to 43 μg/mL), respectively, in comparison with their thiosemicarbazide analogues.

2.6. Morphology of Cell Nuclei

The morphology, shape, and condensation of chromatin is a good indicator of healthy apoptotic and necrotic cells [6]. According to MTT assay data, our synthesized compounds have considerable anticancer activity against Saos-2 cancer cells. With this in mind, the apoptosis induction properties of the samples upon treatment of the Saos-2 bone cancer cell line were investigated by microscopic analysis of DAPI-stained cells. Microscopic images of DAPI-stained cells following 48 h exposure to control media (**A**), ligand **L** (**B**), complex Co^{2+} (**C**), complex Ni^{2+} (**D**), complex Cu^{2+} (**E**), and complex Zn^{2+} (**F**) are shown in Figure 4.

The untreated control Saos-2 cells showed normal nuclei and no necrosis of cancer cells nucleus was observed. Whereas **L** and complex (Co^{2+}, Ni^{2+}, and Cu^{2+})-treated cells showed deformed and fragmented nuclei in their chromatin and consequently, the morphology of the cells was changed. On the other hand, the cancer cells treated with complex Zn^{2+} retained their normal structures, and their morphology and the number of cells were unchanged. Considering the DAPI images of cells treated with **L** and Cu^{2+} derivate, it is evident that the number and the shape of cancer cells decreased and changed drastically, respectively, which is in good agreement with their IC_{50} and higher anticancer activity.

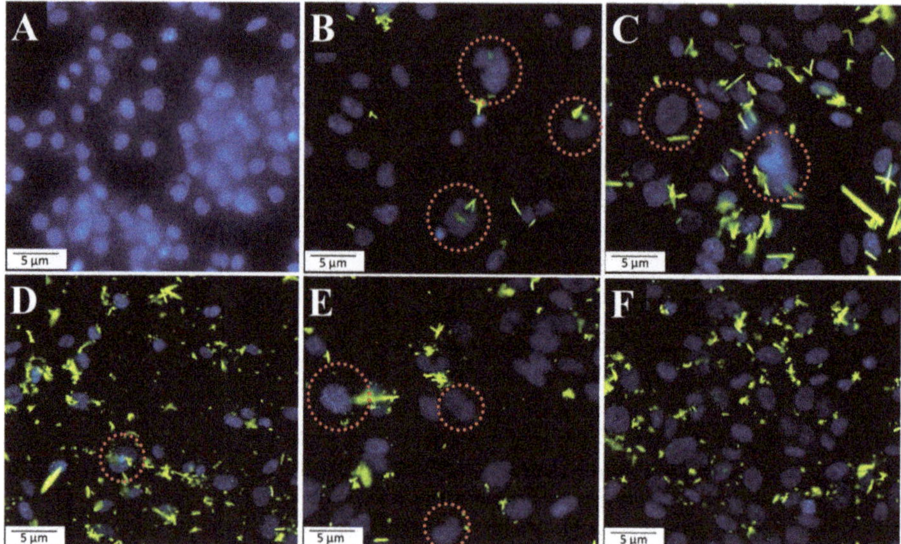

Figure 4. Fluorescent microscopic images of DAPI-stained cells following 48-h treatment with **A** (fresh medium (control)), **B** (ligand **L**), **C** (complex Co^{2+}), **D** (complex Ni^{2+}), **E** (complex Cu^{2+}), and **F** (complex Zn^{2+}). Red circles indicate the Saos-2 nuclei with anomalous morphology after internalization of the fluorescent chemical compounds (**L**, and complexes of Co^{2+}, Ni^{2+}, Cu^{2+}, and Zn^{2+}).

2.7. Biocompatibility Evaluation

In biomedical applications, the effect of synthesized chemical compounds (drugs, drug vehicles, nano-materials) on live tissues and cells (blood cells) is a critical parameter. Different experiments, including the hemolysis assay, coagulation, erythrocytes aggregation, and sedimentation rate, can be used to evaluate the biocompatibility of synthesized materials. In this regard, the hemocompatibility of the **L** and its metal derivatives was investigated by measuring hemolytic activity. To do this, the hemolytic effects of prepared compounds evaluated at the various concentrations in the range 25–1600 µg/mL on HRBCs at physiological conditions (pH 7.4 and 37 °C) and the results are shown in Figure 5. As is clear in Figure 5A, dose-dependent hemolytic effects of all chemical compounds on HRBCs were observed, and the results confirmed that there are only slight hemolytic effects from the synthesized materials even at higher concentrations. For further inspection of the impact of the synthesized compounds on HRBCs, the blood samples were treated with the compounds under investigation at a moderate concentration (200 µg/mL), PBS, and deionized water. The optical microscopy images revealed that the HRBCs shapes show no significant changes after treatment with **L** and its metal complexes for 3 h and they retain their disc-like structure with almost intact erythrocytes membranes. The HRBCs treated with PBS (negative control) and water (positive control) showed diskocytes and lysis forms, respectively (Figure 5B). In general, the results of the hemo-compatibility study confirmed that our synthesized compounds have a very slight lysis effect on HRBCs at lower concentrations and also their hemolytic effects is lower than the permissible range (<4.5%) [58] even at a higher dosage, which would make them suitable candidates for further in vivo investigations.

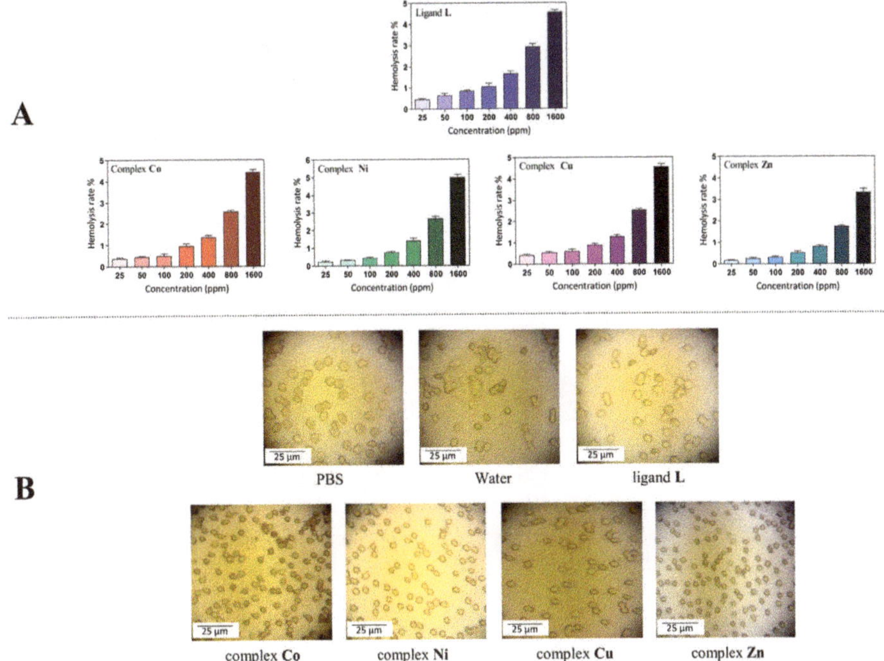

Figure 5. Hemolysis rate of HRBCs in the presence of synthesized compounds at different concentrations (**A**). Optical images of HRBCs treated with synthesized compounds (at the concentration of 200 µg/mL) under light microscopy (**B**).

3. Materials and Methods

3.1. Reagents

All chemical reagents and solvents were purchased from Merck/Aldrich and were used without further purification. Human blood was obtained from the Iranian Blood Transfusion Institute. Mueller Hinton Agar (MHA) and Mueller Hinton Broth (MHB) were purchased from Quelab and Merck, respectively. Roswell Park Memorial Institute 1,640 growth medium (RPMI) was purchased from Gibco BRL Life Technologies. Doxorubicin (DOX) was obtained from Sobhan Pharmaceuticals Co. (Tehran, Iran). Methotrexate (MTX), gentamicin, and nystatin were purchased from Zahravi Pharmaceuticals Co. (Tabriz, Iran). All microorganisms strains, namely, *Staphylococcus aureus* (ATCC® 29213™), *Bacillus subtilis* (ATCC® 6633™), *Escherichia coli* (ATCC® 25922™), *Pseudomonas aeruginosa* (ATCC® 27853™), *Candida albicans* (ATCC® 10231™), and *Candida glabrata* (ATCC® 2001™), were provided from Persian Type Culture Collection (PTCC, Karaj, Iran), and the microbiology department of the Drug Applied Research Centre (Tabriz, Iran).

3.2. Instrumentation

Fourier-transform infrared (FT-IR) spectra were recorded with a Bruker Tensor 27 spectrometer (Karlsruhe, Germany) in the region 4000–500 cm^{-1} using KBr pellets. Proton nuclear magnetic resonance (^{1}H-NMR) spectra were recorded in CDCl$_3$ on a Bruker Spectrospin Avance 400 MHz ultra-shield spectrometer (Karlsruhe, Germany) and Varian 400 MHz (Palo Alto, CA, USA). Mass spectra were recorded on an ion trap instrument Bruker esquire 4000 (Karlsruhe, Germany). An Olympus Bh2-RFCA fluorescence microscope (Tokyo, Japan) was used to evaluate the results of the DAPI staining procedures (vide infra). Microanalyses were carried out using a Heraeus CHN–O–Rapid analyzer (Hanau, Germany). Melting points were measured on an Electrothermal 9100 apparatus (Stone, UK). Crystallographic data were collected on the XRD1 diffraction beam-line of the ELETTRA Synchrotron (Trieste, Italy).

3.3. Synthesis

3.3.1. Preparation of Ligand

Literature procedures for the synthesis of *p*-tert-butylcalix[4]arene (1), 1,3 dinitrile-substituted *p*-tert-butylcalix[4]arene (2), 1,3 diamine-substituted *p*-tert-butylcalix[4]arene (3), and 1,3 diisocyanate-substituted *p*-tert-butylcalix[4]arene (4) were followed [48,59,60]. Bisthiosemicarbazide-substituted *p*-tert-butylcalix[4]arene (5) was prepared as described in our recent paper [49].

Synthesis of 5, 11, 17, 23-tetra-tert-butyl-26, 28-dihydroxy-25, 27-bis(2(2-hydroxybenzeledene) thiosemicarbazonoethoxy) calix[4]arene (**L**)

A solution of salicylaldehyde (0.48 mL, 44 mmol) in ethanol (5 mL) was added to a solution of compound **5** (0.50 g, 5.60 mmol) in ethanol (25 mL). A yellow powder was formed. The mixture was refluxed for 6 h, after which the solution cooled to room temperature and the precipitate was filtered and washed with cold ethanol to remove unreacted aldehyde. Yield 78%. mp: 230 °C (dec.). FTIR (KBr, cm^{-1}); 468, 570, 681, 744, 897, 968, 1035, 1153, 1200, 1266, 1318, 1372, 1479, 1538.7, 1632, 1694, 1944, 2633, 2743, 2955, 3159, 3412, ESI-HRMS: LNa$^+$, [$C_{64}H_{78}N_6O_6S_2$+Na]$^+$ calcd: 1113.5316 *m/z*; found:1113.5315 *m/z*; ^{1}H-NMR: (400 MHz, CDCl$_3$, TMS, 25 °C, δ ppm), 9.74 (s, 2 H; =CH-), 8.98 (s, 2 H; OH), 8.68 (s, 2 H; -NHCS), 7.67 (s, 2 H; OH), 7.33 (s, 2 H; -CSNH), 7.26 (m, 2H, aldehyde ring), 6.99 (d, 2H, aldehyde ring), 6.87 (s, 4 H; ArH), 6.83 (m, 4H, aldehyde ring), 6.71 (s, 4 H; ArH), 4.13 (d, 4 H; ArCH$_2$Ar), 4.36–4.27 (m, 8 H; OCH$_2$CH$_2$NCS), 3.16 (d, 4 H; ArCH$_2$Ar), 1.22 (s, 18H; t-Bu), 0.92 (s, 18H; t-Bu). Anal. Calcd. for $C_{64}H_{78}N_6O_6S_2$ (1091.48): C, 70.43; H, 7.20; N, 7.70; Found: C, 70.13; H, 7.08; N, 7.57.

3.3.2. Preparation of Metal Derivatives

All of the metal derivatives were synthesized according to the following procedure: A solution of metal salt containing Co(NO$_3$)$_2$.6H$_2$O (0.026 g), Ni(NO$_3$)$_2$.6H$_2$O (0.026 g), Cu(NO$_3$)$_2$.3H$_2$O (0.022 g) or Zn(NO$_3$)$_2$.4H$_2$O (0.024 g) in methanol (5 mL) and ligand L (0.1 g) in THF (10 mL) was stirred and refluxed for 24 h. The precipitate was then filtered and the solvent of the filtrate was evaporated under reduced pressure. Finally, the obtained solid was purified by crystallization using THF.

Cobalt Derivative:

Brown, Yield: 84%. mp: 244 °C (dec.). FTIR (KBr, cm^{-1}), 588, 635, 674, 782, 819, 873, 920, 1039, 1120, 1199, 1384, 1482, 1637, 2871, 2958, 3395. ESI-HRMS: LCo(II)Co(III)$^{2+}$, [$C_{64}H_{75}N_6O_6S_2Co_2$]$^{2+}$. calcd: 602.6921 *m/z*; found: 602.6922 *m/z*. Anal. Calcd. for [CoL](NO$_3$)$_2$.2H$_2$O $C_{64}H_{82}N_8O_{14}S_2Co$ (1310.44): C, 58.66; H, 6.31; N, 8.55; Found: C, 58.76; H, 5.97; N, 8.43.

Nickel Derivative:

Orange, Yield: 68%. mp: 214 °C (dec.). FTIR (KBr, cm^{-1}), 587, 685, 747, 817, 873, 922, 1038, 1121, 1197, 1238, 1370, 1481, 1563, 1634, 2095, 2957, 3278. ESI-HRMS: LNi(II)$_2$$^{2+}$, [$C_{64}H_{76}N_6O_6S_2Ni_2$]$^{2+}$ calcd: 603.1972 *m/z*; found: 603.1950 *m/z*. Anal. Calcd. for [NiL](NO$_3$)$_2$ $C_{64}H_{78}N_8O_{12}S_2Ni$ (1274.17): C, 60.33; H, 6.17; N, 8.79; Found: C, 60.32; H, 5.95; N, 8.99.

Copper Derivative:

Green-brown, Yield: 53%. mp: 210 °C (dec.). FTIR (KBr, cm^{-1}), 583, 632, 689, 806, 873, 1034, 1108, 1196, 1303, 1367, 1479, 1576, 1635, 2957, 3046, 3397. ESI-HRMS: LCu(II)$_2$$^{2+}$, [$C_{64}H_{76}N_6O_6S_2Cu_2$]$^{2+}$

calcd: 608.1923 *m/z*; found: 608.1903 *m/z*. Anal. Calcd. for [CuL](NO$_3$)$_2$ C$_{64}$H$_{78}$N$_8$O$_{12}$S$_2$Cu (1279.02): C, 60.10; H, 6.15; N, 8.76; Found: C, 59.72; H, 5.98; N, 8.66.

Zinc Derivative:

Brown, Yield: 91%. mp: 208 °C (dec.). FTIR (KBr, cm^{-1}), 581, 674, 783, 818, 873, 1036, 1115, 1198, 1373, 1481, 1635, 2564, 2958, 3270. ESI-HRMS: for LZn(II)$_2$$^{2+}$, [C$_{64}H_{76}N_6O_6S_2Zn_2$]$^{2+}$ calcd: 610.1905 *m/z*; found: 610.1903 *m/z*. Anal. Calcd. for [CuL](NO$_3$)$_2$ C$_{64}$H$_{78}$N$_8$O$_{12}$S$_2$Zn (1280.87): C, 60.01; H, 6.14; N, 8.75; Found: C, 60.23; H, 6.01; N, 8.64.

3.4. Crystal Structure Determination and Refinement of L

Small single crystal rods were obtained by slow evaporation of a chloroform solution containing **L**. Data collection was carried out at the Macromolecular crystallography XRD1 beamline of the Elettra synchrotron (Trieste, Italy), employing the rotating-crystal method with a Dectris Pilatus 2M area detector. Single crystals were dipped in PEG200 cryoprotectant, mounted on a loop, and immediately flash-frozen under a liquid nitrogen stream at 100 K. The use of synchrotron radiation with crystals cooled to 100 K was necessary to obtain a crystal structure at atomic resolution. The pure diffraction is due to the small crystal size (rods with typical dimensions of 0.1 × 0.05 × 0.02 mm), as well as the significant solvent disorder observed (see below). Diffraction data were indexed and integrated using the XDS package [61] while scaling was carried out with XSCALE [49]. The structure was solved using the SHELXT program [62] and structure refinement was performed with SHELXL-14 by full-matrix least-squares (FMLS) methods on F^2 [63] operating through the WinGX GUI [64].

The crystal structure shows the presence of highly disordered co-crystallized solvent chloroform molecules in voids created by the crystal packing. In particular, in the asymmetric unit, a co-crystallized chloroform solvent molecule disordered over at least 2 positions and a partially occupied co-crystallized chloroform solvent molecule in close proximity to a crystallographic inversion center were found. The PLATON SQUEEZE procedure was used to remove the electron density related to these highly disordered chloroform molecules [65]. The residual electron density of 169 electrons/cell in a total potential solvent area volume of 652.7 (19.1% of the cell volume) can be attributed to 2.9 chloroform solvent molecules (1.45 molecules in the asymmetric unit of the centrosymmetric triclinic crystal).

All non-hydrogen atoms were refined anisotropically. The hydrogen atoms of the water molecule were refined using DFIX and DANG instructions to restrain the OH bond lengths and HOH angle while all other hydrogen atoms were added at the calculated positions and refined using the riding model.

Crystal Data for C$_{64}$H$_{78}$N$_6$O$_6$S$_2$·H$_2$O (M = 1109.46 g/mol): triclinic, space group P-1 (no. 2), a = 12.690(6) Å, b = 15.344(18) Å, c = 19.15(3) Å, α = 104.44(4)°, β = 98.01(4)°, γ = 104.016(10)°, V = 3423(6) Å^3, Z = 2, T = 100(2) K, μ = 0.122 mm^{-1}, D$_{calc}$ = 1.076 g/cm^3, 23853 reflections measured, 6918 unique (R$_{int}$ = 0.1328) which were used in all calculations. The final R$_1$ was 0.0683 (I > 2σ(I)) and wR$_2$ was 0.2108 (all data).

CCDC 1944727 contains supplementary crystallographic data for this paper. These data can be obtained free of charge via http://www.ccdc.cam.ac.uk/conts/retrieving.html (or from the CCDC, 12 Union Road, Cambridge CB2 1EZ, UK; Fax: +44-1223-336033; E-mail: deposit@ccdc.cam.ac.uk)

3.5. Antimicrobial Studies

3.5.1. Preparation of Microorganism's Suspensions

S. aureus (ATCC® 29213™), *B. subtilis* (ATCC® 6633™), *E. coli* (ATCC® 25922™), *P. aeruginosa* (ATCC® 27853™), *C. albicans* (ATCC® 10231™), and *C. glabrata* (ATCC® 2001™) were cultured onto sterile Mueller-Hinton agar plates, and incubated at 37 °C for 24 h. After incubation, a single colony was selected to inoculate a sterile tube containing 2 mL of sterile normal saline to match the turbidity of a 0.5 McFarland standard (10^8 cfu/mL). Further dilution was done with sterile normal saline to 10^5 cfu/mL before incubation.

3.5.2. Broth Microdilution Method

Serial dilution of the synthesized compounds suspension in DMSO (2000–15.62 µg/mL) was done in a sterile 96-well plate. Then, 20 µL of an overnight bacterial and fungal suspension and 180 µL of the compounds suspension in MHB (and 1% glucose for fungal strains) were mixed in each well. The final concentration of DMSO was 10%. After 24 h of incubation at 37 °C, the first concentration with no bacterial growth is regarded as the minimum inhibitory concentration (MIC). Controls (microorganism strain in MHB without chemical and MHB alone) were included in each measurement. Aliquots of 5 µL from wells in which no growth was observed were spotted onto Mueller–Hinton agar plates to determine the minimum bactericidal concentration (MBC) values. The MBC was read as the lowest concentration with no growth after 24 h of incubation [66].

3.6. Cell Culture and MTT Assay

The colorimetric MTT assay, introduced by Mossmann [67], was used for the evaluation of the anti-cancer activity of our synthesized compounds against Saos-2 and MG-63 cell lines (human bone cancer cells). These cells were obtained from the Pasteur Institute of Iran, Tehran, Iran and maintained in RPMI 1640 medium supplemented with 10% FBS and 1% benzylpenicillin/streptomycin. Further, the Saos-2 and MG-63 cell lines were seeded at a density of 1×10^4 cells/well in 96-well plates and maintained at 5% CO_2 in a CO_2 incubator at 37 °C for 24 h, after which the medium was replaced with fresh medium containing different concentrations (200, 100, 50, and 25 µg/mL) of Ligand L and complexes (Co, Ni, Cu, and Zn). The cells without treatment were considered as the control. After 48 h, the medium was removed and replaced with 200 µL of fresh medium containing 100 µg mL^{-1} MTT powder and incubated for a further 4 h. The medium was then withdrawn and 100 µL of DMSO was added to dissolve the formazan crystals inside the cells. The absorbance of solubilized formazan was detected at 570 nm with the reference wavelength of 630 nm using an ELISA plate reader (Stat Fax, 2100, Awareness Technology Inc., Palm City, FL, USA) [68]. The viability of cells in each well was calculated by the following formula:

$$\text{viability of cells } (\%) = \frac{\text{mean absorbance of sample}}{\text{mean absorbance of control}} \times 100. \tag{1}$$

3.7. DAPI Staining

To visually study the apoptotic effects of the synthesized compounds (L and its Co, Ni, Cu, and Zn complexes), the 4′,6-diamidino-2-phenylindole (DAPI) staining assay was performed according to the literature procedure [58]. Briefly, Saos-2 cells were seeded in six-well plates (5×10^4 cells per well) and incubated at 37 °C for 24 h. After the growth of the cells, the culture media were replaced by fresh media containing L and its metal derivatives (sterile filtered through 500 nm) at their half-maximal inhibitory growth concentration (IC$_{50}$). Cells receiving no treatment were considered as the control group. After 48 h, the cells were washed three times with fresh PBS (pH 7.4) to remove any chemical compounds. Subsequently, the cells were fixed with 4 wt% paraformaldehyde for 15 min at room temperature. Afterwards, the cells were washed three times with PBS and permeabilized with Triton X-100 (0.1% *w/v*) for 5 min. The cells were washed again with PBS and stained with 300 ng mL^{-1} DAPI for 15 min. DNA condensation and fragmentation in apoptotic cells were evaluated under a fluorescence microscope.

3.8. Erythrocytes Hemolysis Assay

Ligand **L** and all of the complexes at different concentrations were subjected to a hemolysis assay to assess the toxicities of synthesized compounds to human red blood cells (HRBCs). The erythrocytes hemolysis assay was conducted according to the method of Shafiei et al. [69]. In this regard, fresh human blood was received from IBTI and centrifuged at room temperature (4000 rpm, 10 min). The obtained erythrocytes were washed three times with 10 mL of phosphate-buffered saline (PBS) to remove blood proteins and serum from the RBCs. The total isolated RBCs were diluted 10 times with PBS. In each microtube, 0.5 mL of ligand **L** and all of the complexes at different concentrations (25, 50, 100, 200, 400, 800, 1600 µg/mL) and 0.5 mL of diluted RBCs were mixed and incubated at 37 °C for 3 h in an incubator shaker. Hemolysis was measured by analyzing the absorbance of free hemoglobin leaked out of compromised RBCs in the supernatants at 540 nm. The RBCs incubated with PBS and water (1:1) were used as 0% and 100% hemolysis controls. The hemolysis rate was calculated via the following formula:

$$\text{Hemolysis}(\%) = \frac{A_{\text{Sample}} - A_{\text{PBS}}}{A_{\text{Water}} - A_{\text{PBS}}} \times 100. \tag{2}$$

4. Conclusions

In this study, a novel calix[4]arene-based thiosemicarbazone derivative with multi-dentate chelating properties was synthesized and characterized by NMR, single crystal X-ray structure analysis, and HRMS. Both the 1H-NMR spectra and solid state structure confirmed that the calix[4]arene macrocycle has the cone conformation. In particular, the X-ray structure shows that the cone exhibits a typical open conformation, with two opposite phenyl rings inclined outwards with large angles. The similar gauche conformation of the two ethyl spacers of the alkoxythiosemicarbazone substituents results in a C2 point group symmetry molecule, with the two thiosemicarbazone groups oriented in opposite directions like a two-blade propeller, thereby forming a peculiar chiral helicity. A water molecule has been found, in the center of the two-blade propeller, strongly coordinated through multiple H-bonds. A considerable antibacterial activity against various tested microorganisms (gram-positive bacteria: *S. aureus* and *B. subtilis*; gram-negative bacteria: *E. coli* and *P. aeruginosa*; and fungi: *C. albicans* and *C. glabrata*) was observed for the ligand and for all its metal derivatives (Co^{2+}, Ni^{2+}, Cu^{2+}, and Zn^{2+}). Interesting antifungal activities against yeast (*C. albicans*) were also observed for the ligand and for its Co^{2+} derivative. All compounds showed an increasing cytotoxicity against the tested cancerous cells (Saos-2 and MG-63) with increasing concentration, with the exclusion of Zn^{2+} derivative against Saos-2 cells. Interestingly, for the Saos-2 cell line, the anticancer activity of ligand L was higher than its metal derivatives at all concentrations. Microscopic analysis of DAPI-stained cells showed changes in morphology, with deformation and fragmentation of the nuclei. Finally, the hemo-compatibility study demonstrated that all of the compounds have a very slight lysis activity, which would make them suitable candidates for further in vivo investigations.

Supplementary Materials: The Supplementary Materials are available online.

Author Contributions: E.B.N.: conceptualization and design of the work; acquisition, analysis and interpretation of data; writing, review and editing. B.S.: supervision of the research project. S.G. and N.H.: analysis of structural data; overall interpretation of results; writing, review and editing. P.N.: supervision of the synthesis and interpretation of NMR spectra. H.S.K.: design, acquisition and interpretation of biological experiments. All authors have read and agreed to the published version of the manuscript.

Funding: This study is financially supported by the University of Tabriz (Tabriz, Iran) (94-582A12).

Acknowledgments: We thank the Drug Applied Research Center (DARC), Tabriz University of Medical Sciences, (Tabriz, Iran). All biological experiments in this report were done in DARC. We thank the Elettra Synchrotron (Trieste, Italy) and the staff of the XRD1 beamline for their technical assistance and Fabio Hollan for the HRMS measurements.

Conflicts of Interest: The authors declare no conflict of interest.

References

1. Gutsche, C.D.; Levine, J.A. Calixarenes 6 Synthesis of a Functionalizable Calix[4]arene in a Conformationally Rigid Cone Conformation. *J. Am. Chem. Soc.* **1982**, *104*, 2652–2653. [CrossRef]
2. Erdemir, S.; Tabakci, B.; Tabakci, M. A highly selective fluorescent sensor based on calix [4] arene appended benzothiazole units for Cu^{2+}, S^{2-} and HSO_4^- ions in aqueous solution. *Sens. Actuators B Chem.* **2016**, *228*, 109–116. [CrossRef]

3. Bauer, A.; Jäschke, A.; Azzam, S.S.A.; Glasneck, F.; Ullmann, S.; Kersting, B.; Brendler, V.; Schmeide, K.; Stumpf, T. Multidentate extracting agents based on calix [4] arene scaffold–UVI/EuIII separation studies. *Sep. Purif. Technol.* **2019**, *213*, 246–254. [CrossRef]

4. Shirakawa, S.; Shimizu, S. Inherently Chiral Calix [4] arenes as Supramolecular Catalysts. In *Designed Molecular Space in Material Science and Catalysis*; Springer: Berlin, Germany, 2018; pp. 51–68.

5. De Rosa, M.; La Manna, P.; Soriente, A.; Gaeta, C.; Talotta, C.; Hickey, N.; Geremia, S.; Neri, P. A Simple Tetraminocalix[4]arene as a Highly Efficient Catalyst under "On-Water" Conditions through Hydrophobic Amplification of Weak Hydrogen Bonds. *Chem. Eur. J.* **2017**, *23*, 7142–7151.

6. Rahimi, M.; Karimian, R.; Mostafidi, E.; Noruzi, E.B.; Taghizadeh, S.; Shokouhi, B.; Kafil, H.S. Highly branched amine-functionalized p-sulfonatocalix [4] arene decorated with human plasma proteins as a smart, targeted, and stealthy nano-vehicle for the combination chemotherapy of MCF7 cells. *New J. Chem.* **2018**, *42*, 13010–13024.

7. Rahimi, M.; Karimian, R.; Noruzi, E.B.; Ganbarov, K.; Zarei, M.; Kamounah, F.S.; Yousefi, B.; Bastami, M.; Yousefi, M.; Kafil, H.S. Needle-shaped amphoteric calix [4] arene as a magnetic nanocarrier for simultaneous delivery of anticancer drugs to the breast cancer cells. *Int. J. Nanomed.* **2019**, *14*, 2619–2636. [CrossRef]

8. Böhmer, V.; Rathay, D.; Kämmerer, H. The t-butyl group as a possible protective group in the synthesis of oligo [hydroxy-1, 3-phenylene] methylenes. *Org. Prep. Proc. Int.* **1978**, *10*, 113–121. [CrossRef]

9. Sengupta, A.; Godbole, S.V.; Mohapatra, P.K.; Iqbal, M.; Huskens, J.; Verboom, W. Judd–Ofelt parameters of diglycolamide-functionalized calix [4] arene Eu^{3+} complexes in room temperature ionic liquid for structural analysis: Effects of solvents and ligand stereochemistry. *J. Lumin.* **2014**, *148*, 174–180. [CrossRef]

10. Dessingou, J.; Tabbasum, K.; Mitra, A.; Hinge, V.K.; Rao, C.P. Lower rim 1, 3-di {4-antipyrine} amide conjugate of calix [4] arene: Synthesis, characterization, and selective recognition of Hg^{2+} and its sensitivity toward pyrimidine bases. *J. Org. Chem.* **2012**, *77*, 1406–1413. [CrossRef] [PubMed]

11. Gubbuk, I.H.; Gungor, O.; Alpoguz, H.K.; Ersoz, M.; Yılmaz, M. Kinetic study of mercury (II) transport through a liquid membrane containing calix [4] arene nitrile derivatives as a carrier in chloroform. *Desalination* **2010**, *261*, 157–161. [CrossRef]

12. Bozkurt, S.; Turkmen, M.B.; Soykan, C. Synthesis of new chiral calix [4] arene thiourea derivatives for enantiomeric recognition of carboxylate anions. *J. Incl. Phenom. Macrocyclic Chem.* **2016**, *84*, 35–41. [CrossRef]

13. Sap, A.; Tabakci, B.; Yilmaz, A. Calix [4] arene-based Mannich and Schiff bases as versatile receptors for dichromate anion extraction: Synthesis and comparative studies. *Tetrahedron* **2012**, *68*, 8739–8745. [CrossRef]

14. Iuliano, V.; Talotta, C.; Gaeta, C.; Soriente, A.; De Rosa, M.; Geremia, S.; Hickey, N.; Mennucci, B.; Neri, P. Negative Solvatochromism in a N-Linked p-Pyridiniumcalix[4]arene Derivative. *Org. Lett.* **2019**, *21*, 2704–2707. [CrossRef] [PubMed]

15. Lejeune, M.; Picron, J.-F.; Mattiuzzi, A.; Lascaux, A.; De Cesco, S.; Brugnara, A.; Thiabaud, G.; Darbost, U.; Coquiere, D.; Colasson, B. Ipso-nitration of calix [6] azacryptands: Intriguing effect of the small rim capping pattern on the large rim substitution selectivity. *J. Org. Chem.* **2012**, *77*, 3838–3845. [CrossRef]

16. Kim, S.K.; Sessler, J.L.; Gross, D.E.; Lee, C.-H.; Kim, J.S.; Lynch, V.M.; Delmau, L.H.; Hay, B.P. A calix [4] arene strapped calix [4] pyrrole: An ion-pair receptor displaying three different cesium cation recognition modes. *J. Am. Chem. Soc.* **2010**, *132*, 5827–5836. [CrossRef]

17. Durso, A.; Brancatelli, G.; Hickey, N.; Farnetti, E.; De Zorzi, R.; Bonaccorso, C.; Purrello, R.; Geremia, S. Interactions of a water-soluble calix[4]arene with spermine: Solution and solid-state characterisation. *Supramol. Chem.* **2016**, *28*, 499–505. [CrossRef]

18. Kim, S.K.; Lee, J.; Williams, N.J.; Lynch, V.M.; Hay, B.P.; Moyer, B.A.; Sessler, J.L. Bipyrrole-Strapped Calix [4] pyrroles: Strong Anion Receptors That Extract the Sulfate Anion. *J. Am. Chem. Soc.* **2014**, *136*, 15079–15085. [CrossRef]

19. Teixeira, F.A.; Marcos, P.M.; Ascenso, J.R.; Brancatelli, G.; Hickey, N.; Geremia, S. Selective Binding of Spherical and Linear Anions by Tetraphenyl(thio)urea-Based Dihomooxacalix[4]arene Receptors. *J. Org. Chem.* **2017**, *82*, 11383–11390. [CrossRef]

20. Miranda, A.S.; Serbetci, D.; Marcos, P.M.; Ascenso, J.R.; Berberan-Santos, M.N.; Hickey, N.; Geremia, S. Ditopic Receptors Based on Dihomooxacalix[4]arenes Bearing Phenylurea Moieties with Electron-Withdrawing Groups for Anions and Organic Ion Pairs. *Front. Chem.* **2019**, *7*, 758. [CrossRef]

21. Gattuso, G.; Notti, A.; Parisi, M.F.; Pisagatti, I.; Marcos, P.M.; Ascenso, J.R.; Brancatelli, G.; Geremia, S. Selective recognition of biogenic amine hydrochlorides by heteroditopic dihomooxacalix[4]arenes. *New, J. Chem.* **2015**, *39*, 817–821. [CrossRef]

22. Gao, F.; Cui, L.; Song, Y.; Li, Y.-Z.; Zuo, J.-L. Calix [4] arene-supported mononuclear lanthanide single-molecule magnet. *Inorg. Chem.* **2013**, *53*, 562–567. [CrossRef] [PubMed]

23. Cante-Mota, I.; Moreno-Alcántar, G.; Flores-Alamo, M.; Castillo, I. Benzimidazole-derived calix [4] arenes with polymerizable styrene groups and their Cu (II) complexes. *Inorg. Chim. Acta* **2013**, *407*, 11–18. [CrossRef]

24. Li, L.; Gu, W.-W.; Yan, C.-G. Syntheses, Crystal Structures and Complexing Properties of 1, 3-Distal Calix [4] arene Schiff Bases. *Chem. Res. Chin. Univ.* **2010**, *26*, 38–45.

25. Sgarlata, C.; Brancatelli, G.; Fortuna, C.G.; Sciotto, D.; Geremia, S.; Bonaccorso, C. Three-dimensional network structures based on pyridylcalix[4]arene metal complexes. *Chem. Plus Chem.* **2017**, *82*, 1341–1350.

26. Meninno, S.; Parrella, A.; Brancatelli, G.; Geremia, S.; Gaeta, C.; Talotta, C.; Neri, P.; Lattanzi, A. Polyoxomolybdate-Calix[4]arene Hybrid: A Catalyst for Sulfoxidation Reactions with Hydrogen Peroxide. *Org. Lett.* **2015**, *17*, 5100–5103. [CrossRef]

27. Brancatelli, G.; De Zorzi, R.; Hickey, N.; Siega, P.; Zingone, G.; Geremia, S. New Multicomponent Porous Architecture of Self-Assembled Porphyrins/Calixarenes Driven by Nickel Ions. *Cryst. Growth Des.* **2012**, *12*, 5111–5117. [CrossRef]

28. Casas, J.S.; García-Tasende, M.S.; Sordo, J. Main group metal complexes of semicarbazones and thiosemicarbazones. *A structural review. Coord. Chem. Rev.* **2000**, *209*, 197–261. [CrossRef]

29. De Oliveira, R.B.; de Souza-Fagundes, E.M.; Soares, R.P.P.; Andrade, A.A.; Krettli, A.U.; Zani, C.L. Synthesis and antimalarial activity of semicarbazone and thiosemicarbazone derivatives. *Eur. J. Med. Chem.* **2008**, *43*, 1983–1988.

30. Shn Moorthy, N.; Mfsa Cerqueira, N.; Ramos, M.J.; Fernandes, P.A. Aryl-and heteroaryl-thiosemicarbazone derivatives and their metal complexes: A pharmacological template. *Recent Pat. Anti-Cancer Drug Discov.* **2013**, *8*, 168–182. [CrossRef]

31. Al-Amiery, A.A.; Al-Majedy, Y.K.; Abdulreazak, H.; Abood, H. Synthesis, characterization, theoretical crystal structure, and antibacterial activities of some transition metal complexes of the thiosemicarbazone (Z)-2-(pyrrolidin-2-ylidene) hydrazinecarbothioamide. *Bioinorg. Chem. Appl.* **2011**, *2011*, 483101. [CrossRef]

32. Pathan, A.H.; Bakale, R.P.; Naik, G.N.; Frampton, C.S.; Gudasi, K.B. Synthesis, crystal structure, redox behavior and comprehensive studies on DNA binding and cleavage properties of transition metal complexes of a fluoro substituted thiosemicarbazone derived from ethyl p yruvate. *Polyhedron* **2012**, *34*, 149–156. [CrossRef]

33. Netalkar, P.P.; Netalkar, S.P.; Revankar, V.K. Transition metal complexes of thiosemicarbazone: Synthesis, structures and invitro antimicrobial studies. *Polyhedron* **2015**, *100*, 215–222. [CrossRef]

34. Quiroga, A.G.; Ranninger, C.N. Contribution to the SAR field of metallated and coordination complexes: Studies of the palladium and platinum derivatives with selected thiosemicarbazones as antitumoral drugs. *Coord. Chem. Rev.* **2004**, *248*, 119–133. [CrossRef]

35. Melha, K.S.A. In-vitro antibacterial, antifungal activity of some transition metal complexes of thiosemicarbazone Schiff base (HL) derived from N4-(7'-chloroquinolin-4'-ylamino) thiosemicarbazide. *J. Enzyme Inhib. Med. Chem.* **2008**, *23*, 493–503. [CrossRef]

36. Elsayed, S.A.; El-Hendawy, A.M.; Mostafa, S.I.; Jean-Claude, B.J.; Todorova, M.; Butler, I.S. Antineoplastic activity of new transition metal complexes of 6-methylpyridine-2-carbaldehyde-n (4)-ethylthiosemicarbazone: X-Ray crystal structures of [VO$_2$ (mpETSC)] and [Pt (mpETSC) Cl]. *Bioinorg. Chem. Appl.* **2010**, *2010*, 149149. [CrossRef]

37. Santini, C.; Pellei, M.; Gandin, V.; Porchia, M.; Tisato, F.; Marzano, C. Advances in copper complexes as anticancer agents. *Chem. Rev.* **2013**, *114*, 815–862. [CrossRef]

38. Glisoni, R.J.; Cuestas, M.L.; Mathet, V.L.; Oubiña, J.R.; Moglioni, A.G.; Sosnik, A. Antiviral activity against the hepatitis C virus (HCV) of 1-indanone thiosemicarbazones and their inclusion complexes with hydroxypropyl-β-cyclodextrin. *Eur. J. Pharm. Sci.* **2012**, *47*, 596–603. [CrossRef]

39. Lessa, J.A.; Soares, M.A.; Dos Santos, R.G.; Mendes, I.C.; Salum, L.B.; Daghestani, H.N.; Andricopulo, A.D.; Day, B.W.; Vogt, A.; Beraldo, H. Gallium (III) complexes with 2-acetylpyridine-derived thiosemicarbazones: Antimicrobial and cytotoxic effects and investigation on the interactions with tubulin. *Biometals* **2013**, *26*, 151–165. [CrossRef]

40. Kalaivani, P.; Prabhakaran, R.; Ramachandran, E.; Dallemer, F.; Paramaguru, G.; Renganathan, R.; Poornima, P.; Padma, V.V.; Natarajan, K. Influence of terminal substitution on structural, DNA, protein binding, anticancer and antibacterial activities of palladium (II) complexes containing 3-methoxy salicylaldehyde-4 (N) substituted thiosemicarbazones. *Dalton Trans.* **2012**, *41*, 2486–2499. [CrossRef]

41. Li, M.X.; Chen, C.L.; Zhang, D.; Niu, J.Y.; Ji, B.S. Mn (II), Co (II) and Zn (II) complexes with heterocyclic substituted thiosemicarbazones: Synthesis, characterization, X-ray crystal structures and antitumor comparison. *Eur. J. Med. Chem.* **2010**, *45*, 3169–3177. [CrossRef]

42. Chawla, H.M.; Sahu, S.N.; Shrivastava, R.; Kumar, S. Calix [4] arene-based ditopic receptors for simultaneous recognition of fluoride and cobalt (II) ions. *Tetrahedron Lett.* **2012**, *53*, 2244–2247. [CrossRef]

43. Chawla, H.M.; Singh, S.P. Calix [4] arene based neutral receptor for dihydrogen phosphate anion. *Tetrahedron* **2008**, *64*, 741–748. [CrossRef]

44. Liu, Z.; Liu, D.; Wang, J.; Sun, J.; Yan, C. Synthesis and crystal structure of p-tert-butylcalix [4] arene 1, 3-distal and monosubstituted semicarbazones and thiosemicarbazones. *Chem. Res. Chin. Univ.* **2014**, *30*, 415–419. [CrossRef]

45. Rao, C.N.R.; Venkataraghavan, R. The C= S stretching frequency and the "-NC= S bands" in the infrared. *Spectrochim. Acta* **1962**, *18*, 541–547. [CrossRef]

46. Mostafa, M.M. Spectroscopic studies of some thiosemicarbazide compounds derived from Girard's T and P. *Spectrochim. Acta A* **2007**, *66*, 480–486. [CrossRef] [PubMed]

47. Viñuelas-Zahínos, E.; Luna-Giles, F.; Torres-García, P.; Fernández-Calderón, M.C. Co (III), Ni (II), Zn (II) and Cd (II) complexes with 2-acetyl-2-thiazoline thiosemicarbazone: Synthesis, characterization, X-ray structures and antibacterial activity. *Eur. J. Med. Chem.* **2011**, *46*, 150–159. [CrossRef]

48. Quiroga-Campano, C.; Gomez-Machuca, H.; Moris, S.; Jara, P.; De la Fuente, J.; Pessoa-Mahana, H.; Jullian, C.; Saitz, C. Synthesis of bifunctional receptor for fluoride and cadmium based on calix [4] arene with thiourea moieties. *J. Mol. Struct.* **2017**, *1141*, 133–141. [CrossRef]

49. Bahojb Noruzi, E.; Kheirkhahi, M.; Shaabani, B.; Geremia, S.; Hickey, N.; Asaro, F.; Nitti, P.; Kafil, H.S. Design of a thiosemicarbazide functionalized calix [4] arene ligand and related transition metal complexes: Synthesis, characterization and biological studies. *Front. Chem.* **2019**, *7*, 663. [CrossRef]

50. Zhong, Z.; Aotegen, B.; Xu, H.; Zhao, S. Structure and antimicrobial activities of benzoyl phenyl-thiosemicarbazone-chitosans. *Int. J. Biol. Macromol.* **2012**, *50*, 1169–1174. [CrossRef]

51. Kovala-Demertzi, D.; Demertzis, M.A.; Miller, J.R.; Papadopoulou, C.; Dodorou, C.; Filousis, G. Platinum(II) complexes with 2-acetyl pyridine thiosemicarbazone: Synthesis, crystal structure, spectral properties, antimicrobial and antitumour activity. *J. Inorg. Biochem.* **2001**, *86*, 555–563. [CrossRef]

52. Hall, I.H.; Lackey, C.B.; Kistler, T.D.; Durham, R.W.; Jouad, E.M.; Khan, M.; Thanh, X.D.; Djebbar-Sid, S.; Benali-Baitich, O.; Bouet, G.M. Cytotoxicity of copper and cobalt complexes of furfural semicarbazone and thiosemicarbazone derivatives in murine and human tumor cell lines. *Pharmazie* **2000**, *55*, 937–941. [PubMed]

53. Baldini, M.; Belicchi-Ferrari, M.; Bisceglie, F.; Pelosi, G.; Pinelli, S.; Tarasconi, P. Cu(II) Complexes with Heterocyclic Substituted Thiosemicarbazones: The Case of 5-Formyluracil. Synthesis, Characterization, X-ray Structures, DNA Interaction Studies, and Biological Activity. *Inorg. Chem.* **2003**, *42*, 2049–2055. [CrossRef]

54. Das, D.; Rameshbabu, A.P.; Ghosh, P.; Patra, P.; Dhara, S.; Pal, S. Biocompatible nanogel derived from functionalized dextrin for targeted delivery of doxorubicin hydrochloride to MG 63 cancer cells. *Carbohydr. Polym.* **2017**, *171*, 27–38. [CrossRef] [PubMed]

55. Chai, D.; Hao, B.; Hu, R.; Zhang, F.; Yan, J.; Sun, Y.; Huang, X.; Zhang, Q.; Jiang, H. Delivery of Oridonin and Methotrexate via PEGylated Graphene Oxide. *ACS Appl. Mater. Interfaces* **2019**, *11*, 22915–22924. [CrossRef] [PubMed]

56. Meshkini, A.; Oveisi, H. Methotrexate-F127 conjugated mesoporous zinc hydroxyapatite as an efficient drug delivery system for overcoming chemotherapy resistance in osteosarcoma cells. *Colloids Surf. B* **2017**, *158*, 319–330. [CrossRef] [PubMed]

57. Soares, P.I.P.; Sousa, A.I.; Silva, J.C.; Ferreira, I.M.M.; Novo, C.M.M.; Borges, J.P. Chitosan-based nanoparticles as drug delivery systems for doxorubicin: Optimization and modelling. *Carbohydr. Polym.* **2016**, *147*, 304–312. [CrossRef]

58. Rahimi, M.; Shojaei, S.; Safa, K.D.; Ghasemi, Z.; Salehi, R.; Yousefi, B.; Shafiei-Irannejad, V. Biocompatible magnetic tris (2-aminoethyl) amine functionalized nanocrystalline cellulose as a novel nanocarrier for anticancer drug delivery of methotrexate. *New J. Chem.* **2017**, *41*, 2160–2168. [CrossRef]

59. Gutsche, C.D.; Iqbal, M.; Stewart, D. Calixarenes 18. Syntheses procedures for p-tert-butylcalix [4] arene. *J. Org. Chem.* **1986**, *51*, 742–745. [CrossRef]

60. Zhang, W.-C.; Huang, Z.-T. Synthesis of 4-tert-butylcalix [4] arenes bearing two schiff-base units at the lower rim. *Synthesis* **1997**, *1997*, 1073–1076. [CrossRef]

61. Kabsch, W. XDS. *Acta Crystallogr. Sect. D Biol. Crystallogr.* **2010**, *66*, 125–132. [CrossRef]

62. Sheldrick, G.M. SHELXT–Integrated space-group and crystal-structure determination. *Acta Crystallog. Sect. A Found. Crystallogr.* **2015**, *71*, 3–8. [CrossRef] [PubMed]

63. Sheldrick, G.M. A short history of SHELX. *Acta Crystallog. Sect. A Found. Crystallogr.* **2008**, *64*, 112–122. [CrossRef] [PubMed]

64. Farrugia, L.J. WinGX and ORTEP for Windows: An update. *J. Appl. Crystallog.* **2012**, *45*, 849–854. [CrossRef]

65. Spek, A.L. Structure validation in chemical crystallography. *Acta Crystallogr. Sect. D Biol. Crystallogr.* **2009**, *65*, 148–155. [CrossRef] [PubMed]

66. Skovgaard, S.; Larsen, M.H.; Nielsen, L.N.; Skov, R.L.; Wong, C.; Westh, H.; Ingmer, H. Recently introduced qacA/B genes in Staphylococcus epidermidis do not increase chlorhexidine MIC/MBC. *J. Antimicrob. Chemother.* **2013**, *68*, 2226–2233. [CrossRef]

67. Mosmann, T. Rapid colorimetric assay for cellular growth and survival: Application to proliferation and cytotoxicity assays. *J. Immunol. Methods* **1983**, *65*, 55–63. [CrossRef]

68. Rahimi, M.; Shafiei-Irannejad, V.; Safa, K.D.; Salehi, R. Multi-branched ionic liquid-chitosan as a smart and biocompatible nano-vehicle for combination chemotherapy with stealth and targeted properties. *Carbohydr. Polym.* **2018**, *196*, 299–312. [CrossRef] [PubMed]

69. Shafiei-Irannejad, V.; Rahimi, M.; Zarei, M.; Dinparast-isaleh, R.; Bahrambeigi, S.; Alihemmati, A.; Shojaei, S.; Ghasemi, Z.; Yousefi, B. Polyelectrolyte Carboxymethyl Cellulose for Enhanced Delivery of Doxorubicin in MCF7 Breast Cancer Cells: Toxicological Evaluations in Mice Model. *Pharm. Res.* **2019**, *36*, 68. [CrossRef]

Sample Availability: Samples of the compound L are available from the authors.

 *molecules*

Review

Antimicrobial Activity of Calixarenes and Related Macrocycles

Dmitriy N. Shurpik [1], **Pavel L. Padnya** [1], **Ivan I. Stoikov** [1] and **Peter J. Cragg** [2,*]

[1] A.M. Butlerov Chemical Institute, Kazan Federal University, Kremlyovskaya St, 18, 420008 Kazan, Russia; dnshurpik@mail.ru (D.N.S.); padnya.ksu@gmail.com (P.L.P.); ivan.stoikov@mail.ru (I.I.S.)

[2] School of Pharmacy and Biomolecular Sciences, University of Brighton, Huxley Building, Moulsecoomb, Brighton, East Sussex BN2 4GJ, UK

* Correspondence: P.J.Cragg@brighton.ac.uk

Academic Editors: Mario Berberan-Santos and Paula M. Marcos

Received: 30 September 2020; Accepted: 1 November 2020; Published: 5 November 2020

Abstract: Calixarenes and related macrocycles have been shown to have antimicrobial effects since the 1950s. This review highlights the antimicrobial properties of almost 200 calixarenes, resorcinarenes, and pillararenes acting as prodrugs, drug delivery agents, and inhibitors of biofilm formation. A particularly important development in recent years has been the use of macrocycles with substituents terminating in sugars as biofilm inhibitors through their interactions with lectins. Although many examples exist where calixarenes encapsulate, or incorporate, antimicrobial drugs, one of the main factors to emerge is the ability of functionalized macrocycles to engage in multivalent interactions with proteins, and thus inhibit cellular aggregation.

Keywords: calixarene; resorcinarene; pillararene; antibiotic; fungicide; biofilm inhibition

---

1. Introduction

Calixarenes have found an enormous range of applications since their first discovery and later exploitation by researchers worldwide [1]. Their ability to complex small molecules, either within each individual macrocycle or through aggregated nanostructures, has opened up avenues for drug delivery. Functionalization, though upper and lower rims, offers up the possibility of their use as inhibitors of biochemical processes or as prodrugs with the active substituents released in response to external stimuli. Some useful summaries of the broader biological and biochemical effects of water-soluble calixarenes [2,3] and pillararenes [4] have been published.

The use of calixarenes and related macrocycles as antimicrobial agents dates back to the 1950s and Cornforth's work on *Macrocyclon* (Figure 1) [5]. Since his original antitubercular study, the biological effects of calixarenes and their relatives, the resorcinarenes and pillararenes, have become more widely appreciated. The antimicrobial impact of calixarenes was first assessed in 2002 by Regnouf-de-Vains [6] and has been followed by subsequent reviews [7–9]. Herein, we update the field of antimicrobial calixarenes and related macrocycles with the most recent advances.

Before embarking on a survey of these compounds' antimicrobial effects, several points need to be considered. There are three main modes of action by which macrocycles can induce a biological response. Individual calixarenes may do so through, for example, insertion in a cell membrane to destroy its integrity, or by acting as a prodrug, incorporating a substituent that is cleaved from its parent macrocycle to become biologically active. Macrocycles, or aggregates of macrocycles, can deliver single drug molecules by transporting them in their central cavities, releasing them in response to an external stimulus. Finally, there is the possibility of disrupting the biofilms formed by colonies of bacteria when adhering to surfaces.

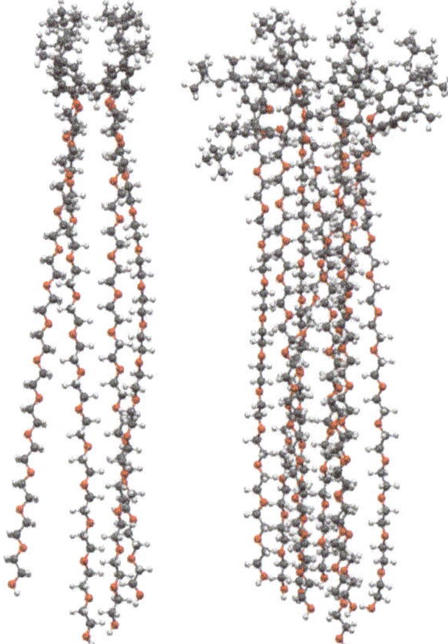

Figure 1. *Macrocyclon* as originally envisaged by Cornforth (**left**) and as later determined (**right**).

In all these cases, the solubility of the calixarene, or its complex, must be considered. If it is very hydrophobic, its application, except perhaps as a part of a topical cream, will be limited. Lack of aqueous solubility will make administration difficult. Aqueous solubility is one aspect of Lipinski's 'rule of five', which states that an orally active drug should have no more than five hydrogen bond donors and 10 hydrogen bond acceptors, a molecular mass below 500 Da, and an octanol-water partition coefficient below five [10]. In practice, not all of these parameters need to be met, but it is a useful rule of thumb, particularly when considering macrocycles with high molecular masses.

Once a calixarene or its complex has been shown to be 'druggable', it needs to reach its target. A water-soluble calix[4]arene with quaternary ammonium groups on the upper rim and a fluorophore on the lower rim, shown in Figure 2, was used by Matthews, Mueller, and colleagues to probe calixarene transport into cells [11]. Using fluorescence confocal microscopy, it was possible to show that the molecules were not transported across the membrane by endocytosis, nor did they appear to traverse the membrane between lipid raft domains. The authors posit that other, non-specific processes must be responsible as the macrocycles were found to be retained in the cytoplasm and unable to enter the cell nuclei.

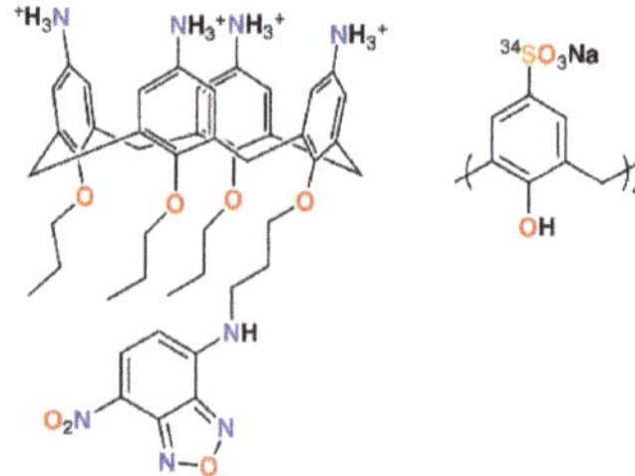

Figure 2. Matthews' fluorescent calixarene (**left**) and Coleman's calixarene radiotracer (**right**).

One final thought must be given to toxicity. An ideal antimicrobial will display specific activity against one species of bacteria, while having no effects on any other organisms, whether they be humans being treated for an infection or the diversity of animal and plant life that may exist in a field where crops are sprayed. An encouraging study into the toxicity and biodistribution of *p*-sulfonatocalix[4]arene in mice by Coleman and colleagues using a ^{34}S-labelled derivative found no toxic effects in mice up to 100 mg kg^{-1} of body weight (Figure 2) [12]. One animal was given a dose of 400 mg kg^{-1} and died within an hour, whereas 14 others, receiving doses of 40 or 100 mg kg^{-1}, were outwardly unaffected. Biodistribution was found in all major organs, but particularly the lungs, as well as in blood and plasma. At both 40 and 100 mg kg^{-1}, very little was found in muscle tissue and none was detected in the brain. A majority of the 40 mg kg^{-1} dose had been cleared through the urine within an hour.

Taken together, the toxicity data and macrocycles' cellular destination indicate that a mechanism, yet to be determined, exists by which they can enter cells; however, they are not in themselves harmful.

2. Molecular Prodrugs and Drug Delivery Agents

Phenol-based macrocycles provide a rigid scaffold to which numerous substituents can be appended through lower rim substitution reactions. The regiochemistry of these reactions can be controlled by numerous well-known techniques to maximize the distance between substituents, in alternate conformations, or make use of the macrocyclic cavity to preorganize the spatial relationships between substituents in a cone conformer. If the substituents are pharmacologically active in their own right, the macrocycle becomes the agent that delivers the prodrug to a destination within the target microbe. Once at the destination, local conditions, typically pH or enzyme action, initiate bond cleavage to release the active drug while the macrocycle is either chemically degraded or excreted. Alternatively, the calixarene essentially acts as a vector by delivering its drug-derived substituents to their destination while remaining attached throughout. The benefits of such an approach are that the solubility of the macrocycle can enhance the bioavailability of the drug and that multiple copies of that drug can be delivered to the same cellular site. This latter effect is an example of multivalency and is important in other antimicrobial applications of calixarenes, notably in biofilm inhibition.

It is often only through conjugation to a carrier molecule that a drug can be delivered to its target either because of solubility issues, which are addressed by tuning the carrier's physical properties to those of its biological target, or through the affinity of the carrier for a particular feature of the target such as a phospholipid membrane or protein recognition site. This 'magic bullet' approach has long been the goal of medicinal chemists, particularly when the therapeutic agents involved are highly toxic [13].

2.1. Incorporation of Antibiotic Motifs

The group of Regnouf-de-Vains has investigated the effects of introducing classical antibiotic motifs into calixarenes over research covering two decades. Calix[4]arenes have been used as a platform for drug delivery through the incorporation of penicillin core moieties as lower rim substituents (Figure 3). The lower rim of 4-*t*-butylcalix[4]arene was functionalised in the 1,3-positions with acid groups, and then further reacted with *N*-hydroxysuccinamide in the presence of dicyclohexylcarbodiimide to form a 1,3-diester. The β-lactams' carboxylic acids were protected as pivavoyloxymethyl esters and reacted with the calixarene diester to form diamide **1** [14]. Although antimicrobial data were not published, this route led the way to an analogue, **2**, in which penicillin V was appended to the calixarene [15] and tested against Gram-positive and Gram-negative bacteria [16]. The group also prepared a nalidixic acid delivering prodrug, **3**. Nalidixic acid is a bacteriostatic agent that works by interfering in the DNA replication mechanism rather than as a biocide, which kills the host. Reaction of the sodium salt of nalidixic acid with dibromopropane followed by *O*-alkylation of 4-*t*-butylcalix[4]arene in the 1,3-positions gave the disubstituted calixarene in 65% yield [17]. A fourth derivative incorporating nalidixic acid and penicillin V substituents on opposite rings, **4**, was prepared and tested for antimicrobial activity [16].

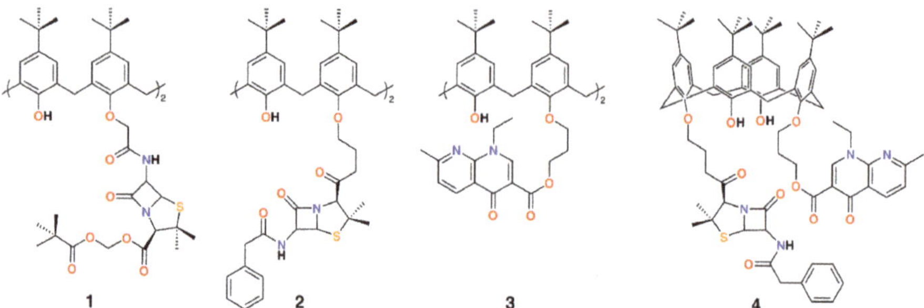

Figure 3. Regnouf-de-Vains' calixarene prodrugs.

Given the poor aqueous solubilities of these 4-*t*-butylcalix[4]arene derivatives, antimicrobial disc-diffusion tests were done in dimethyl sulfoxide with the pure solvent as a control. Only Gram-negative reference strain *Pseudomonas aeruginosa* ATCC 27853 appeared to be affected by DMSO and, as the same degree of diffusion was seen for all compounds, it was determined that the antimicrobial effects in all cases were due to the solvent alone. No significant activity was found for Gram-positive reference strains, *Staphylococcus aureus* ATCC 29213 and *Enterococcus faecalis* ATCC 29212, as well as against Gram-negative reference strain *E. coli* ATCC 25922. Only calixarenes **2** and **4**, with two penicillin V or one penicillin V and one nalidixic acid subunit, had any significant inhibitory effect, and then only against *S. aureus* ATCC 25923 [11].

Considering the solubility issues with 4-*t*-butylcalix[4]arene, the group also explored a 4-guanidinoethylcalix[4]arene derivative, **5**, prepared through the reaction of Boc-triflylguanidine with 4-ethylaminocalix[4]arene (Figure 4) [18,19]. A monomeric analogue of the calixarene was synthesized by the same method to assess the importance of the macrocyclic scaffold. To determine mammalian toxicity, MTT assays were performed for both compounds using human embryonic lung fibroblast (MRC-5) cells. Selectivity indices of IC$_{50}$/minimum inhibitory concentration (MIC) showed that the monomer was as toxic as it was active, whereas the calixarene had remarkably low toxicity. Both disc diffusion and minimum inhibitory concentration tests were carried out on Gram-negative *E. coli* ATCC 25922 and *P. aeruginosa* ATCC 27853, and Gram-positive *S. aureus* ATCC 25923 and *E. faecalis* ATCC 29212 bacteria. At 18 h, the monomer required concentrations of 512 µg mL^{-1} or higher to have any activity at all, whereas the calix[4]arene was active at 16 µg mL^{-1} against *E. coli*, *S. aureus*, and *E. faecalis*, and at 64 µg mL^{-1} against *P. aeruginosa* [20]. A later study extended the range of bacteria tested to include a second *S. aureus* strain, ATCC 29213, which gave identical results to the ATCC 25923 strain, and clinical isolates of penicillinase-producing *E. coli*, methicillin-sensitive *S. aureus* (MRSA) (mecA gene), *E. faecium* (vanA gene) and (vanB gene), and *P. aeruginosa* (overexpression of efflux pumps) [21]. The effects of reference antibiotics amoxicillin, oxacillin, vancomycin, and ticarcillin were also determined for comparison. The calixarene's efficacy was unchanged against the clinical isolates, indicating that its mode of action was unaffected by the resistance mechanisms commonly employed by antibiotic-resistant bacteria. The authors comment that, in presenting its four cationic groups co-facially, the calixarene mimics the cytoplasmic perturbation mechanism observed for cation-rich polypeptides. This is an example of multivalency, which is also key to the activities of biofilm-inhibiting calixarenes, discussed later. In 2010, the group published a more comprehensive assessment of the antimicrobial activity of the cationic calixarene against 69 clinical isolates compared with two antiseptic cationic compounds, chlorhexidine and hexamidine [22]. Interestingly, in another study, the group found a 4-*t*-butylcalix[4]arene with *n*-propylguanidinum lower rim substituents, **6**, was most effective in the *1,3-alt* conformation [23]. Tests against Gram-positive *E. faecalis* and two strains of *S. aureus* and Gram-negative *E. coli* and *P. aeruginosa*, together with reference and isoniazid (INH)-resistant *Mycobacterium tuberculosis*, showed that the *cone* conformer had MICs of 8 µg mL^{-1} and 9.5 µg mL^{-1}

against *E. faecalis* and INH-resistant *M. tuberculosis*, respectively, but otherwise had MICs between 32 and 128 µg mL^{-1}. By contrast, the *1,3-alt* conformation had values below 10 µg mL^{-1}, apart from *P. aeruginosa*, for which it was 16 µg mL^{-1}. This conformer was particularly effective against INH-resistant *M. tuberculosis*, with an MIC of 1.2 µg mL^{-1}.

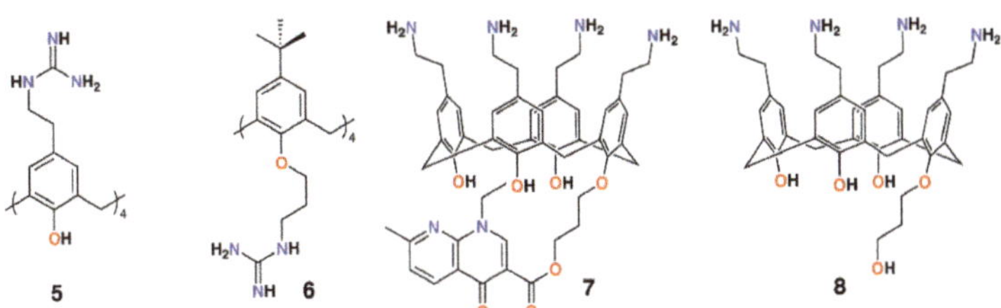

Figure 4. Regnouf-de-Vains' soluble guanidinium-functionalized calixarenes.

Another water-soluble derivative, an ethylaminocalix[4]arene with a single nalidixic substituent on the lower rim, **7**, was prepared through *tert*-butyloxycarbonyl-protection of the amines followed by alkylation with bromopropylnalidixate and upper rim deprotection (Figure 4) [24]. In biological media, the prodrug decomposes and releases about 30% of the nalidixic acid over the first 24 h, as assessed by HPLC. An inhibitory effect on the growth of Gram-negative *E. coli* ATCC 25922 (MIC 25 µg mL^{-1}) and Gram-positive *S. aureus* ATCC 25923 and 29213 (MIC 70 µg mL^{-1} for both) was found, but Gram-negative *P. aeruginosa* ATCC 29212 and Gram-positive *E. faecalis* ATCC 27853 (MIC 140 µg mL^{-1} for both) were relatively unaffected. An analogue without the nalidixic acid, **8**, was broadly ineffective. The parent ethylaminocalix[4]arene was also tested against the same strains and shown to have activity against *E. coli* (MIC 4 µg mL^{-1}) and both *S. aureus* strains (MIC 8 µg mL^{-1}) and, to some degree, *E. faecalis* and *P. aeruginosa* (MIC 32 µg mL^{-1}) [25]. This is an apparent improvement over the activity of the naldixic acid derivative, but similar to that of 4-guanidinoethylcalix[4]arene.

Pur and Dillmaghani introduced 6-aminopenicillanic acid substituents to calix[4]arene via esterification at either the lower rim, **9**, or upper rim, **10**, to produce 'calixpenams', followed by oxidation to their sulfoxide derivatives and finally through ring expansion and sulfur insertion to cephalosporin-containing analogues, or 'calixcephems', **11** and **12** (Figure 5). The calixpenams' MICs against *S. pyogenes* ATCC 19615, *Streptococcus agalactiae* ATCC 12386, and *Streptococcus pneumoniae* ATCC 49619 were found to be five to six times lower than penicillin V or X [26]. Tests against five strains of β-lactamase-producing or -non-producing methicillin-sensitive *S. aureus* were determined [27]. The calixpenams were more effective against methicillin-sensitive *S. aureus* (MSSA) β-lactamase (−) than MSSA β-lactamase (+), whereas the calixcephems showed broad antibiotic activity. In all cases, the calixarenes were between 6 and 10 times as effective as their penicillin or cephalosporin analogues. The synergistic effect of the four substituents was suggested as one reason for the enhanced activity.

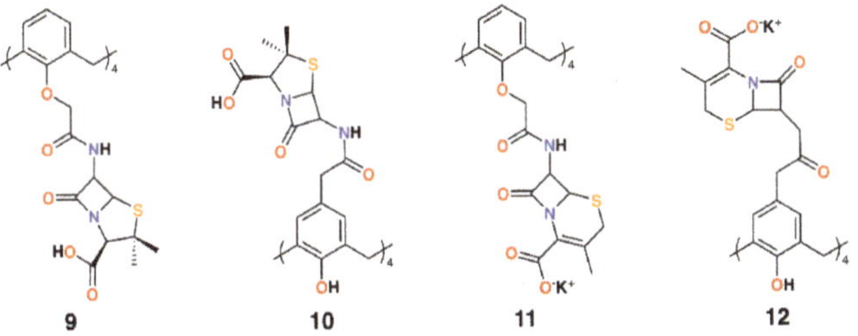

Figure 5. Dillmaghani's calixpenams and calixcephems.

2.2. Incorporation of Oxazole, Thiadiazole, and Bithiazole Motifs

Menon and colleagues prepared a library of nine 4-*t*-butylcalix[4]arenes, **13** to **21**, with 1,3,4-oxadiazole and 1,3,4-thiadiazole derivatives of isoniazid, nicotinic acid, benzoic acid, and *cis*-cinnamic acid (Figure 6) [28]. These, along with the unattached oxadiazole and thiadiazole derivatives, were assessed for activity against *S. aureus* MTCC 96, *S. pyogenes* MTCC 442, *E. coli* MTCC 443, *P. aeruginosa* MTCC 1688, and fungal species *Candida albicans* MTCC 227 and *Aspergillus clavatus* MTCC 1323. While the zones of inhibition for all compounds in DMSO were similar (8–19 mm), marginally better than ampicillin, there was more variability against the two fungal species. The oxadiazole and thiadiazole substituents were less active (*C. albicans* 6.5 to 16.2 mm, *A. clavatus* 11.5 to 16.4 mm) than when appended to calixarenes (*C. albicans* 9.0 to 23.3 mm, *A. clavatus* 11.3 to 24.3 mm), but none are quite as effective as griseofulvin (24 mm). Separately, the authors assessed the compounds' inhibitory effects against *M. tuberculosis* H37Rv and found that calixarenes with isoniazid substituents, **13** and **15**, displayed greater than 90% inhibition, significantly higher than the 71 to 82% range for the substituents alone.

Figure 6. Menon's' calixarene prodrugs.

Gezelbash and Dilmaghani subsequently reported four 4-*t*-butylcalix[4]arene-based aryl-oxadiazole derivatives, **22** to **25** (Figure 7), in which the substituents were linked by ethanethioate moieties in the 1- and 3-rings [29]. These, along with 14, previously reported by Menon [21], were assessed for their effects against *E. coli* and *Aspergillus fumigatus*. The furyl-containing derivative was found to be the most effective against *A. fumigatus*, whereas the 3-nitrophenyl derivative had broad spectrum activity.

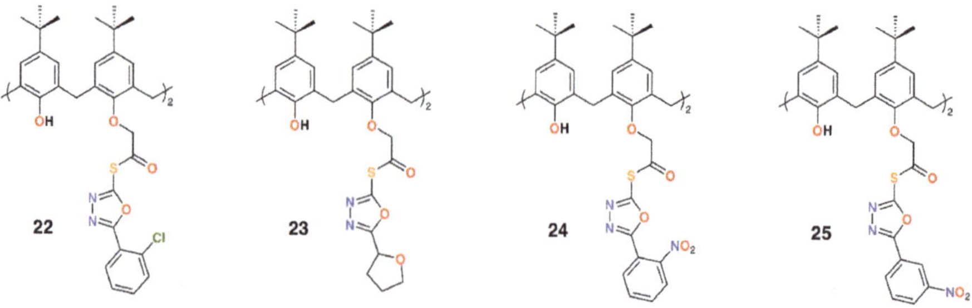

Figure 7. Dillmaghani's aryl-oxadiazole derivatives.

A family of six water-soluble calixarenes, **26** to **31** shown in Figure 8, incorporating 4-methyl-2,2′-bithiazole lower rim substituents on the 1- and 3-rings, was reported by Regnouf-de-Vains and colleagues [30]. Their effects against cells infected with HIV were determined and, compared with three parent derivatives with sulfonate, **32**; carboxylate, **33**; and phosphonate, **34**, upper rim substituents.

While none appeared cytotoxic to uninfected cells, only a 4-sulfonatocalix[4]arene derivative displayed any significant anti-HIV activity.

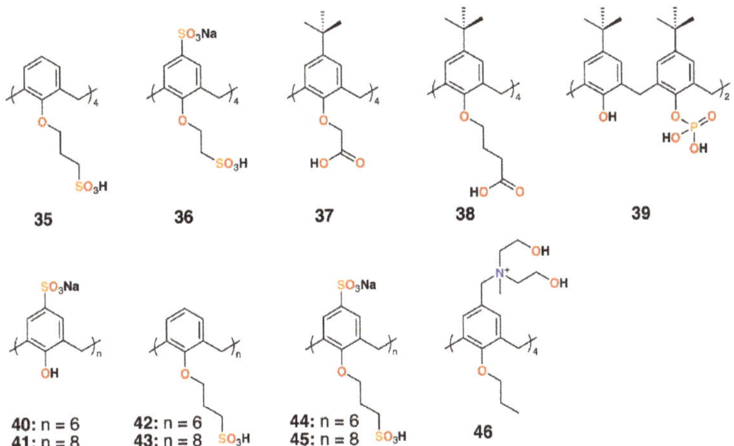

Figure 8. Regnouf-de-Vains' anti-HIV calixarene prodrugs.

2.3. Functionalized Nanoparticles

An alternative approach to delivering drugs as macrocyclic substituents is to utilize the calixarenes' central cavity to act as a controlled release site for drug guests. Coleman and colleagues used six water-soluble calixarenes, **32** and **35** to **39** (Figure 9), to cap silver nanoparticles (AgNPs) and subsequently to bind chlorhexidine and gentamycin with the presumption that the surface decorated AgNPs could be used to deliver the drugs in high concentrations [31]. In related work, a further six 4-sulfonatocalixarene derivatives, **40** to **45**, were used to cap AgNPs and to inhibit the growth of *Bacillus subtilis* 168 and *E. coli* [32]. No inhibition was observed for *E. coli*, however, four derivatives were effective against *B. subtilis*. The parent sulfonatocalix[4]-, -[6]-, and -[8]arenes together with sulfontocalix[6]arene with *O*-propylsulfonate lower rim substituents all retarded the growth to a maximum optical density from 16 to 18 h.

Figure 9. Coleman's AgNP-capping calixarenes and Nostro's drug-delivering calixarene.

2.4. Drug-Delivering Calixarenes

Wheate and colleagues investigated the potential for 4-sufonatocalix[8]arene, **41** (Figure 9), to act as a vehicle for the antibiotics isoniazid and ciprofloxacin [33]. In human kidney cells (HEK-293), the isoniazid complex was shown to increase uptake owing to the solubilizing effects of the macrocycle by 35%, although the ciprofloxacin complex showed no effects up to its solubility limit of 185 µg mL^{-1}. The MICs of the complexes against *E. coli*, *S. aureus*, *E. faecalis*, and *P. aeruginosa* were unchanged from the values of the drugs alone, however, the authors note that the potential for the large calix[8]arenes to bind two different drugs simultaneously warranted further investigation.

A polycationic calix[4]arene derivative, **46** (Figure 9), was prepared by Nostro and colleagues, through the reaction of *N*-methyldiethanolamine on tetra-propoxy-4-chloromethylcalix[4]arene, to deliver ofloxacin, chloramphenicol, and tetracycline [34]. The calixarene alone had a similar profile to chloramphenicol against *S. aureus* ATCC 6538, but its MIC was lower by a factor of two against MRSA 15, MRSE 17, and *P. aeruginosa* isolate 1. It was lower by a factor of four against *S. epidermidis* ATCC 35984, but performed less well against *P. aeruginosa* ATCC 9027. The authors speculate that the polycationic nature of the macrocycle may have a disruptive effect on the membrane structure of Gram-negative bacteria in a similar manner to the guanidinium derivatives discussed above.

2.5. Metal-Binding Calixarenes

The use of calixarenes as cation-releasing platforms was investigated by Memon and colleagues, who used a diamide calixarene derivative, **47** (Figure 10), to complex iron(III) [35] and copper(II) [36]. Iron(III) was found to bind in a 1:1 ratio to the calixarene by a spectroscopic Job's plot and copper(II) in a 2:1 ratio by the same method. Tests against bacteria and fungi found that the calixarene had MIC values between 1.5 and 3 µg mL^{-1}, whereas the iron(III) complex had a value of 0.37 µg mL^{-1} against *S. albus*, *E. coli*, and the fungal strain *Rhizopus stolonifera*. The copper(II) complex had the same MIC for *E. coli*, but was slightly less active against *S. albus* and *R. stolonifera*, with an MIC of 0.75 µg mL^{-1}.

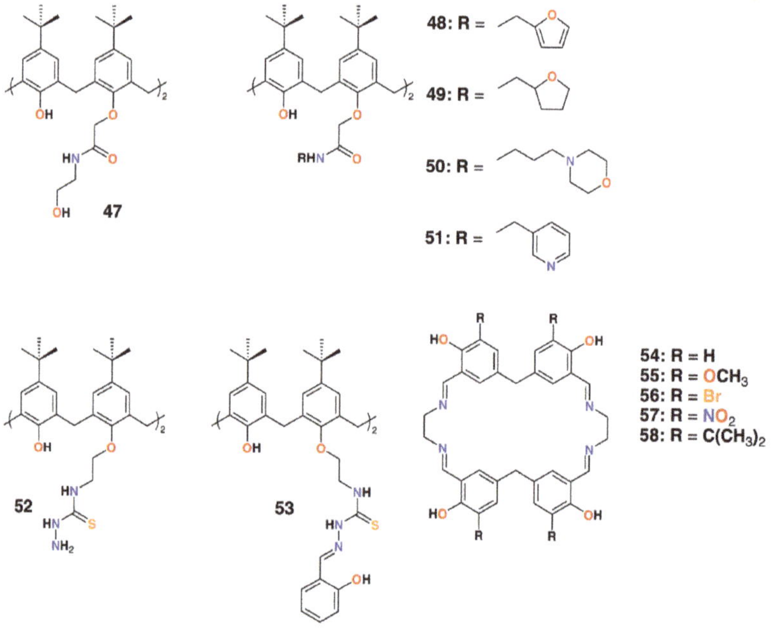

Figure 10. Cation-releasing calixarenes from Memon and Yilmaz with Desai's corates.

Yilmaz and colleagues prepared a number of calixarene diamides, **48** to **51**, designed to bind copper with a 1:1 stoichiometry (Figure 10) [37]. The authors propose that copper binds as Cu^{2+} in a slightly distorted square planar geometry, based on spectroscopic data, although whether the heteroatoms in the lower rim substituents are involved in this was not considered. Their antibacterial and antifungal properties were assessed against four Gram-positive, four Gram-negative, and two fungal strains. Quite surprisingly, given the antimicrobial activities of other calixarene-metal complexes, no antimicrobial activity was seen for Cu·**50** and Cu·**51**, even at 10,000 µg mL^{-1}. Metal-free macrocycles **48** and **50** displayed some activity against *B. subtilis*, *B. cereus*, and *E. coli*, but at best, this was in the range of 40 to 160 µg mL^{-1}. This suggests that it is the nature of the substituent, and not the bound copper, that imparts antimicrobial activity.

Shaabani and colleagues prepared a thiosemicarbazide functionalized calixarene, **52** (Figure 10), and investigated its effects, and those of its transition metal complexes, on *S. aureus* ATCC 29213, *B. subtilis* ATCC 6633, *E. coli* ATCC 25922, *P. aeruginosa* ATCC 27853, *C. albicans* ATCC 10231, and *C. glabrata* ATCC 2001 [38]. The tests revealed a higher antibacterial activity against Gram-positive *B. subtilis*, with an MIC of 31.25 µg mL^{-1}, than against Gram-negative *E. coli*, with an MIC of 250 µg mL^{-1}, and *P. aeruginosa*, with an MIC of 62.5 µg mL^{-1}. The metal complexes generally showed enhanced activity against *E. coli*, with the greatest effects seen for the nickel(II) and zinc(II) complexes, with MICs of 62.5 µg mL^{-1} and 31.25 µg mL^{-1}, respectively.

The Shaabani group extended the thiosemicarbazide motif through reaction with salicylaldehyde to give 2-hydroxybenzeledene-thiosemicarbazone **53** (Figure 10) [39]. Reaction with the appropriate metal nitrate salts gave Co·**53**, Cu·**53**, Ni·**53**, and Zn·**53**, which, along with the parent calixarene, were assessed for their antibacterial effects on *S. aureus* ATCC 29213, *B. subtilis* ATCC 6633, *E. coli* ATCC 25922, and *P. aeruginosa* ATCC 27853, together with their antifungal activity on *C. albicans* ATCC 10231 and *Candida glabrata* ATCC 2001. All had MICs of 31.25 µg mL^{-1} against *B. subtilis*, *E. coli*, and *P. aeruginosa*, but only Cu·**53** had any activity against *S. aureus*, with an MIC of 31.25 µg mL^{-1}. Only **53** and Co·**53** had any antifungal effects, and then only against *C. albicans*, with MICs of 31.25 µg mL^{-1} and 31.25 µg mL^{-1}, respectively. Nevertheless, the authors note that this level of activity is between two- and eightfold better than their analogs prepared from **52**.

Ray, Deolalkar, and Desai reported the synthesis of calixarene-salen hybrid corand macrocycles **54–58** together with their silver complexes (Figure 10) [40]. The antibacterial properties of the silver complexes were assessed against *E. coli* MTCC 433, *P. aeruginosa* MTCC 1688, *Salmonella typhi* MTCC 98, *S. aureus* MTCC 96, *S. pyogenus* MTCC 442, and *B. subtilis* MTCC 441. Their antifungal effects on *C. albicans* MTCC 227 were also investigated. These tests showed similar antibacterial effects to ampicillin, however, Ag·**56** and Ag·**57** were much more effective against *S. typhi* (MIC of 80.7 µg mL^{-1} and 48.8 µg mL^{-1}, respectively, vs. 100 µg mL^{-1} for ampicillin), and all were better than ampicillin against *S. pyogenes*. Ag·**58** was significantly more potent than ampicillin against all strains tested. Of particular note was its activity against *E. coli* (MIC of 49.5 µg mL^{-1} vs. 100 µg mL^{-1} for ampicillin and 24.8 µg mL^{-1} for ciproflaxin), *S. aureus* (MIC of 49.5 µg mL^{-1} vs. 250 µg mL^{-1} for ampicillin or 50.0 µg mL^{-1} for ciproflaxin), and *S. pyogenus* (MIC of 79.3 µg mL^{-1} vs. 100 µg mL^{-1} for ampicillin and ciproflaxin). All silver complexes were better fungicides than griseofulvin when tested against *C. albicans*. Ag·**55** and Ag·**56** were the most active, with minimal fungicidal concentrations (MFCs) of 384.1 µg mL^{-1} and 405.2 µg mL^{-1}, respectively, versus 500 µg mL^{-1} for griseofulvin.

2.6. Sulfonamide-Containing Calixarenes

The use of sulfonamides as bacteriostatic antibiotics predates treatment with penicillin by a decade, which makes them potentially effective when introduced as functional groups to calixarenes. Hamid and colleagues used upper rim diazotization to introduce sulfonamides, **59** and **60**, and other potentially therapeutic moieties, **61** to **63**, to calix[4]arenes (Figure 11) [41]. The compounds were screened against *B. subtilis*, *S. aureus*, MRSA, *S. epidermidis*, *E. faecalis*, *E. coli*, *P. aeruginosa*, *C. albicans*, and *S. cerevisiae*. Compound **59** showed good inhibition against *S. epidermidis* and *S. aureus*, both at 7.8 µg mL^{-1}, as well as MRSA and *B. subtilis*, both at 15.6 µg mL^{-1}. It was also active against Gram-negative *P. aeruginosa*, with an MIC of 15.6 µg mL^{-1}, and the fungal strain *C. albicans*, with an MIC of 62.5 µg mL^{-1}. Activity was greatest against Gram-positive strains, with **60** having the lowest MIC values against MRSA of 0.97 µg mL^{-1} and *B. subtilis* of 0.97 µg mL^{-1}. It also showed good inhibition against *S. aureus* with an MIC of 3.9 µg mL^{-1}, *S. epidermidis* at 15.6 µg mL^{-1}, and *E. faecalis* at 3.9 µg mL^{-1}. The tetrasubstituted derivatives **62** and **63** had limited inhibitory activity, which was explained by docking studies on neuraminidase- and penicillin-binding protein receptors. A single substituent was able to bind to the active site, while the macrocycle interacted with a number of neighboring protein side chains. By comparison, the tetrasubstituted derivatives were too sterically crowded for a single substituent to approach the binding sites.

Figure 11. Hamid's sulfonamides and related compounds.

2.7. Antibiotic Pillar[5]arenes

Moving away from the calixarene framework, Notti and colleagues utilized the water-soluble deca-carboxylatopillar[5]arene, **64** (Figure 12), as a carrier for the aminoglycosidic antibiotic amikacin and demonstrated that, where the macrocycle:drug ratio was 2:1 or greater, no antibiotic effects were observed against *S. aureus* ATCC 29213 based on the density of colony forming units (CFUs) [42]. At a ratio of 0.5:1, the complex was as potent as the drug alone and was also shown to release amikacin upon addition of acid as the protonated pillar[5]arene precipitates. Amikacin is active against both Gram-positive and Gram-negative bacteria, but is required at high concentrations to be of therapeutic use. Therapeutic concentrations can cause side-effects including nephrotoxicity and ototoxicity, so a method of targeting delivery such as pH-activated release would improve its therapeutic profile [43].

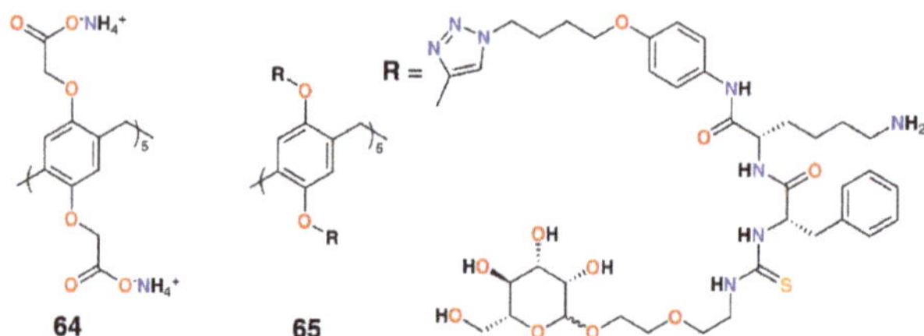

Figure 12. Drug-delivering pillararenes from Notti and He.

A more complex pillar[5]arene, **65**, was prepared by He and colleagues to deliver vancomycin to MRSA within cells (Figure 12) [44]. The pillar[5]arene was decasubstituted with substituents incorporating a hydrophobic region attached by click chemistry to an acid-sensitive lysine-phenylalanine region, which was linked in turn by thiourea to a terminal mannose. Upon addition of vancomycin, self-assembly resulted in the formation of vesicles with externally facing mannose groups able to bind to mannose receptors prior to endocytosis. The pillar[5]arene-vancomycin vesicles were found not to affect bacterial growth, but once inside cells, could be degraded by cathepsin A or low pH to release vancomycin. Pillar[5]arene vesicles formed in the presence of the fluorescent dye propidium iodide were used to demonstrate that vesicle release occurred within macrophages. MTT assays on three cell lines found no cytotoxic effects. To assess the ability of vancomycin-containing vesicles to target intracellular MRSA, RAW264.7 cells, originally from leukaemia-transformed murine cells, were infected with the WHO-2 strain of MRSA. Vancomycin released into cells was measured by HPLC and was consistent with the incubating concentrations of the drug. The impact on cells was determined by observing the number of CFUs from initial administration of the vancomycin-containing vesicles, which revealed a decrease of two orders of magnitude over 25 h.

A summary of the antibiotic activities of the most active macrocyclic derivatives is given in Table 1.

Table 1. Summary of macrocycles with significant antibiotic activities. MIC, minimum inhibitory concentration.

Compound	Target	MIC/μg mL^{-1}	Ref
5	*E. coli*	16	[20]
	S. aureus	16	[20]
	E. faecium	16	[20]
6 (*cone*)	*E. faecalis*	8	[23]
	M. tuberculosis	9.5	[23]
6 (*1,3-alt*)	*M. tuberculosis*	1.2	[23]
8	*E. coli*	4	[25]
	S. aureus	8	[25]
Fe·**47**	*E. coli*	0.37	[35]
	S. albus	0.37	[35]
	R. stolonifera	0.37	[35]
Cu$_2$·**47**	*E. coli*	0.37	[36]
	S. albus	0.37	[36]
	R. stolonifera	0.75	[36]
59	*S. aureus*	7.8	[41]
	S. epidermidis	7.8	[41]
	MRSA	15.6	[41]
	B. subtilis	15.6	[41]
	P. aeruginosa	15.6	[41]
60	*S. aureus*	3.9	[41]
	S. epidermidis	15.6	[41]
	MRSA	0.97	[41]
	B. subtilis	0.97	[41]

3. Cell Destruction

As an alternative mode of action to drug delivery or release of toxic metal ions, the amphiphilic properties of macrocycles can be utilized to disrupt bacterial and fungal membranes, leading to rupture and cell death. Where the compounds are charged, there may be an even greater affinity for biological membranes and, potentially, different responses to Gram-positive and Gram-negative bacteria, and to fungi, which have a different membrane composition.

3.1. Macrocyclon

In 1951, Cornforth and colleagues reported on the suppressive effects of the non-ionic surfactant Triton A20 on tuberculosis in mice [45]. The surfactant consists of a 4-(2,4,4-trimethylpentan-2-yl)phenol head group, or 'octylphenol', linked to polyether substituents of varying lengths formed through reaction with ethylene oxide. In a later paper, reaction of these compounds with their 2,6-bis(hydroxymethyl) analogues resulted in complex mixtures from which linear polymers comprising an odd number of octylphenyl units could be isolated [5]. No cyclic products were found, so the group used the Zinke method to prepare a macrocycle from octylphenol and formaldehyde [46]. The condensation product was believed to be a cyclotetramer, or *cyclotetra-m-benzylene*, and subsequently reacted with ethylene oxide to give *Macrocyclon*, **66**, a macrocyclic analogue of the authors' linear polymer shown in Figures 1 and 13. It was essentially non-toxic to mammals and, unlike its linear analogue, found to be more potent than streptomycin in the treatment of the human virulent strain of *M. tuberculosis*, H37Rv. The greatest effects were seen where the polyether comprised 15 to 20 repeat units, but when this extended to 45 units, the compound became pro-tuberculous. None of the derivatives inhibited growth of tubercular bacilli and it was suggested that the activity was due to their surfactant effects. In a later paper, D'Arcy Hart and colleagues revisited these compounds and compared *Macrocyclon*, with an average of 12.5 -CH_2CH_2O- repeat units and terminating in a hydroxy group, with 'HOC-60', **67**, an analogue with 60 repeat units [47]. At this point, it was also appreciated that the compounds were calix[8]- rather than calix[4]arene derivatives. It was observed that growth of *M. tuberculosis* inside macrophages was inhibited by *Macrocyclon*, but stimulated by HOC-60. Similarly, lipase activity was inhibited by *Macrocyclon* and stimulated by HOC-60, suggesting that lipids and lipid metabolism were affected by the macrocycles.

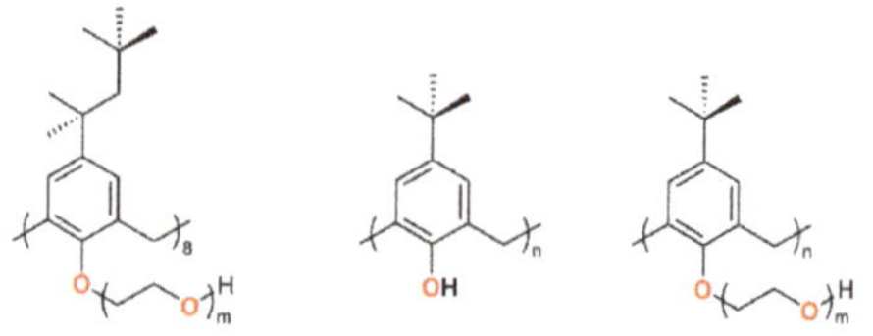

66: m = 12.5, *Macrocyclon*
67: m = 60, HOC-60
68: n = 6
69: n = 8
70: n = 6, m = 6
71: n = 8, m = 6
72: n = 8, m = 12

Figure 13. Cornforth's *Macrocyclon* and HOC-60 with Tascon's derivatives.

Almost 50 years after the original report, *Macrocyclon* was reinvestigated by Tascon and colleagues, who confirmed the original findings [48]. In these later experiments, both macrophages and live mice were infected with *M. tuberculosis* and the range of calixarenes, **68** to **72** (Figure 13), increased.

Treatment with *Macrocyclon* supported the novel therapeutic pathway proposed by D'Arcy Hart, as the macrocycle enhanced the innate defense mechanisms in the murine macrophages. This work was followed up by a much more extensive screen of 25 calixarenes, **41**, **66**, and **68** to **91**, against *M. tuberculosis* by Hailes and colleagues (Figure 14) [49]. Derivatives based on calix[4], -[6], -[7], and -[8]arenes with *t*-butyl, phenyl, and sulfonate upper rim substituents, and a range of lower rim, largely ethylene glycol substituents, were assessed alongside *Macrocyclon*. The parent 4-*t*-butylcalix[8]arene and 4-phenylcalix[8]arene were more active than their smaller homologues and the 4-sulfonatocalix[8]arene, being water soluble, had activity approaching that of *Macrocyclon*. The addition of polyethylene glycol substituents enhanced the calixarenes' anti-mycobacterial properties, with complete substitution having greater effects than partial substitution. Longer substituents were required for 4-phenylcalix[7]arene derivatives to be effective, presumably owing to the parent compound's lower solubility compared with the 4-*t*-butylcalixarenes. Lower rim acetate groups elicited pro-tubercular activity and other substituents, such as cyanopropoxy groups, had little effect.

3.2. Charged Calixarenes

Regnouf-de-Vains and colleagues also investigated the anti-mycobacterial activities of 17 charged calix[4]arenes [50–52]. In addition to parent compounds **5**, **32** to **34**, and **92**, derivatives with bithiazolyl lower rim subunits, **26**, **29**, and **31**, previously tested as antibiotics [28], and bipyridyl analogues with sulfonate, carboxylate, phosphonate, ethylamine, or ethylguanidinium upper rim termini, **92** to **104** (Figure 15), were also assessed for activity against *M. tuberculosis* H37Rv. None of the anionic species exhibited activity, but the parent 4-guanidinoethylcalix[4]arene, **5**, and its derivative with two lower rim bipyridyl substituents, **96**, provided almost 100% inhibition at concentrations of 1.22 µg mL^{-1} and 1.89 µg mL^{-1}, respectively.

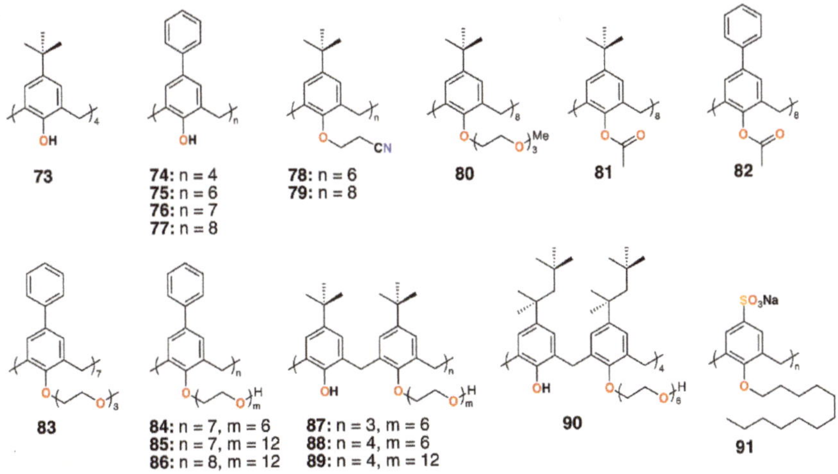

Figure 14. Hailes' *Macrocyclon* analogues.

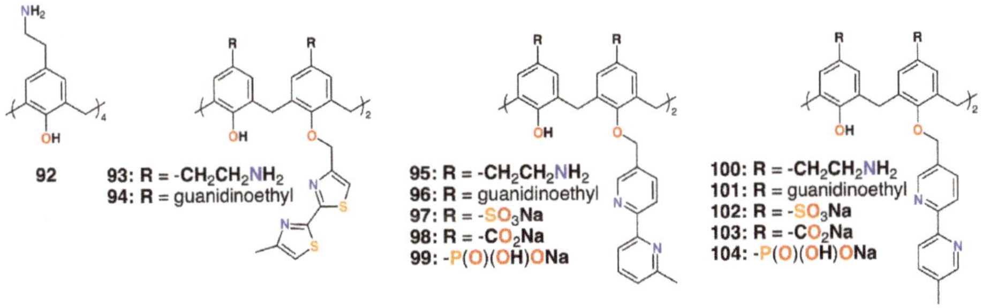

Figure 15. Regnouf-de-Vains' charged calixarenes.

Independently, the groups of Yushchenko [53] and Loftsson [54] focused on cationic calix[4]arenes with trimethylammonium, *N*-(2-hydroxyethyl)-*N*,*N*-dimethylammonium, or *N*-(2-aminoethyl-*N*,*N*-dimethylammonium groups linked to the upper rims by methylene groups, and O-methyl, O-propyl, or O-octyl lower rim substituents, **105** to **108** (Figure 16). Derivatives **105** and **106** showed increasing hemolysis from 0.71 µg mL^{-1} and 0.89 µg mL^{-1}, respectively. The antimicrobial activity of 2-hydroxyethyl derivatives **105** and **108**, together with N-methylimidazolium derivatives **109** and **110** (Figure 16) against *S. aureus* 209 P and *E. coli* M-17 bacteria, was studied. Compounds **105** and **109** had pronounced activity with MICs of 1.95 µg mL^{-1} for *S. aureus* 209 P, while **107** and **108** had no such activity and all compounds were inactive against the Gram-negative *E. coli* M-17. The authors concluded that the antibacterial activity depended on the size and conformational rigidity of the macrocycle and the length of the alkyl substituents on the lower rim of the calixarene. Conformationally rigid macrocycles **106** and **109** were active, while conformationally flexible **108** was inactive. In addition, with an increase in the length of the lower rim alkyl substituents from propyl to octyl, the antibacterial activity decreases. Tests on *S. aureus* ATCC 29213 and *E. coli* ATCC 25922 and M-17 showed that **106** and **107** had very similar profiles to the guanidinoethylcalix[4]arenes reported by Regnouf-de-Vains.

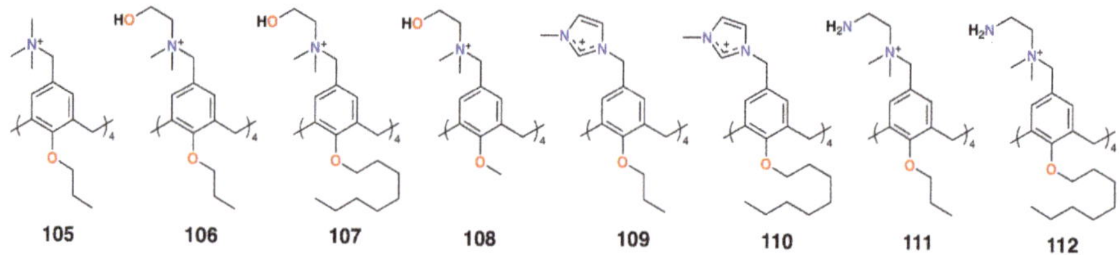

Figure 16. Charged calixarenes reported by Loftsson, Yushchenko, and Melzhyk.

Melezhyk and colleagues also investigated **106** and **107** together with the four cationic derivatives, **109** to **112** (Figure 16) [55]. Tests against *P. aeruginosa* ATCC 9027, *S. aureus* ATCC 6538, *K. pneumonia* ATCC 10031, and *E. coli* ATCC 25922 found that **105**, **109**, and **110** had MICs of 10 µM against *S. aureus* ATCC 6538, but the remaining calixarenes were ineffective. Other than **109**, with an MIC of 1.11 µg mL^{-1}, none of the compounds were effective against *K. pneumonia* ATCC 10031 or *E. coli* ATCC 25922, and only **111** had biofilm inhibiting activity, but this was seen for all strains except *P. aeruginosa* ATCC 9027.

Consoli and colleagues designed calixarene, **113**, which binds a nitric oxide (NO)-releasing guest in response to irradiation at 400 nm (Figure 17) [56]. Switching between 'dark' and 'light' conditions allowed NO release to be followed electrochemically. The calixarene itself had no observable antimicrobial activity against either *S. aureus* ATCC 6538 or *P. aeruginosa* ATCC 9027, but when combined with the otherwise insoluble guest, approached 100% inhibition for both after 20 min of illumination. The group developed a more soluble derivative of the calixarene, **114** (Figure 17), which incorporated aspects of the NO-releasing species to determine if this would impart antibiotic activity [57]. Experiments on human skin fibroblasts showed negligible antiproliferative activity, however, disc diffusion experiments and observation of CFUs demonstrated a 98.9% reduction of *S. aureus* ATCC 6538 CFUs in the dark and a 99.95% reduction after 20 min irradiation. *E. coli* ATCC 10536 was unaffected in the dark, but after 30 min irradiation, it saw a 93.5% reduction of CFUs. The difference was ascribed to the combination of broad bactericidal NO release, which affected both *S. aureus* and *E. coli*, and the quaternary ammonium groups in the calixarene framework, which had a greater disruptive effect on the cell membranes of Gram-positive *S. aureus*.

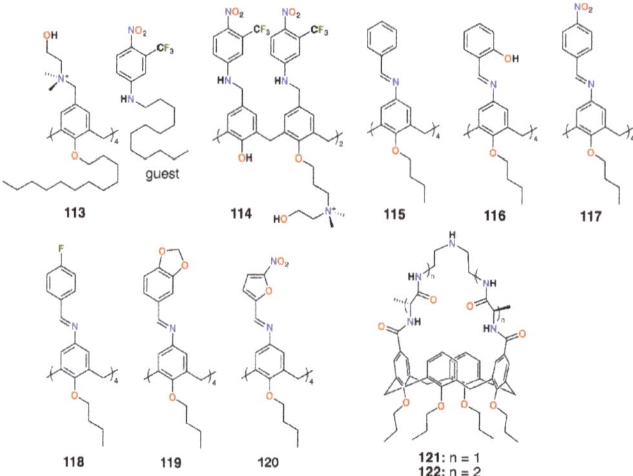

Figure 17. Antibiotic and antifungal calixarenes of Consoli and Di Fatima with Ungaro's vancomycin mimics.

Di Fatima and colleagues prepared a family of iminocalixarenes, **115** to **120** (Figure 17), and investigated their antifungal properties compared with their monomeric analogues [58]. MIC tests against *C. albicans* ATCC 18804, *Candida krusei* ATCC 20298, *Candida tropicalis* ATCC 750, *Candida parapsilosis* ATCC 22019, *Candida glabrata* ATCC 90030, and *Candida dubliniensis* CBS 7987 showed the calixarenes to have up to a 36-fold enhancement active against *Candida* strains compared with the monomers. The antifungal response of **115** was generally comparable to that of fluconazole and twice as active against *C. krusei*.

3.3. Vancomycin Mimicking Calixarenes

A different approach to imparting antibiotic activity to calixarenes was pioneered by Ungaro and colleagues in 1996 [59]. Vancomycin mimics were prepared by bridging two opposite rings in *O*-propyl calix[4]arene through their upper rims with two, **121**, or four alanine residues, **122**, linked by diethylenetriamine, as shown in Figure 17. The compounds' effects on three strains of *S. aureus* (penicillin-sensitive 663, penicillin-resistant 853, and methicillin-resistant 1131) together with *Bacillus cereus, Saccharomyces cerevisiae, Acholeplasma laidlawii, S. epidermidis, E. coli* 1852, and *C. albicans* were determined and compared to that of vancomycin. Significant activity against *S. epidermidis* and all strains of *S. aureus*, with MICs between 4 μg mL^{-1} and 16 μg mL^{-1}, were found for the -D-Ala-NHCH$_2$CH$_2$NHCH$_2$CH$_2$-D-Ala- and its L-Ala analogue. These compared favorably with 2 μg mL^{-1} seen for vancomycin. Neither vancomycin nor the alanine-bridged calixarenes showed activity against *S. cerevisiae, A. laidlawii*, or *C. albicans*.

3.4. Other Calixarenes

Jain and colleagues prepared a number of oxacalix[4]arenes, including **123** and **124** (Figure 18), which exhibited a range of activities against bacterial and fungal strains [60]. Compound **123** was particularly effective against *E. coli, P. aeruginosa*, and *S. aureus*, with MICs below 2 μg mL^{-1}, whereas **124** was additionally effective against the fungal strains *C. albicans* and *Aspergillus clavatus*, with MICs of 12.5 μg mL^{-1} in both cases.

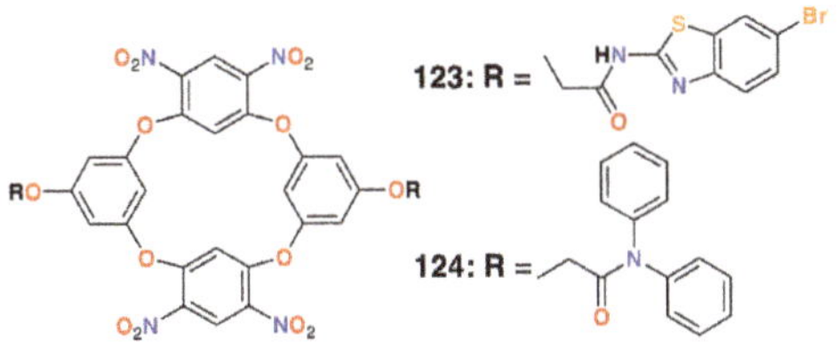

Figure 18. Jain's oxacalixarenes.

In 2012, Memon and colleagues tested the antibacterial and fungicidal activity of a calixarene with morpholine groups introduced to the upper rim, **125** (Figure 19), based on the antifungal properties of morpholine derivatives [61]. Using the disc diffusion method, *Staphylococcus albus* ATCC 10231, *Streptococcus viridans* ATCC 12392, *Bacillus procynous* ATCC 51189, *Enterobacter aerogenes* ATCC 13048, *Klebsiella aerogenous* ATCC 10031, *E. coli* ATCC 8739 *Sallmonella* ATCC 6017, *A. niger* ATCC 16404 *Aspergillus fumagatus* ATCC 90906, and *Penicillium* ATCC 32333 were used as a range of organisms that covered Gram-positive, Gram-negative, and fungal strains. Calixarene **125** exhibited excellent activity against all strains except *S. albus*, with MICs between 4 and 8 mg mL^{-1}. Continuing with this theme, the group synthesized the compound's pyrrolidine analogue, **126** (Figure 19) [62]. The MIC values were in the range from 1170 to 2340 μg mL^{-1} for *S. aureus, S. viridans*, and *E. coli*, and from 580 to 2340 μg mL^{-1} for fungal strains *A. niger, A. flavus*, and *C. albicans*. The group also investigated the effect of 4-nitrocalix[4]arene, **127** (Figure 19), on a similar range of strains [63]. Disc diffusion experiments using *S. aureus, S. viridans, E. coli, A. niger, Aspergillus flavus*, and *C. albicans* showed good antibacterial activity against *S. aureus* and *S. viridans*, with MICs below 5000 μg mL^{-1}, together with *A. flavus* and *C. albicans*, at 2300 μg mL^{-1}, but the best results were against *E. coli* and the fungal strain *A. niger*, with an MIC of 580 μg mL^{-1} for both. In support of their findings, the authors state that phenol-containing compounds with electron-withdrawing groups in the 4-position increase activity.

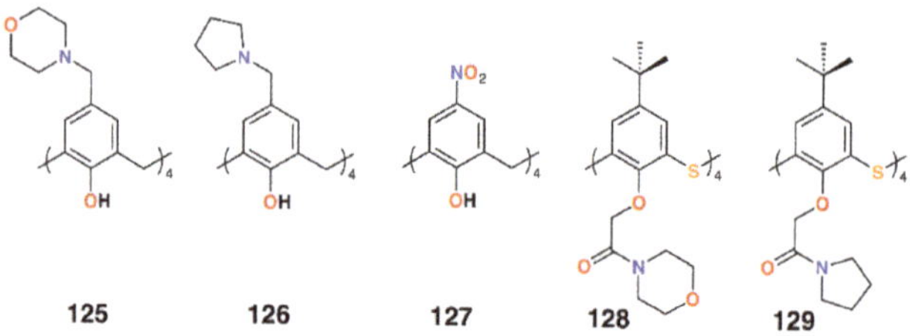

125 **126** **127** **128** **129**

Figure 19. Morpholine and pyrrolidine calixarene derivatives from Memon and Galitskaya.

Morpholine and pyrrolidine groups were chosen as lower rim substituents by Galitskaya and colleagues [64]. These two 4-*t*-butylthiacalix[4]arene derivatives, **128** and **129** (Figure 19), formed nanocomposites with silver cations that were used to treat polyethylene films. The *cone* conformer was found to be the most active, totally inhibiting the growth of *P. aeruginosa* on the film's surface after 24 h; however, no effects were seen for *Bacillus pumilus*.

3.5. Pore-Forming Pillar[5]arenes

Pillar[5]arenes can be modified to incorporate short amino acid substituents that allow them to penetrate phospholipid membranes. Hou and colleagues prepared pillar[5]arenes **130** to **133** (Figure 20) with substituents comprising one to four amino acids (-L-Trp-CO_2H, -D-Leu-L-Trp-CO_2H, -L-Leu-D-Leu-L-Trp-CO_2H and -D-Leu-L-Leu-D-Leu-L-Trp-CO_2H) based on the terminus of gramicidin A, a natural pore-forming protein [65]. Tests on *E. coli*, *S. epidermidis*, *S. aureus*, and *B. subtilis* showed significant growth inhibition by all compounds at concentrations of 10 μM (from 3.05 μg mL^{-1} for **130** up to 6.64 μg mL^{-1} for **133**) on the Gram-positive bacteria, but had no effect on *E. coli*.

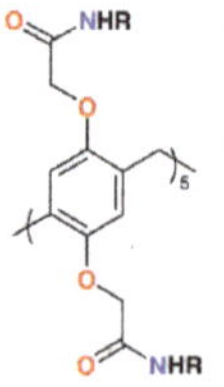

130: R = -Trp-CO_2H
131: R = -D-Leu-Trp-CO_2H
132: R = -Leu-D-Leu-Trp-CO_2H
133: R = -D-Leu-Leu-D-Leu-Trp-CO_2H

Figure 20. Hou's pore-forming pillar[5]arenes.

Xin, Dong, Chen, and colleagues undertook similar experiments with five pillar[5]arenes, **134** to **138** (Figure 21), with two helical peptides joined to the same ring [66]. Here, the peptides were between 8 and 16 residues long and also tested against *E. coli*, *S. epidermidis*, *S. aureus*, and *B. subtilis*. An interesting correlation was observed between activity and length of peptide chain for the Gram-positive bacteria and the 16-residue derivative approached the IC_{50} of gramicidin A. Much lower activity was seen against *E. coli* with 30% of bacteria surviving even at concentrations of 500 μM (equating to 138 μg mL^{-1} for **134** up to 235 μg mL^{-1} for **135**).

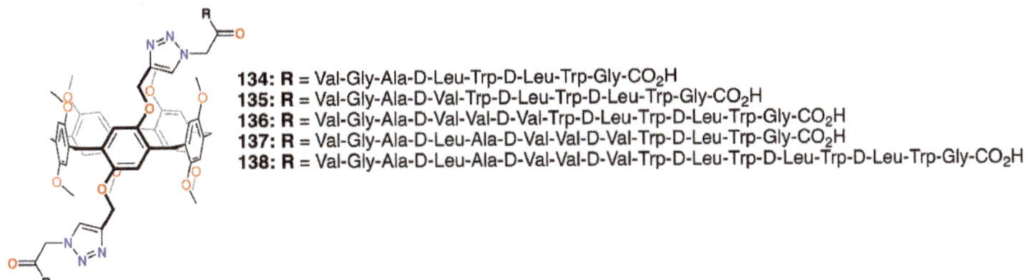

Figure 21. Xin's pore-forming pillar[5]arenes.

A summary of the cytotoxic activities of the most active macrocyclic derivatives is given in Table 2.

Table 2. Summary of macrocycles with significant cytotoxic effects on bacterial and fungal species.

Compound	Target	MIC/µg mL^{-1}	Ref
5	*M. tuberculosis*	1.22	[50]
96	*M. tuberculosis*	1.89	[50]
105	*S. aureus*	1.95	[53]
109	*S. aureus*	1.95	[55]
122	*S. aureus*	4	[59]
	S. epidermidis	16	[59]
130	*S. aureus*	3.05	[65]
	S. epidermidis	3.05	[65]
	B. subtilis	3.05	[65]
131	*S. aureus*	4.32	[65]
	S. epidermidis	4.32	[65]
	B. subtilis	4.32	[65]
132	*S. aureus*	5.31	[65]
	S. epidermidis	5.31	[65]
	B. subtilis	5.31	[65]
133	*S. aureus*	6.64	[65]
	S. epidermidis	6.64	[65]
	B. subtilis	6.64	[65]

4. Biofilm Inhibition

Direct delivery of a drug or prodrug to individual cells is one antimicrobial strategy, however, disruption of bacterial colonies growing on a surface is another approach. Once individual bacteria are able to bind to a surface, initially through hydrophobic and other weak interactions, they can act as anchors for further cell deposition. Biofilms then form when the bacteria are able to deposit an extracellular matrix of polymeric material, largely composed of polysaccharides, which allows them to extend the colony [67]. The polysaccharides are recognized by lectins, carbohydrate-binding proteins present on cell surfaces, leading to the formation of a cellular matrix. These complex, surface-bound films accumulate more bacteria, facilitating chemical communication between them. One consequence of this is increased resistance to antibiotics and surfactant-based detergents as the outer layers form a protective barrier for the remainder of the colony. While the biofilm may comprise a single species, it is possible for synergistic ecosystems to emerge in which molecules generated by one species are metabolized by others. As different species of bacteria have different susceptibilities to antibiotics, biofilm heterogeneity can also confer a level of overall protection.

While antibiotics and surfactants are widely used against biofilms, inhibition of their formation and growth has been approached using multivalent macrocycles. Several groups have developed derivatives with extended substituents terminating in saccharides such as mannose [68,69]. The sugar termini bind to receptor sites on bacterial proteins. These sites would usually interact with polysaccharide groups in the extracellular matrix, so blocking them inhibits adhesion. Calixarenes, along with related macrocycles such as pillararenes, incorporating several substituents have been developed as inhibitors. A single macrocycle with several sugar groups has two benefits. Its multivalency maximizes the chances that it will block binding sites and extended substituents are able to span between several receptor sites on the same protein. In doing so, they can form a physical barrier between the protein surface and the extracellular matrix. Consequently, the conformation of the calixarene is often a vital factor in the compound's effectiveness. The *cone* conformer of calix[4]arene is able to bind four sites on the same protein face and is essentially tetravalent, whereas the same compound in the *1,3-alt* conformer would have a greater span between binding sites, but would only be divalent.

Lectin binding by macrocyclic glycoconjugates can also reduce infection by stopping the cell adhesion process from the very beginning. With carbohydrate recognition sites blocked, bacteria have to rely on weak protein–surface interactions to aggregate and these are easily reversed. To determine the efficacy of the macrocyclic agents, two key factors need to be quantified: affinity and inhibition. Experiments undertaken in the presence of different lectins will reveal specificity and affinity, while biofilm inhibition properties are evaluated through cell culture experiments.

4.1. Calixsugars

In 2009, Imberty, Matthews, Vidal, and colleagues prepared a family of 4-*t*-butylcalix[4]arenes that were partially *O*-alkylated with propyl bromide and then *O*-propargylated at the remaining lower rim positions [70]. Acetylated mannose or galactose moieties were attached by a triethylene glycol tether using click chemistry. Galactose derivatives with one, two, or three substituents were isolated and deprotected to give **139** to **142** (Figure 22). The fully substituted derivatives were synthesized from the tetra-*O*-propargyl precursor, which gave rise to *cone*, *1,3-alt*, and *partial cone* products **143** to **145** (Figure 22). The tetrasubstituted mannose analogue, **146** (Figure 22), was also prepared, in the *cone* conformer, as a comparator. Isothermal titration calorimetry was used to show binding to Lec A (galactophilic *P. aeruginosa* first lectin, or PA-IL) and that the tetravalent galactose ligands bound most strongly. Enhanced binding of the *partial cone* and *1,3-alt* conformers demonstrated the importance of geometric alignment when binding to the protein surface. Surface plasmon resonance (SPR), and later atomic force microscopy (AFM) and computer modelling studies [71], supported these findings. Imberty, Vidal, and colleagues then investigated the influence of the linker arm between the calixarene and sugar terminus [72]. Linkers with increasing rigidity, but similar lengths, shown in Figure 22 and incorporating ethylene glycol, **147**; diethylene glycol acetamide, **148**; ethylene glycol acetamidoacetamide, **149** and **150**; and phenyl acetamidoamide, **151** to **153**, were compared with the original triethylene glycol linker. Binding to PA-IL was assessed by SPR, isothermal titration microcalorimetry (ITC), hemagglutination (HIA), and enzyme-linked lectin assays (ELLA). The cost of increasing rigidity was lower solubility, nevertheless, the researchers found that the calixarene with four diethylene glycol acetamide linkers was more potent than the original trethylene glycol-containing compounds, but only when in the *1,3-alt* conformation.

139: **R** = **R'** = **R"** = *n*-propyl
140: **R** = **R'** = *n*-propyl, **R"** = **V**
141: **R** = **R"** = *n*-propyl, **R'** = **V**
142: **R** = *n*-propyl, **R'** = **R"** = **V**
143: **R** = **R'** = **R"** = **V** (*cone*)
144: **R** = **R'** = **R"** = **V** (*1,3-alt*)
145: **R** = **R'** = **R"** = **V** (*partial cone*)

146 (*cone*)

147: **R** = **W** (*1,3-alt*)
148: **R** = **X** (*1,3-alt*)
149: **R** = **Y** (*1,3-alt*)
150: **R** = **Y** (*partial cone*)
151: **R** = **Z** (*cone*)
152: **R** = **Z** (*1,3-alt*)
153: **R** = **Z** (*partial cone*)

Figure 22. Imberty's glycosylated calixarenes.

Using a similar, but simpler approach, Consoli, Geraci, and colleagues synthesized a tetra-*O*-propyl *cone*-calix[4]arene, **154** (Figure 23), with fucose groups on the upper rim by an acetamidoacetamide linker [73]. A planktonic antimicrobial susceptibility test was performed with wild-type *P. aeruginosa* PAO1 using the microdilution broth method and no antimicrobial activity was seen up to 32.7 µg mL^{-1}. Biofilm inhibition was observed, however, with a dose-dependent response seen over the range from 2.0 µg mL^{-1} to 32.7 µg mL^{-1}, with concomitant inhibition rates rising from 35% to 73%. The precursor calixarene without the sugar groups, terminating in four amines, also actively inhibited biofilm formation, but to a lesser extent. The authors ascribe this to interactions between the bacteria and the formation of positively charged ammonium groups.

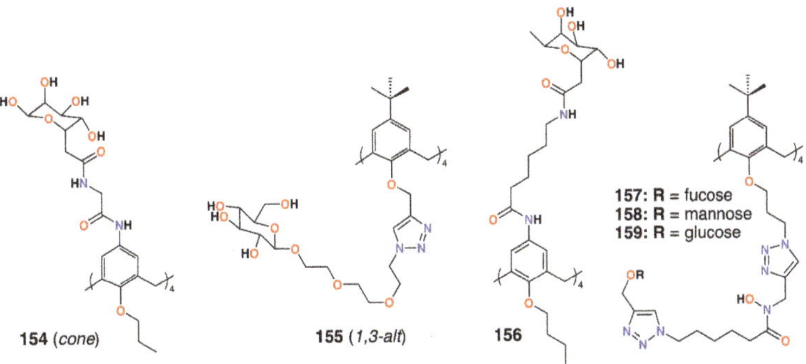

154 (*cone*)   **155** (*1,3-alt*)   **156**   **157:** **R** = fucose
158: **R** = mannose
159: **R** = glucose

Figure 23. Glycosylated calixarenes from Geraci, Vidal, and Benazza.

Returning to the triethylene glycol spacer motif, Vidal and colleagues extended the family of calixarenes to include those with four fucose termini, **155** (Figure 23) [74]. With glucose, galactose, mannose, and fucose analogues in hand, their effects on bacterial aggregation, cell adhesion, and biofilm inhibition could be compared. As LecA is selective for galactose and LecB (fucophilic *P. aeruginosa* second lectin, or PA-IIL), discrimination between the two should prove possible. ITC was used to determine affinities of the compounds for LecA and LecB. A monovalent fucosylated linker, prepared as a control, bound to LecA with an affinity in the nanomolar range, in agreement with similar literature examples. The data suggested that a *1,3-alt* calixarene with four fucose-terminated substituents appeared to utilise three of them when binding and, as expected, its affinity for LecB was 50 times lower. The glucosylated analogue was not recognised by either lectin. *P. aeruginosa* wild-type PAO1 strains and those overproducing either LecA or LecB were used in aggregation experiments. Little aggregation was observed with galactosylated of fucosylated calixarenes for wild-type *P. aeruginosa*, but the results from PAO1Δ*lecA* and PAO1Δ*lecB* implied that clustering was LecA-dependent and LecB-independent. Cell adhesion experiments using A549 lung epithelial cells showed dose-dependent inhibition of between 70% and 90%, regardless of which calixarene was tested, but this was significantly better than the linkers alone. Inhibition of LecA- and LecB-dependent biofilms after 24 h was observed at calixarene concentrations of 1.1 μg mL^{-1} or higher, regardless of the terminal sugar, but did not affect bacterial growth.

In later work, Geraci and colleagues altered the structure of the upper rim slightly to give a more extended reach for the calixarene's fucose termini, **156** (Figure 23) [75]. Biofilm inhibition tests with *P. aeruginosa* (PAO1) and *S. epidermidis* ATCC 35984 showed 35.4% inhibition of PAO1 at 27.9 μg mL^{-1} and complete inhibition between 55.7 μg mL^{-1} and 111.5 μg mL^{-1} with similar results for *S. epidermidis*. Monomeric analogues had some effect against PAO 1, but enhanced *S. epidermis* biofilm formation. The authors note that their results for a fucose-terminated calixarene in the *cone* conformation were orders of magnitude better than those obtained by Vidal for **155** in the *1,3-alt* conformation [64].

Benazza and colleagues prepared fuco-, **157**; manno-, **158**; and glucocluster, **159**, *cone*-calix[4]arene derivatives, shown in Figures 23 and 24, and introduced an iron chelating hydroxamic acid region in each linker [76]. It was proposed that, in addition to binding to lectins, these compounds could sequester soluble iron, an essential nutrient for bacteria, to enhance their antibacterial activities. No antibiotic activity against wild-type *P. aeruginosa* PAO1 was seen for the fucose or mannose terminated derivatives, so additional experiments investigated the effects of iron chelation. Under iron-depleted conditions, production of the bacteria's main siderophore, pyoverdine-I (Pvd-I), increased to counteract the sequestering effects of the calixarenes. When a siderophore-deficient strain of *P. aeruginosa* was used, the mannosylated calixarene significantly disrupted growth. Binding assays using fluorescence polarization revealed an unexpected interaction between LecB and the mannosylated calixarene over 200 times stronger than a monomeric analogue. Computer modelling indicated that the calixarene could bind simultaneously to four sites on the LecB protein, which may explain this effect. *P. aeruginosa* PAO1 biofilm growth studies at 5.4 μg mL^{-1} found that not only did the mannosylated-derivative inhibit biofilm formation by 84%, but the fucoslyated-derivative did so by 72%. Even more surprising was the effect of the glucosylated-derivative, used as a negative control, which gave 92% inhibition. The authors concluded that other inhibitory mechanisms could be involved such as the possibility that nitric oxide was being released from the iron-binding site.

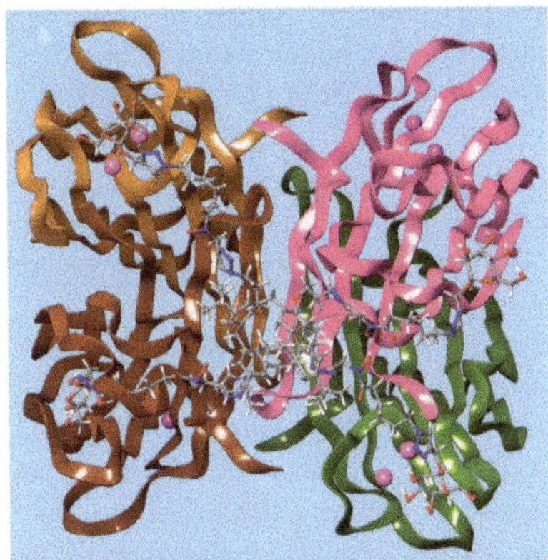

Figure 24. A glycosylated calixarene binding to a lectin. Reprinted with permission from [76]. Copyright © 2019 American Chemical Society.

4.2. Biofilm Inhibiting Resorcinarenes

A much simpler approach to biofilm inhibition was taken by Guildford and colleagues, who assessed the effects of a polyethylene glycol-bound resorcinarene catheter coating on biofilms of *E. coli* NCTC 10418 and *Proteus mirabilis* NCTC 11938 [77]. Although the macrocycle's solubility is often an issue, this can be improved through functionalisation with polyethylene glycol, **160** (Figure 25). The incorporation of upper rim silane groups facilitates its attachment to silicone coated devices [78]. The study found that the coating was able to significantly increase biofilm inhibition of *P. mirabilis* over 10 days, but was less effective against *E. coli*. Resorcin[4]arenes had also been used by Vidal, Matthews, and colleagues as tetravalent ligands for LecA and the lactose-specific lectin, galectin-1 (Gal-1) [79]. Resorcin[4]arene derivatives terminating in either galactose or lactose groups were synthesized in both chair (*rctt*) and boat (*rccc*) stereoisomers, **161** to **164**, as shown in Figure 25. Solubility issues resulted in only certain combinations of compounds and lectins being assessed for binding. Compared with a monomeric analogue, the galactose-terminated resorcin[4]arenes had a lower IC_{50} by a factor of 240 to 300, but no significant differences arose as a result of their topology. Molecular modelling suggested that two resorcin[4]arene substituents bound to adjacent LecA monomers in a similar manner to calix[4]arene-based analogues.

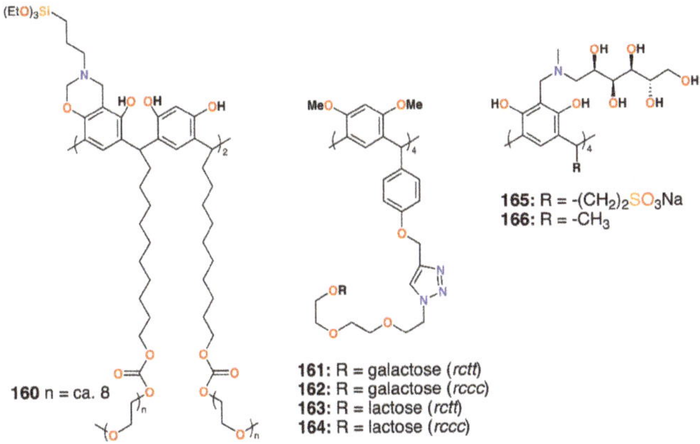

Figure 25. Resorcinarene derivatives of Guildford, Matthews, and Kashapov.

Calix[4]resorcinarenes were also used by Kashapov and colleagues [80]. Two derivatives, **165** and **166** (Figure 25), incorporating *N*-methyl-D-glucosamine at the upper rim were prepared and assessed for their activities against *S. aureus* 209 P, *Bacillus cereus* 8035, *P. aeruginosa* 9027, *E. coli* F-50, *Trichophyton* mentagrophytes var. gypseum 1773, *Aspergillus niger* BKMF-1119, and *C. albicans* 885-653. Both compounds were selective for *S. aureus* 209 P, and **166** had activity against *Bacillus cereus* 8035, with MICs ranging from 17.2 µg mL^{-1} to 137.2 µg mL^{-1}. The authors suggest that both compounds can form aggregates, which was supported by NMR diffusion spectroscopy, but that differences in their structures occur due to the more hydrophilic nature of **166**.

Several other groups have also employed resorcinarenes. Utomo and colleagues reported on a C-4-methoxyphenyl[4]resorcinarene with hexadecyltrimethylammonium substituents, **167** (Figure 26), which had a higher activity than the unsubstituted parent compound against *S. aureus*, but was ineffective against *E. coli* [81]. Yadav, Kumari, and colleagues used C-methyl[4]resorcinarene, **168** (Figure 26), to complex the antibiotic gatifloxacin and deliver it to *S. aureus* subsp. aureus and the human lung pathogen *Legionella pneumophila* subsp. pneumophila ATCC 33152 [82]. The complex had a lower MIC against *S. aureus* than the antibiotic alone, 0.16 µg mL^{-1} against 0.195 µg mL^{-1}, whereas with *L. pneumophila*, an MIC of 0.025 µg mL^{-1} was found whether the drug was complexed or not. Interestingly, the macrocycle itself had no effects on *S. aureus* below 25 µg mL^{-1}, but an MIC of 0.78 µg mL^{-1} against *L. pneumophila*. Yamin and colleagues investigated the activity of C-5-bromo-2-hydroxyphenyl-2-methyl-[4]resorcinarene, **169** (Figure 26), against MRSA, *S. aureus*, *E. faecalis*, *Enterobacter aerogenes*, and *P. aeruginosa* [83]. The macrocycle had little effect against *E. aerogenes* and *P. aeruginosa*, but had an MIC of 6.25 mg mL^{-1} against *S. aureus* and *E. faecalis* and an excellent MIC of 1.56 mg mL^{-1} against MRSA.

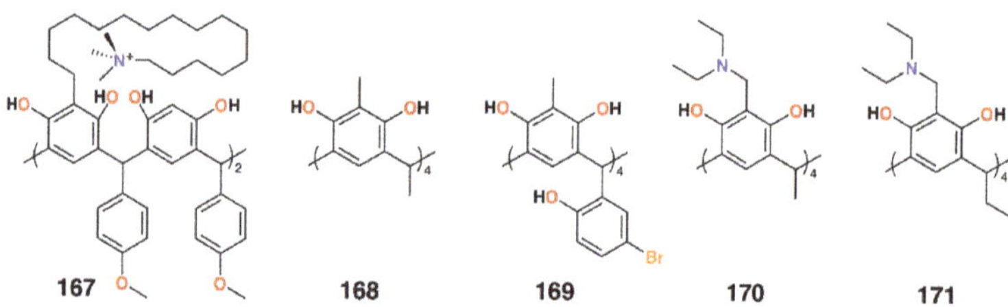

Figure 26. Resorcinarenes from the groups of Utomu, Yadav, Jumari, Yamin, and Vagapova.

Not all [4]resorcinarenes have antimicrobial activity. Vagapova and colleagues synthesized aminomethylated [4]resorcinarenes, **170** and **171** (Figure 26), which were found to have no in vitro antimicrobial activity against *S. aureus* ATCC 209 P, *B. cereus* ATCC 8035, *E. coli* CDC F-50, *P. aeruginosa* ATCC 9027, *A. niger* BKMF- 1119, *T. mentagrophytes* 1773, or *C. albicans* 855-653 over the concentration range of 0.97 to 500 µg mL^{-1} [84].

4.3. Biofilm Disruption through Drug Delivery

Shah and colleagues used a drug-delivery approach to biofilm disruption by encapsulating the broad spectrum antibiotic clarithromycin within self-assembling nanostructures formed by the amphiphilic *O*-decyl sulfonatocalix[6]arene, **172** (Figure 27) [85]. The nanostructures were in the region of 135 nm in diameter and had a calixarene/drug composition of 5:1, which represented over 50% drug loading efficiency. The drug-filled nanostructures had a 40% lower MIC and IC$_{50}$ against *Streptococcus pneumoniae* ATCC 6303 (antibiotic sensitive strain) than the drug alone, with a similar improvement observed when tested against *S. pneumoniae* ATCC 700669 (antibiotic resistant strain). Biofilm inhibition was similarly improved with minimum biofilm inhibition concentrations for clarithromycin of 19.69 µg mL^{-1} and 44.35 µg mL^{-1} with *S. pneumoniae* ATCC 6303 and ATCC 700669, respectively. When the drug-containing nanostructures were, used these values reduced to

13.69 μg mL^{-1} and 15.97 μg mL^{-1}, respectively. Drug-free nanostructures had no effect on antibiotic activity or biofilm inhibition. Efficacy was ascribed to cell penetration by the nanostructures and the inhibition of the normal drug efflux systems due to its encapsulation.

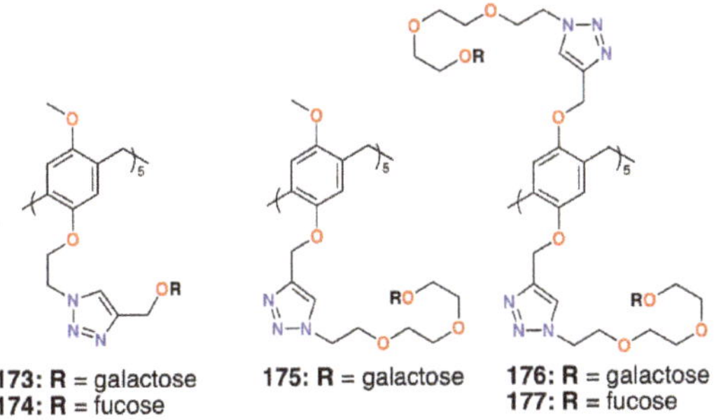

172

Figure 27. Shah's nanostructure-forming calixarene.

4.4. Biofilm-Inhibiting Pillar[n]arenes

The growth in the interest in pillararenes has led to a number of groups to exploit their symmetric, tubular structures and introduce extended substituents. This feature has resulted in their potential as biofilm inhibitors. Imberty, Vidal, and colleagues prepared asymmetric pillar[5]arenes with the intent to galactose and fucose termini, **173** to **175** (Figure 28), connected by two different linkers, on one face and methoxy groups on the other [86]. Cyclisation of the starting monomers 1-(2-bromoethoxy)-4-methoxybenzene or 1-methoxy-4-(2-propynyloxy)benzene led to a mixture of diastereoisomers, which could not be readily separated; thus, to ensure multivalency, symmetric, decasubstituted analogues, **176** and **177** (Figure 28), were prepared as comparators. HIA, ELLA, ITC, and SPR were used to examine the binding properties of the pillararenes with LecA; LecB; and a fucose-selective lectin, BambL, from *Burkholderia ambifaria*. The linker length of the galactose-functionalised pillar[5]arenes was the most important factor. The longer chain derivative displayed a much lower IC$_{50}$, in the micromolar range, and MIC. This was shown to be due to the latter's ability to interact with five lectin monomers, whereas the former was sterically crowded, owing to short linkers between the sugars and the macrocycle, and it could only interact with three monomers. As expected, the galactosylated derivatives bound LecA better than monomeric analogues and the fucosylated derivatives bound to the fucose-specific lectins LecB and BambL. The asymmetric derivative was specific for BambL, with low nanomolar affinity, as it bound no better to LecB than a monovalent ligand. Although not tested on biofilms, this specificity would undoubtedly impart inhibition where BambL-expressing bacteria were involved. The authors note that, as pillararenes have inherent chirality, both diastereoisomers must be present in a racemic mixture and the effects are an average of the two, but whether or not the chirality is important for recognition is unknown.

173: R = galactose
174: R = fucose

175: R = galactose

176: R = galactose
177: R = fucose

Figure 28. Vidal's glycosylated pillar[5]arenes.

Nierengarten and colleagues investigated the effects of increasing the numbers of sugars by incorporating *O*-alkyl substituents with both one and two monosaccharides [87]. Derivatives with different lengths of linkers, **178** to **183** (Figure 29), were also prepared to assess their effects. As expected, the galactosylated derivatives bound to LecA with a longer substituent, giving better results. Binding by the fucosylated derivatives to LecB increased with linker length to a point as the use of two triethylene glycol spacers gave poor results. The derivatives with 20 fucose groups were also superior to their analogues with just 10. Despite its poor result with LecB, the fucose derivative with the longest linkers and 20 sugars gave an IC$_{50}$ value in the picomolar range against BambL, over seven orders of magnitude better than the monovalent control.

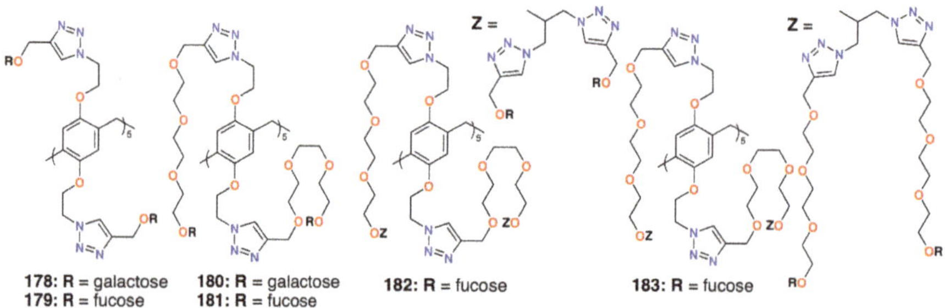

Figure 29. Nierengarten's glycosylated pillararenes.

In an attempt to incorporate both fucose and galactose termini, the Nierengarten group prepared a number of rotaxanes from pillar[5]arenes, **184** to **187** (Figure 30) [88]. Reaction between the azido-functionalized pillar[5]arene and acetylated fucose or galactose followed by diacylation gave the shuttle and an alkyl thread, terminating in either of those sugars, was formed by the same click methods. The combination of ten galactose units and a difucose thread was highly effective against both LecA and LecB, as it contains complementary sugars to both lectins.

Figure 30. Nierengarten's pillar[5]arene rotaxanes.

The mannose pillar[5]arene analogue, **188** (Figure 30), had previously been prepared and hemagglutination assays with the pillar[5]arene and an acyclic control showed an almost sevenfold enhancement in inhibition due to the presence of the glycocluster [89]. Later work on the same compound found some inhibition against a bacterial liposaccharide heptosyltransferase, WaaC, involved in lipopolysaccharide biosynthesis [90]. Shortly after Nierengarten's first report, Huang and colleagues published the synthesis of galactose-appended **189** and asymmetric derivative **190** (Figure 31) [91]. No cellular agglutination was observed for the symmetric pillar[5]arene or an acyclic control, owing to the presence of the mannose groups, but an asymmetric derivative with mannose and *n*-decyl substituents appeared to facilitate bacterial adhesion.

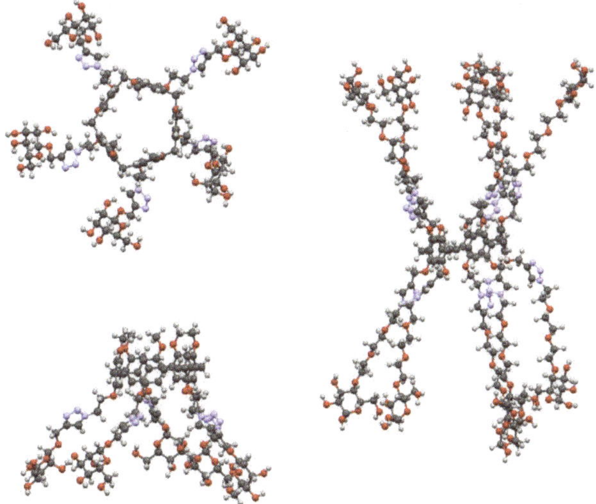

Figure 31. Huang's glycosylated pillar[5]arenes.

The multivalent potential of the sugar-terminated pillar[5]arenes can be seen in Figure 32. Their fivefold symmetry is shown in the top view of asymmetric pillar[5]arene **173**, with one of its four conformers illustrated, and the effect of extending the polyether linkers to increase the compound's span can be seen in **176**.

Figure 32. Glycosylated pillar[5]arenes **173** (**top** and **side**) and **176** (**side**) illustrating their multivalent binding potential.

While most of the previous examples use sugar-lectin interactions to inhibit biofilm formation, Cohen and colleagues investigated symmetric pillar[5]arenes, **191** to **196** (Figure 33), with ethyl- and methylphosphonium and ethyl- and methylammonium substituents alongside monomeric analogues [92]. Tests against methicillin-resistant *S. aureus* ATCC 33592 and *E. faecalis* ATCC 29212 showed that the monomers were ineffective, but the pillararenes had excellent inhibiting properties. MBIC$_{50}$ values of the phosphonium derivatives were 0.17 µg mL^{-1} (**191**) and 0.47 µg mL^{-1} (**192**) for both bacteria, as were the ammonium analogues (0.17 µg mL^{-1} for **193** and 0.47 µg mL^{-1} for **194**), indicating that it is the positive charges that are essential for biofilm inhibition. The group expanded their investigation to include imidazolium groups, **197** (Figure 32), and varied the counterions and the size of the pillararene, **198** (Figure 32) [93]. Biofilm inhibitory tests against Gram-positive *S. aureus* subsp. *aureus Rosenbach* ATCC 33592, *S. aureus* ATCC 29213, *S. aureus* BAA/043, *E. faecalis* ATCC 29212, *S. epidermidis* RP62A, and *S. mutans* ATCC 700610 showed pillar[5]- and -[6]arenes with trimethylammonium termini were the most potent, with MBIC$_{50}$ values of 0.20 µg mL^{-1} to 0.79 µg mL^{-1} across all species for pillar[5]arene **195** and 0.11 µg mL^{-1} to 0.79 µg mL^{-1} for pillar[6]arene **198**. By contrast, the negatively charged pillar[5]arene carboxylate derivative **64** had little effect on biofilm formation. Tests of the cationic pillararenes against Gram-negative *E. coli* ATCC 25922 and *P. aeruginosa* PAO1 showed no inhibition.

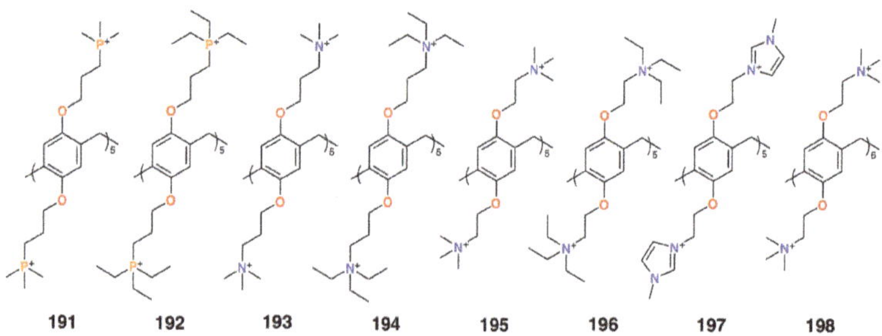

Figure 33. Cohen's charged pillar[*n*]arenes.

A summary of the biofilm inhibition demonstrated by the most active macrocyclic derivatives is given in Table 3.

Table 3. Summary of macrocycles demonstrating significant biofilm inhibition.

Compound	Target	MBIC/µg mL^{-1}	Ref
154	*P. aeruginosa*	2	[73]
155	*P. aeruginosa*	1.1	[74]
158	*P. aeruginosa*	5.4	[76]
165	*S. aureus*	17.2	[80]
168	*L. pneumophila*	0.78	[82]
168·gatifloxacin	*S. aureus*	0.16	[82]
169	*S. aureus*	6.25	[83]
	E. faecalis	6.25	[83]
	MRSA	1.56	[83]
172·clarithromycin	*S. pneumoniae*	13.69	[85]
191	*S. aureus*	0.17	[92]
	E. faecalis	0.47	[92]
192	*S. aureus*	0.17	[92]
	E. faecalis	0.47	[92]
193	*S. aureus*	0.17	[92]
	E. faecalis	0.47	[92]
194	*S. aureus*	0.17	[92]
	E. faecalis	0.47	[92]
195	*S. aureus*	0.20	[92]
	E. faecalis	0.20	[92]
	S. mutans	0.50	[92]
198	*S. aureus*	0.11	[92]
	E. faecalis	0.11	[92]
	S. epidermidis	0.23	[92]

5. Conclusions

From Cornforth's *O*-alkylation of calixarenes with cell membrane-disrupting polyethers, to Ungaro's vancomycin mimics through to Regnouf-de-Vains' nalidixic acid-appended calixarene prodrugs, and Matthews' multivalent lectin-binding calixarenes, calixarenes and their related macrocycles have been demonstrating antimicrobial activity for almost 70 years. During this time, the modes of action have evolved from simple surfactant activity through to drug delivery and, most recently, the use of multivalent macrocycles to bind proteins to block their aggregation. Indeed, it is the chemist's ability to prepare complex multifunctional derivatives with precise regioisomerism, which is propelling the most recent advances in antimicrobial macrocycles. Control of molecular recognition over extended distances is key in this endeavor and yet it relies on some of the first aspects of calixarene chemistry to be exploited chemically: lower rim substitution and conformer control.

Several themes are beginning to emerge: calixarenes incorporating drug moieties such as penicillin are more effective when coupled to ammonium or guanidinium substituents; metal chelating and releasing calixarenes are not as effective as might be expected; and the ability to use macrocycles as multivalent ligands capable of binding in spatially remote locations has been shown to be a highly effective biofilm-disrupting strategy.

Future development will, no doubt, revolve around substituents with greater specificity for target binding sites, such that the compounds are effective at vanishingly small doses. When those discoveries are made, calixarenes may well become the pharmacist's magic bullets.

Author Contributions: Writing—original draft preparation, P.J.C.; Writing—review & editing, all authors; Identification of papers for incorporation in the manuscript, all authors; Assistance with Russian language papers, I.I.S., P.L.P. and D.N.S. All authors have read and agreed to the published version of the manuscript.

Funding: The work was supported by the grant of the President of the Russian Federation for state support of leading scientific schools of the Russian Federation (NSh-2499.2020.3) for Dmitriy N. Shurpik, Pavel L. Padnya, and Ivan I. Stoikov.

Conflicts of Interest: The authors declare no conflict of interest.

References

1. Neri, P.; Sessler, J.L.; Wang, M.-X. (Eds.) *Calixarenes and Beyond*; Springer International Publishing: Cham, Switzerland, 2016.
2. Perret, F.; Lazar, A.N.; Coleman, A.W. Biochemistry of the para-sulfonato-calix[n]arenes. *Chem. Commun.* **2006**, 2425–2438. [CrossRef] [PubMed]
3. Perret, F.; Coleman, A.W. Biochemistry of anionic calix[n]arenes. *Chem. Commun.* **2011**, *47*, 7303–7319. [CrossRef] [PubMed]
4. Cragg, P.J. Pillar[n]arenes at the chemistry-biology interface. *Isr. J. Chem.* **2018**, *58*, 1158–1172. [CrossRef]
5. Cornforth, J.W.; D'Arcy Hart, P.; Nicholls, G.A.; Rees, R.J.W.; Stock, J.A. Antituberculous effects of certain surface-active polyoxyethylene ethers. *Br. J. Pharmacol.* **1955**, *10*, 73–86. [CrossRef] [PubMed]
6. Lamartine, R.; Tsukada, M.; Wilson, D.; Shirata, A. Antimicrobial activity of calixarenes. *Comptes Rendus Chim.* **2002**, *5*, 163–169. [CrossRef]
7. Naseer, M.M.; Ahmed, M.; Hameed, S. Functionalized calix[4]arenes as potential therapeutic agents. *Chem. Biol. Drug Des.* **2017**, *89*, 243–256. [CrossRef]
8. Rodik, R.; Poberezhnyk, M.; Kalchenko, V. Calixarene derivatives for (nano)biotechnologies. *Macroheterocycles* **2017**, *10*, 421–431. [CrossRef]
9. Rodik, R.V. Антимікробна та Противірусна АктивністьКаліксаренів. *J. Org. Pharmaceut. Chem.* **2015**, *13*, 67–78. [CrossRef]
10. Lipinski, C.A.; Lombardo, F.; Dominy, B.W.; Feeny, P.J. Experimental and computational approaches to estimate solubility and permeability in drug discovery and development settings. *Adv. Drug Deliv. Rev.* **1997**, *23*, 3–25. [CrossRef]
11. Lalor, R.; Baillie-Johnson, H.; Redshaw, C.; Matthews, S.E.; Mueller, A. Cellular uptake of a fluorescent calix[4]arene derivative. *J. Am. Chem. Soc.* **2008**, *130*, 2892–2893. [CrossRef] [PubMed]

12. Coleman, A.W.; Jebors, S.; Cecillon, C.; Perret, P.; Garin, D.; Marti-Battle, D.; Moulin, M. Toxicity and biodistribution of para-sulfonato-calix[4]arene in mice. *New J. Chem.* **2008**, *32*, 780–782. [CrossRef]

13. Strebhardt, K.; Ullrich, A. Paul Ehrlich's magic bullet concept: 100 years of progress. *Nat. Rev. Cancer* **2008**, *8*, 473–480. [CrossRef]

14. Ben Salem, A.; Regnouf-de-Vains, J.-B. Synthesis and characterisation of a new podand based on a calixarene and a β-lactam. *Tetrahedron Lett.* **2001**, *42*, 7033–7036. [CrossRef]

15. Korchowiec, B.; Ben Salem, A.; Corvis, Y.; Regnouf-de-Vains, J.-B.; Korchowiec, J.; Rogalska, E. Calixarenes in a membrane environment: A monolayer study on the miscibility of three *p-tert*-butylcalix[4]arene *b*-lactam derivatives with 1,2-dimyristoyl-sn-glycero-3-phosphoethanolamine. *J. Phys. Chem. B* **2007**, *111*, 13231–13242. [CrossRef]

16. Ben Salem, A.; Sautrey, G.; Fontanay, S.; Duval, R.E.; Regnouf-de-Vains, J.-B. Molecular drug-organiser: Synthesis, characterization and biological evaluation of penicillin V and/or nalidixic acid calixarene-based podands. *Bioorg. Med. Chem.* **2011**, *19*, 7534–7540. [CrossRef]

17. Ben Salem, A.; Regnouf-de-Vains, J.-B. Towards a new family of calixarene-based podands incorporating quinolone arms. An example using nalidixic acid. *Tetrahedron Lett.* **2003**, *44*, 6769–6771. [CrossRef]

18. Gutsche, C.D.; Nam, K.C. Calixarenes. 22. Synthesis, properties, and metal complexation of aminocalixarenes. *J. Am. Chem. Soc.* **1988**, *110*, 6153–6162. [CrossRef]

19. Baker, T.J.; Tomioka, M.; Goodman, M.; Mergott, D.G.; Roush, W.R. Preparation and use of *N,N'*-di-BOC-*N''*-triflylguanidine. *Org. Synth.* **2000**, *78*, 91.

20. Mourer, M.; Duval, R.E.; Finance, C.; Regnouf-de-Vains, J.-B. Functional organisation and gain of activity: The case of the antibacterial tetra-para-guanidinoethyl-calix[4]arene. *Bioorg. Med. Chem. Lett.* **2006**, *16*, 2960–2963. [CrossRef]

21. Grare, M.; Mourer, M.; Fontanay, S.; Regnouf-de-Vains, J.-B.; Finance, C.; Duval, R.E. In vitro activity of *para*-guanidinoethylcalix[4]arene against susceptible and antibiotic-resistant Gram-negative and Gram-positive bacteria. *J. Antimicrob. Chemother.* **2007**, *60*, 575–581. [CrossRef]

22. Grare, M.; Massimba Dibama, H.; Lafosse, S.; Ribon, A.; Mourer, M.; Regnouf-de-Vains, J.-B.; Duval, R.E. Cationic compounds with activity against multidrug-resistant bacteria: Interest of a new compound compared with two older antiseptics, hexamidine and chlorhexidine. *Clin. Microbiol. Infect.* **2010**, *16*, 432–438. [CrossRef]

23. Mourer, M.; Duval, R.E.; Constant, P.; Daffé, M.; Regnouf-de-Vains, J.-B. Impact of tetracationic calix[4]arene conformation-from conic structure to expanded bolaform-on their antibacterial and antimycobacterial activities. *ChemBioChem* **2019**, *20*, 911–921. [CrossRef]

24. Massimba Dibama, H.; Clarot, I.; Fontanay, S.; Ben Salem, A.; Mourer, M.; Finance, C.; Duval, R.E.; Regnouf-de-Vains, J.-B. Towards calixarene-based prodrugs: Drug release and antibacterial behaviour of a water-soluble nalidixic acid/calix[4]arene ester adduct. *Bioorg. Med. Chem. Lett.* **2009**, *19*, 2679–2682. [CrossRef]

25. Grare, M.; Mourer, M.; Regnouf-de-Vains, J.-B.; Finance, C.; Duval, R.E. Vers de nouvelles molécules antibactériennes. Intérêt du para-guanidinoéthylcalix[4]arène. *Pathol. Biol.* **2006**, *54*, 470–476. [CrossRef]

26. Pur, F.N.; Dilmaghani, K.A. Calixpenams: Synthesis, characterization, and biological evaluation of penicillins V and X clustered by calixarene scaffold. *Turk. J. Chem.* **2014**, *38*, 288–296.

27. Pur, F.N.; Dilmaghani, K.A. Calixcephems: Clustered cephalosporins analogous to calixpenams as novel potential anti-MRSA agents. *Turk. J. Chem.* **2014**, *38*, 850–858.

28. Patel, M.B.; Modi, N.R.; Raval, J.P.; Menon, S.K. Calix[4]arene based 1,3,4-oxadiazole and thiadiazole derivatives: Design, synthesis, and biological evaluation. *Org. Biomol. Chem.* **2012**, *10*, 1785–1794. [CrossRef] [PubMed]

29. Gezelbash, Z.D.; Dilmaghani, K.A. Synthesis, antifungal and antibacterial activity of calix[4]arene-based 1,3,4-oxadiazole derivatives. *J. Chin. Chem. Soc.* **2020**, *67*, 1446–1452. [CrossRef]

30. Mourer, M.; Psychogios, N.; Laumond, G.; Aubertin, A.-M.; Regnouf-de-Vains, J.-B. Synthesis and anti-HIV evaluation of water-soluble calixarene-based bithiazolyl podands. *Bioorg. Med. Chem.* **2010**, *18*, 36–45. [CrossRef]

31. Perret, F.; Tauran, Y.; Suwinska, K.; Kim, B.; Chassain-Nely, C.; Boulet, M.; Coleman, A.W. Molecular recognition and transport of active pharmaceutical ingredients on anionic calix[4]arene-capped silver nanoparticles. *J. Chem.* **2013**, *2013*. [CrossRef]

32. Boudebbouze, S.; Coleman, A.W.; Tauran, Y.; Mkaouar, H.; Perret, F.; Garnier, A.; Brioude, A.; Kim, B.; EMaguina, E.; Rhimi, M. Discriminatory antibacterial effects of calix[n]arene capped silver nanoparticles with regard to Gram positive and Gram negative bacteria. *Chem. Commun.* **2013**, *49*, 7150–7152. [CrossRef] [PubMed]

33. Moussa, Y.E.; Ong, Y.Q.E.; Perry, J.D.; Cheng, Z.; Kayser, V.; Cruz, E.; Kim, R.R.; Sciortino, N.; Wheate, N.J. Demonstration of in vitro host-guest complex formation and safety of *para*-sulfonatocalix[8]arene as a delivery vehicle for two antibiotic drugs. *J. Pharmaceut. Sci.* **2018**, *107*, 3105–3111. [CrossRef] [PubMed]

34. Consoli, G.M.L.; Granata, G.; Picciotto, R.; Blanco, A.R.; Geraci, C.; Marino, A.; Nostro, A. Design, synthesis and antibacterial evaluation of a polycationic calix[4]arene derivative alone and in combination with antibiotics. *Med. Chem. Commun.* **2018**, *9*, 160–164. [CrossRef]

35. Memon, S.; Chandio, A.A.; Memon, A.A.; Nizamani, S.M.; Bhatti, A.A.; Brohi, N.A. Synthesis, characterization, and exploration of antimicrobial activity of copper complex of diamide derivative of *p-tert*-butylcalix[4]arene. *Polycycl. Aromat. Compd.* **2017**, *37*, 362–374. [CrossRef]

36. Chandio, A.A.; Memon, A.A.; Memon, S.; Memon, F.N.; Panhwar, Q.K.; Durmaz, F.; Nizamani, S.M.; Brohi, N.A. Synthesis and antimicrobial assessment of Fe^{3+} inclusion complex of *p-tert*-butylcalix[4]arene diamide derivative. *J. Chem.* **2019**, *2019*. [CrossRef]

37. Özkan, Ş.Ç.; Yilmaz, A.; Arslan, E.; Açık, L.; Sayın, Ü.; Mutlu, E.G. Novel copper (II) complexes of p-tert-butylcalix [4] arene diamide derivatives: Synthesis, antimicrobial and DNA cleavage activities. *Supramol. Chem.* **2015**, *27*, 255–267. [CrossRef]

38. Noruzi, E.B.; Kheirkhahi, M.; Shaabani, B.; Geremia, S.; Hickey, N.; Asaro, F.; Nitti, P.; Kafil, H.S. Design of a thiosemicarbazide functionalized calix [4] arene ligand and related transition metal complexes: Synthesis, characterization and biological studies. *Front. Chem.* **2019**, *7*, 663. [CrossRef]

39. Noruzi, E.B.; Shaabani, B.; Geremia, S.; Hickey, N.; Nitti, P.; Kafil, H.S. Synthesis, crystal structure, and biological activity of a multidentate calix[4]arene ligand doubly functionalized by 2-hydroxybenzeledene-thiosemicarbazone. *Molecules* **2020**, *25*, 370. [CrossRef]

40. Roy, H.; Deolalkar, M.; Desai, A.S. Synthesis of calix-salen silver corates for evaluation of their antimicrobial and anticancer activities. *ACS Omega* **2019**, *4*, 21346–21352. [CrossRef]

41. Ali, Y.; Bunnori, N.M.; Susanti, D.; Alhassan, A.M.; Hamid, S.A. Synthesis, in vitro and in silico studies of azo-based calix [4] arenes as antibacterial agent and neuraminidase inhibitor: A new look into an old scaffold. *Front. Chem.* **2018**, *6*, 210. [CrossRef]

42. Barbera, L.; Franco, D.; De Plano, L.M.; Gattuso, G.; Guglielmino, S.P.P.; Lentini, G.; Manganaro, N.; Marino, N.; Pappalardo, S.; Parisi, M.F.; et al. A water-soluble pillar[5]arene as a new carrier for an old drug. *Org. Biomol. Chem.* **2017**, *15*, 3192–3195. [CrossRef] [PubMed]

43. Rybak, M.J.; Abate, B.J.; Kang, S.L.; Ruffing, M.J.; Lerner, S.A.; Drusano, G.L. Prospective evaluation of the effect of an aminoglycoside dosing regimen on rates of observed nephrotoxicity and ototoxicity. *Antimicrob. Agents Chemother.* **1999**, *43*, 1549–1555. [CrossRef]

44. Peng, H.; Xie, B.; Yang, X.; Dai, J.; Wei, G.; He, Y. Pillar[5]arene-based, dual pH and enzyme responsive supramolecular vesicles for targeted antibiotic delivery against intracellular MRSA. *Chem. Commun.* **2020**, *56*, 8115–8118. [CrossRef] [PubMed]

45. Cornforth, J.W.; D'Arcy Hart, P.; Rees, R.J.W.; Stock, J.A. Antituberculous effect of certain surface-active polyoxyethylene ethers in mice. *Nature* **1951**, *168*, 150–153. [CrossRef]

46. Zinke, A.; Zigeuner, G.; Hössinger, K.; Hoffmann, G. Zur Kenntnis des Härtungsprozesses von Phenol-Formaldehyd-Harzen. XVIII., vorläufige Mitteilung: Über cyclische Mehrkernphenole. *Monatsh. Chem.* **1948**, *79*, 438–439. [CrossRef]

47. D'Arcy Hart, P.; Armstrong, J.A.; Brodaty, E. Calixarenes with host-mediated potency in experimental tuberculosis: Further evidence that macrophage lipids are involved in their mechanism of action. *Infect. Immun.* **1996**, *64*, 1491–1493. [CrossRef] [PubMed]

48. Colston, M.J.; Hailes, H.C.; Stavropoulos, E.; Hervé, A.C.; Hervé, G.; Goodworth, K.J.; Hill, A.M.; Jenner, P.; D'Arcy Hart, P.; Tascon, R.E. Antimycobacterial calixarenes enhance innate defense mechanisms in murine macrophages and induce control of *Mycobacterium tuberculosis* infection in mice. *Infect. Immun.* **2004**, *72*, 6318–6323. [CrossRef]

49. Goodworth, K.J.; Hervé, A.C.; Stavropoulos, E.; Hervé, G.; Casades, I.; Hill, A.M.; Weingarten, G.G.; Tascon, R.E.; Colston, M.J.; Hailes, H.C. Synthesis and in vivo biological activity of large-ringed calixarenes against *Mycobacterium tuberculosis*. *Tetrahedron* **2011**, *67*, 373–382. [CrossRef]

50. Mourer, M.; Massimba Dibama, H.; Constant, P.; Daffé, M.; Regnouf-de-Vains, J.-B. Anti-mycobacterial activities of some cationic and anionic calix[4]arene derivatives. *Bioorg. Med. Chem.* **2012**, *20*, 2035–2041. [CrossRef]

51. Mourer, M.; Massimba Dibama, H.; Fontanay, S.; Grare, M.; Duval, R.E.; Finance, C.; Regnouf-de-Vains, J.-B. *p*-Guanidinoethyl calixarene and parent phenol derivatives exhibiting antibacterial activities. Synthesis and biological evaluation. *Bioorg. Med. Chem.* **2009**, *17*, 5496–5509. [CrossRef]

52. Mourer, M.; Fontanay, S.; Duval, R.E.; Regnouf-de-Vains, J.-B. Synthesis, characterization, and biological evaluation as antibacterial agents of water-soluble calix[4]arenes and phenol derivatives incorporating carboxylate groups. *Helv. Chim. Acta* **2012**, *95*, 1373–1386. [CrossRef]

53. Yushchenko, T.I.; Germanyuk, T.A.; Chornoknyzhny, S.I.; Zaichko, N.V.; Korol, A.P.; Prokopchuk, Z.M.; Rodik, R.V.; Cheshun, E.A. Antibacterial and antiplatelet activity of calix[4,6]arene tetraalkylamines. *Pharmacol. Drug Toxicol.* **2012**, *5*, 79–88.

54. Ukhatskaya, E.V.; Kurkov, S.V.; Hjálmarsdóttir, M.A.; Karginov, V.A.; Matthews, S.E.; Rodik, R.V.; Kalchenko, V.I.; Loftsson, T. Cationic quaternized aminocalix[4]arenes: Cytotoxicity, haemolytic and antibacterial activities. *Int. J. Pharm.* **2013**, *458*, 25–30. [CrossRef] [PubMed]

55. Melezhyk, I.O.; Rodik, R.V.; Iavorska, N.V.; Klymchenko, A.S.; Mely, Y.; Shepelevych, V.V.; Skivka, L.M.; Kalchenko, V.I. Antibacterial properties of tetraalkylammonium and imidazolium tetraalkoxycalix[4]arene derivatives. *Anti-Infect. Agents* **2015**, *13*, 87–94. [CrossRef]

56. Di Bari, A.; Picciotto, R.; Granata, G.; Blanco, A.R.; Consoli, G.M.L.; Sortino, S. A bactericidal calix[4]arene-based nanoconstruct with amplified NO photorelease. *Org. Biomol. Chem.* **2016**, *14*, 8047–8052. [CrossRef] [PubMed]

57. Consoli, G.M.L.; Di Bari, A.; Blanco, A.R.; Nostro, A.; D'Arrigo, M.; Pistarà, V.; Sortino, S. Design, synthesis, and antibacterial activity of a multivalent polycationic calix[4]arene-NO photodonor conjugate. *ACS Med. Chem. Lett.* **2017**, *8*, 881–885. [CrossRef]

58. da Silva, C.M.; da Silva, D.L.; Magalhães, T.F.F.; Alves, R.B.; de Resende-Stoianoff, M.A.; Martins, F.T.; de Fátima, A. Iminecalix[4]arenes: Microwave-assisted synthesis, X-ray crystal structures, and anticandidal activity. *Arab. J. Chem.* **2019**, *12*, 4365–4376. [CrossRef]

59. Casnati, A.; Fabbi, M.; Pelizzi, N.; Pochini, A.; Sansone, F.; Ungaro, R. Synthesis, antimicrobial activity and binding properties of calix[4]arene based vancomycin mimics. *Bioorg. Med. Chem. Lett.* **1996**, *6*, 2699–2704. [CrossRef]

60. Mehta, V.; Athar, M.; Jha, P.C.; Panchal, M.; Modi, K.; Jain, V.K. Efficiently functionalized oxacalix[4]arenes: Synthesis, characterization and exploration of their biological profile as novel HDAC inhibitors. *Bioorg. Med. Chem. Lett.* **2016**, *26*, 1005–1010. [CrossRef]

61. Soomro, A.M.; Oad, R.J.; Memon, S.; Qureshi, I. Bioactivity assessment of water soluble calix[4]arene derivative. *Pak. J. Anal. Environ. Chem.* **2012**, *13*, 36–39.

62. Muneer, S.; Memon, S.; Panhwar, Q.K.; Bhatti, A.A.; Khokhar, T.S. Synthesis and investigation of antimicrobial properties of pyrrolidine appended calix[4]arene. *J. Anal. Sci. Technol.* **2017**, *8*, 3. [CrossRef]

63. Muneer, S.; Memon, S.; Panhwar, Q.K.; Khushik, F.; Khokhar, T.S.; Noor, A.A. Synthesis and antimicrobial activity of p- tetranitrocalix[4]arene derivative. *Polycycl. Aromat. Compd.* **2016**, *36*, 554–563. [CrossRef]

64. Galitskaya, P.; Fomin, V.; Stiokov, I.; Andreyko, E.; Selivanovskaya, S. Antimicrobial activity of nanoparticles from solid phase supramolecular assemblies based on stereoisomers of *p-tert*-butylthiacalix[4]arene with silver cations. *Int. J. Pharm. Technol.* **2016**, *8*, 15048–15053.

65. Zhang, M.; Zhu, P.-P.; Xin, P.; Si, W.; Li, Z.-T.; Hou, J.-L. Synthetic channel specifically inserts into the lipid bilayer of gram- positive bacteria but not that of mammalian erythrocytes. *Angew. Chem. Int. Ed.* **2017**, *56*, 2999–3003. [CrossRef] [PubMed]

66. Xin, P.; Sun, Y.; Kong, H.; Wang, Y.; Tan, S.; Guo, J.; Jiang, T.; Dong, W.; Chen, C.-P. A unimolecular channel formed by dual helical peptide modified pillar[5]arene: Correlating transmembrane transport properties with antimicrobial activity and haemolytic toxicity. *Chem. Commun.* **2017**, *53*, 11492–11495. [CrossRef]

67. Flemming, H.-C.; Wingender, J. The biofilm matrix. *Nat. Rev. Microbiol.* **2010**, *8*, 623–633. [CrossRef]

68. Matthews, S.E. Calixsugars: Finally reaching their potential? In *Calixarenes and Beyond*; Neri, P., Sessler, J.L., Wang, M.-X., Eds.; Springer International Publishing: Cham, Switzerland, 2016; pp. 559–600.

69. Baldini, L.; Casnati, A.; Sansone, F.; Ungaro, R. Peptido- and glycocalixarenes. In *Calixarenes in the Nanoworld*; Vicens, J., Harrowfield, J., Baklouti, L., Eds.; Springer: Dordrecht, The Netherlands, 2007; pp. 233–257.

70. Cecioni, S.; Lalor, R.; Blanchard, B.; Praly, J.-P.; Imberty, A.; Matthews, S.E.; Vidal, S. Achieving high affinity towards a bacterial lectin through multivalent topological isomers of calix[4]arene glycoconjugates. *Chem. Eur. J.* **2009**, *15*, 13232–13240. [CrossRef]

71. Sicard, D.; Cecioni, S.; Iazykov, M.; Chevolot, Y.; Matthews, S.E.; Praly, J.-P.; Souteyrand, E.; Imberty, A.; Vidal, S.; Phaner-Goutorbe, M. AFM investigation of *Pseudomonas aeruginosa* lectin LecA (PA-IL) filaments induced by multivalent glycoclusters. *Chem. Commun.* **2011**, *47*, 9483–9485. [CrossRef]

72. Cecioni, S.; Praly, J.-P.; Matthews, S.E.; Wimmerová, M.; Imberty, A.; Vidal, S. Rational design and synthesis of optimized glycoclusters for multivalent lectin–carbohydrate interactions: Influence of the linker arm. *Chem. Eur. J.* **2012**, *18*, 6250–6263. [CrossRef] [PubMed]

73. Consoli, G.M.L.; Granata, G.; Cafiso, V.; Stefani, S.; Geraci, C. Multivalent calixarene-based C-fucosyl derivative: A new *Pseudomonas aeruginosa* biofilm inhibitor. *Tetrahedron Lett.* **2011**, *52*, 5831–5834. [CrossRef]

74. Boukerb, A.M.; Rousset, A.; Galanos, N.; Meár, J.-B.; Thépaut, M.; Grandjean, T.; Gillon, E.; Cecioni, S.; Abderrahmen, C.; Faure, K.; et al. Antiadhesive properties of glycoclusters against *Pseudomonas aeruginosa* lung infection. *J. Med. Chem.* **2014**, *57*, 10275–10289. [CrossRef]

75. Granata, G.; Stracquadanio, S.; Consoli, G.M.L.; Cafiso, V.; Stefani, S.; Geraci, C. Synthesis of a calix[4]arene derivative exposing multiple units of fucose and preliminary investigation as a potential broad-spectrum antibiofilm agent. *Carb. Res.* **2019**, *476*, 60–64. [CrossRef]

76. Taouai, M.; Chakroun, K.; Sommer, R.; Michaud, G.; Giacalone, D.; Ben Maaouia, M.A.; Vallin-Butruille, A.; Mathiron, D.; Abidi, R.; Darbre, T.; et al. Glycocluster tetrahydroxamic acids exhibiting unprecedented inhibition of *Pseudomonas aeruginosa* biofilms. *J. Med. Chem.* **2019**, *62*, 7722–7738. [CrossRef] [PubMed]

77. Guildford, A.; Morris, C.; Kitt, O.; Cooper, I. The effect of urinary Foley catheter substrate material on the antimicrobial potential of calixerene-based molecules. *J. Appl. Microbiol.* **2017**, *124*, 1047–1059. [CrossRef]

78. Barlow, I.J.; Williams, N.H.; Stirling, C.J.M. Medical Devices and Coatings. Patent No. WO 2013/104916A2, 18 July 2013.

79. Soomro, Z.H.; Cecioni, S.; Blanchard, H.; Praly, J.-P.; Imberty, A.; Vidal, S.; Matthews, S.E. CuAAC synthesis of resorcin[4]arene-based glycoclusters as multivalent ligands of lectins. *Org. Biomol. Chem.* **2011**, *9*, 6587–6597. [CrossRef] [PubMed]

80. Kashapov, R.R.; Razuvayeva, Y.S.; Ziganshina, A.Y.; Mukhitova, R.K.; Sapunova, A.S.; Voloshina, A.D.; Syakaev, V.V.; Latypov, S.K.; Nizameev, I.R.; Kadirov, M.K.; et al. *N*-Methyl-*d*-glucamine-calix[4]resorcinarene conjugates: Self-assembly and biological properties. *Molecules* **2019**, *24*, 1939. [CrossRef]

81. Utomo, S.B.; Fujiyanti, M.; Lestari, W.P.; Mulyani, S. Antibacterial activity test of the C-4-methoxyphenylcalix[4]resorcinarene compound modified by hexadecyltrimethylammonium-bromide against *Staphylococcus aureus* and *Escherichia coli* bacteria. *JKPK* **2018**, *3*, 201–209. [CrossRef]

82. Dawn, A.; Chandra, H.; Ade-Browne, C.; Yadav, J.; Kumari, K. Multifaceted supramolecular interactions from C-methylresorcin[4]arene lead to an enhancement in in vitro antibacterial activity of gatifloxacin. *Chem. Eur. J.* **2017**, *23*, 18171–18179. [CrossRef]

83. Abosadiya, H.M.; Hasbullah, S.A.; Mackeen, M.M.; Low, S.C.; Ibrahim, N.; Koketsu, M.; Yamin, B.M. Synthesis, characterization, X-ray structure and biological activities of C-5-bromo-2-hydroxyphenylcalix[4]-2-methyl resorcinarene. *Molecules* **2013**, *18*, 13369–13384. [CrossRef]

84. Vagapova, L.; Nasirova, Z.; Burilova, E.; Zobov, V.; Burilov, A.; Amirov, R.; Pudovik, M. New salt structures based on aminomethylated calix[4]resorcinarenes and (1-hydroxyethane-1,1-diyl) bisphosphonic acid. *Russ. J. Org. Chem.* **2017**, *53*, 312–314. [CrossRef]

85. Ali, I.; Imran, M.; Saifullah, S.; Tian, H.-W.; Guo, D.-S.; Shah, M.R. Amphiphilic *p*-sulfonatocalix[6]arene based self-assembled nanostructures for enhanced clarithromycin activity against resistant *Streptococcus pneumoniae*. *Colloids Surf. B* **2020**, *186*, 110676. [CrossRef] [PubMed]

86. Galanos, N.; Gillon, E.; Imberty, A.; Matthews, S.E.; Vidal, S. Pentavalent pillar[5]arene-based glycoclusters and their multivalent binding to pathogenic bacterial lectins. *Org. Biomol. Chem.* **2016**, *14*, 3476–3481. [CrossRef]

87. Buffet, K.; Nierengarten, I.; Galanos, N.; Gillon, E.; Holler, M.; Imberty, A.; Matthews, S.E.; Vidal, S.; Vincent, S.P.; Nierengarten, J.-F. Pillar[5]arene-based glycoclusters: Synthesis and multivalent binding to pathogenic bacterial lectins. *Chem. Eur. J.* **2016**, *22*, 2955–2963. [CrossRef]

88. Vincent, S.P.; Buffet, K.; Nierengarten, I.; Imberty, A.; Nierengarten, J.-F. Biologically active heteroglycoclusters constructed on a pillar[5]arene-containing [2]rotaxane scaffold. *Chem. Eur. J.* **2016**, *22*, 88–92. [CrossRef]

89. Nierengarten, I.; Buffet, K.; Holler, M.; Vincent, S.; Nierengarten, J.-F. A mannosylated pillar[5]arene derivative: Chiral information transfer and antiadhesive properties against uropathogenic bacteria. *Tetrahedron Lett.* **2013**, *54*, 2398–2402. [CrossRef]

90. Tikad, A.; Fu, H.; Sevrain, C.M.; Laurent, S.; Nierengarten, J.-F.; Vincent, S.P. Mechanistic insight into heptosyltransferase inhibition by using Kdo multivalent glycoclusters. *Chem. Eur. J.* **2016**, *22*, 13147–13155. [CrossRef]

91. Yu, G.; Ma, Y.; Han, C.; Yao, Y.; Tang, G.; Mao, Z.; Gao, C.; Huang, F. A sugar-functionalized amphiphilic pillar[5]arene: Synthesis, self- assembly in water, and application in bacterial cell agglutination. *J. Am. Chem. Soc.* **2013**, *135*, 10310–10313. [CrossRef]

92. Joseph, R.; Kaizerman, D.; Herzog, I.M.; Hadar, M.; Feldman, M.; Fridman, M.; Cohen, Y. Phosphonium pillar[5]arenes as a new class of efficient biofilm inhibitors: Importance of charge cooperativity and the pillar platform. *Chem. Commun.* **2016**, *52*, 10656–10659. [CrossRef]

93. Joseph, R.; Naugolny, A.; Feldman, M.; Herzog, I.M.; Fridman, M.; Cohen, Y. Cationic pillararenes potently inhibit biofilm formation without affecting bacterial growth and viability. *J. Am. Chem. Soc.* **2016**, *138*, 754–757. [CrossRef]

Review

Calix[3]arene-Analogous Metacyclophanes: Synthesis, Structures and Properties with Infinite Potential

Md. Monarul Islam [1], Paris E. Georghiou [2,*], Shofiur Rahman [3] and Takehiko Yamato [4]

[1] Chemical Research Division, Bangladesh Council of Scientific and Industrial Research (BCSIR), Dhanmondi, Dhaka 1205, Bangladesh; mmipavel@yahoo.com
[2] Department of Chemistry, Memorial University of Newfoundland, St. John's, NL A1B 3X7, Canada
[3] Aramco Laboratory for Applied Sensing Research, King Abdullah Institute for Nanotechnology, King Saud University, Riyadh 11451, Saudi Arabia; msrahman@mun.ca
[4] Department of Applied Chemistry, Faculty of Science and Engineering, Saga University, Honjo-Machi 1, Saga 840-8502, Japan; yamatot@cc.saga-u.ac.jp
* Correspondence: parisg@mun.ca

Academic Editors: Mario Berberan-Santos and Paula M. Marcos
Received: 24 August 2020; Accepted: 10 September 2020; Published: 14 September 2020

Abstract: Calixarene-analogous metacyclophanes (CAMs) are a special class of cyclophanes that are cyclic polyaromatic hydrocarbons containing three or more aromatic rings linked by one or more methylene bridging groups. They can be considered to be analogues of calixarenes, since, in both types of molecules, the component aromatic rings are linked by methylene groups, which are meta to each other. Since the prototype or classical calix[4]arene consists of four benzene rings each linked by methylene bridges, which are also meta to each other, it can be considered to be an example of a functionalized [1.1.1.1]metacyclophane. A metacyclophane (MCP) that consists of three individual hydroxyl-group functionalized aromatic rings linked by methylene groups, e.g., a trihydroxy[1.1.1]MCP may therefore, by analogy, be termed in the broadest sense as a "calix[3]arene" or a "calix[3]arene-analogous metacyclophane". Most of the CAMs reported have been synthesized by fragment coupling approaches. The design, synthesis and development of functionalized CAMs, MCPs, calixarenes and calixarene analogues has been an area of great activity in the past few decades, due their potential applications as molecular receptors, sensors and ligands for metal binding, and for theoretical studies, etc. In this review article, we focus mainly on the synthesis, structure and conformational properties of [1.1.1]CAMs, i.e., "calix[3]arenes" and their analogues, which contain three functionalized aromatic rings and which provide new scaffolds for further explorations in supramolecular and sensor chemistry.

Keywords: calix[3]arenes; metacyclophanes; calixarene-analogous metacyclophanes; inherent chirality; host-guest chemistry

---

1. Introduction

Cyclophanes are an important class of bridged aromatic hydrocarbons, consisting of one or more aromatic units linked by methylene group chains in such a way that the aromatic units form part of a macrocycle. A generic synthetic structure of cyclophanes in which two or more aromatic groups are linked via their *ortho, meta* or *para* positions by variably sized methylene groups is shown in Figure 1. These molecules have been extensively pursued and studied for a variety of different considerations [1,2]. Cyclophane chemistry can be considered to have been initiated by Pellegrin [3] who, in 1899, reported the synthesis of [2.2]metacyclophane (i.e., [2,2]MCP) **1** by Wurtz coupling of 1,3-bis(bromomethyl)benzene. The real development of the chemistry of cyclophanes, however, is well-acknowledged to have occurred after Cram and Steinberg's synthesis of [2.2]paracyclophane **2** in 1951 [4].

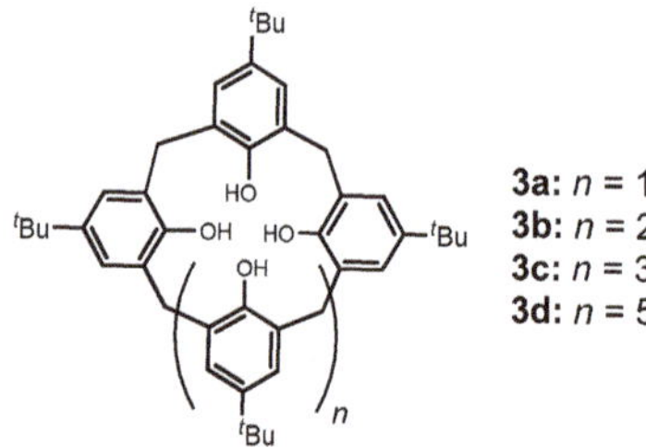

Figure 1. Top: Generic cyclophane structure with the *ortho-*, *meta-* or *para-*substitution of the phenyl rings and where *m*, *n* and *z* are variable. Compound **1** is [2.2]metacyclophane and **2** is [2.2]paracyclophane, where *m* = *n* = 2 and *z* = 1.

On the other hand, calixarenes such as **3a–d** (Figure 2), which can be considered to be examples of [1$_n$]metacyclophanes (where *n* = 4, 5, 6 or 8, respectively), have also become an important class of compounds since Zincke and Zeigler's first report in 1944 of a cyclic tetrameric compound (later confirmed to be **3a**) from the base-induced reaction of *para*-substituted *tert*-butylphenol with formaldehyde [5]. The importance of this class of compounds derives from the fact that calixarenes have "basket"- or "cup"-like 3-D topologies, with the hydroxyl groups forming a narrower or "lower" rim due to the strong intermolecular hydrogen bonds. The opposite rim is termed the "upper" rim and is wider due to the steric repulsion of the *para*-substituents and the orientation of the aromatic rings although for the larger calix[*n*]arenes (*n* > 6); however, their shapes or conformations can vary or be more complex. As a result, these molecules can be selectively functionalized at either rim and they can serve as building blocks for a variety of applications, such as, for example, ionic or molecular receptors. Gutsche and coworkers pioneered and developed methodologies for synthesizing *p-tert*-butylcalix[4]arene **3a** [6] (also commonly and inaccurately referred to simply as "calix[4]arene", which should strictly be confined to the de-*tert*-butylated derivative of **3a**), *p-tert*-butylcalix[5]arene **3b** [7], *p-tert*-butylcalix[6]arene **3c** [8], and *p-tert*-butylcalix[8]arene **3d** [9] (recently, Haase reported a highly selective and high-yield synthesis for the production of calix[8]arenes [10]) as major products. Gutsche's syntheses were achieved using *p-tert*-butylphenol and formaldehyde under different but reproducible reaction conditions, which have allowed these compounds, and derivatives thereof, to become easily accessible to be widely studied. Derivatives of **3a** are well-known to be able to exist in four different major conformationally immobile conformers known as *cone*, *partial-cone*, *1,3-alternate*, and *1,2-alternate* conformers (Figure 3) and which can be selectively generated and isolated by introducing different functional groups onto the lower rim of **3a** [11].

Figure 2. Basic structural motifs of calix[*n*]arenes **3a–d**.

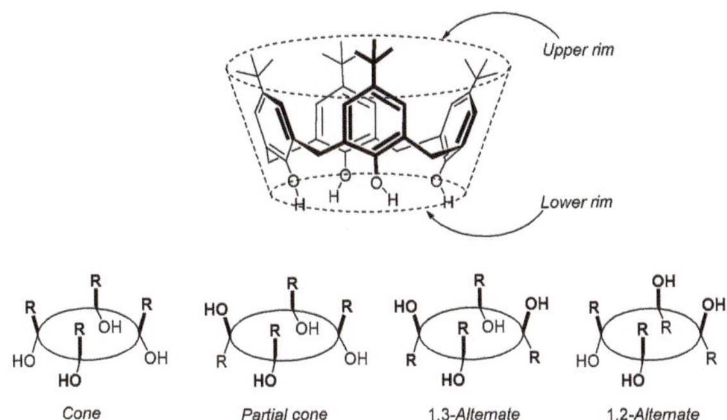

Figure 3. General structure of p-substituted calix[4]arene (e.g., R = *t*-Bu) and schematic representations of its formal conformational isomers.

Many calixarene-based molecular receptors have been synthesized and studied, but calix[4]arenes and its derivatives with amide, ketone and ester functionalities at the lower rim, which have shown significant cation affinity [12,13], have received the most attention to date. In general, modifications of calixarenes have been via functionalization of their lower or upper rims. The spirodienone route [14] and Lhtoak's mercuration [15] route, however, are notable exceptions for effecting the direct modifications to the basic calix[4]arene skeleton not specifically involving the functionalization of the upper and/or lower rims, or via convergent fragment-based approaches. Since there are many excellent reviews and monographs that have been published on calixarenes, and their more typical analogues, including homocalixarenes [16], homooxacalixarenes [17,18], azacalixarenes [19] and hexahomoazacalixarenes [20], and other types [21,22] the reader is referred to these so will not be further elaborated upon here. Instead, for this review, we have chosen to emphasize the MCP skeleton that has been extensively explored as a versatile and stable platform, or scaffold, for functionalization and study. In particular, we present a specific focus upon those functionalized MCPs, which in the broadest sense, can be considered to be directly analogous with calix[3]arenes in general. A review of the syntheses and properties of cyclophanes incorporating three aromatic units is presented below.

2. Calixarene-Analogous MCPs Containing Three Aromatic Rings: Calix[3]arene Analogues

Many reports during the past few decades have dealt with the modification and studies of the properties of calix[*n*]arenes, particularly those in which *n* = 4,5,6 or 8, but not many have concerned calix[3]arenes containing only three aromatic rings, i.e., calix[3]arenes or [1.1.1]MCPs (or [1$_n$]MCPs), with the exception of the hexahomoxacalix[3]arenes in which the bridging groups are one or more -CH$_2$OCH$_2$- groups [17,18]. [1.1.1]MCPs cyclophanes, which incorporate only three aromatic rings, have been a major focus of our group in recent years and provide useful molecular platforms, particularly for their synthesis and study of their conformational properties and molecular strain. In 1982, Moshfegh and co-workers reported the first synthesis of a series of functionalized [1.1.1]MCPs, i.e., p-halocalix[3]-arenes **4a–c** [23] (Figure 4).

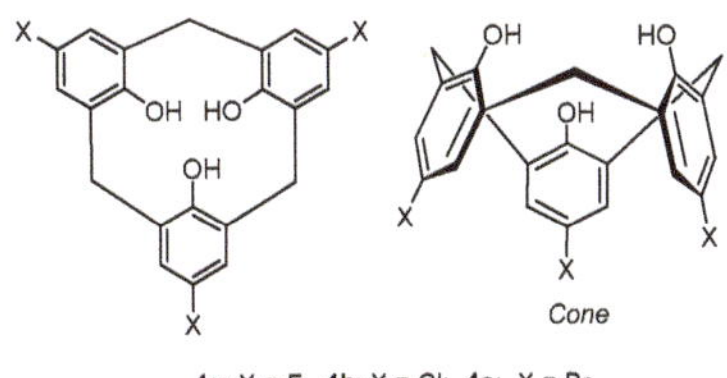

4a: X = F; **4b**: X = Cl; **4c**: X = Br

Figure 4. General structures of the [1.1.1]metacyclophane (MCP) calixarene analogues **4a–c**.

These compounds which contained intra-annular hydroxy groups and halogens on the *para* positions of one, two or all three of the phenyl groups and which adopted *cone* type conformations, were obtained in 69–90% yields by the cyclocondensation(s) of the corresponding precursor mono- or dihalo-2,2′-dihydroxydiphenyl-methane with 2,6-bis(hydroxymethyl)phenol, or with 4-halo-2,6-bis(hydroxymethyl)phenols [23]. The antimicrobial properties of these compounds which were named as "phloroglucide" analogues were reported but surprisingly, to date, no other studies have been reported with these particular compounds.

3. Homocalixarene-Analogous MCPs Containing Three Aromatic Rings: Homocalix[3]arene Analogues

Expanded macrocyclic calixarene-analogous metacyclophanes (CAMs) containing polymethylene bridging groups such as [3.1.1]MCP 7 which contains a single trimethylene ($-(CH_2)_3-$) and two methylene bridges can be regarded as an example of an unsymmetrical homocalix[3]arene analogous MCP. Thus, 7 which can be considered to be an example of a [3.1.1]dihomocalix[3]arene-analogous MCP was synthesized using the Nafion-H catalyzed cyclobenzylation of 5 and *tert*-butylphenol 6 [24,25]. The trimethoxy and trihydroxy derivatives of **7** were also reported. The room temperature ^{1}H-NMR spectrum of dimethoxy **7** revealed two sets of doublets for the methylene protons which implied that it is in an asymmetric "*2-partial-cone*" conformation, in which the two aromatic rings linked by the trimethylene group are anti to each other as shown in Scheme 1. However, at 80 °C in CDBr$_3$, coalescence of the methylene protons indicates that there is conformational ring flipping above this temperature. A mechanism was proposed in which two possible structures results from the alternating hydrogen bonding formed between one of the methoxy groups and the hydroxyl group. The triol however was fixed in a typical *cone* conformation with all three hydroxyl groups syn to one another.

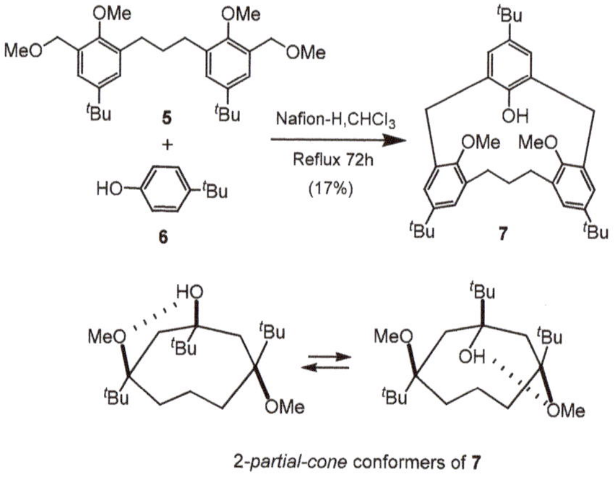

Scheme 1. Synthesis of hydroxy[3.1.1]MCP calixarene analogue **7** and schematic representation of interconverting *2-partial-cone* conformer structures.

Tashiro and coworkers reported the synthesis and conformational properties of the dihomocalix[3]arene containing one methylene and two dimethylene ($-(CH_2)_2-$) bridges, i.e., trimethoxy[2.2.1]MCP **8** and several of its derivatives. [26] The synthesis of **8**, which can also be considered to be a dihomocalix[3]arene analogue, was found to be a 2:1 mixture of asymmetric and symmetrical atropisoisomers **8a** and **8b**, respectively (Scheme 2). The synthesis was accomplished via the corresponding dithia[3.3.1]MCP precursor, which was transformed by the oxidation and sulfone pyrolysis-extrusion methodology under vacuum pyrolysis at 450 °C [27,28]. Based upon their ^{1}H-NMR spectra, and with the use of Pirkle's reagent, these molecules were shown to exist in alternate conformations designated as *2,1-alternate* **8a** and *2,2-alternate* **8b**, which could undergo rapid interconversion.

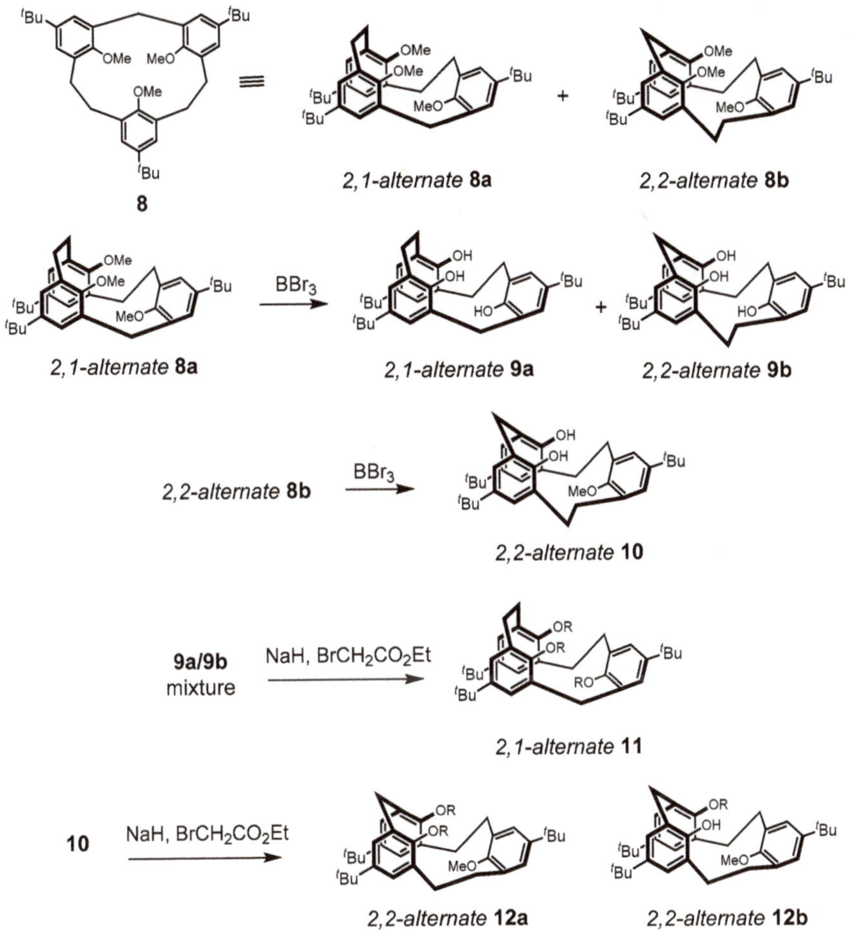

Scheme 2. Conformers of [2.2.1]MCP dihomocalix[3]arene analogues **8–12** (R = CH$_2$CO$_2$Et).

Trihydroxy[2.2.1]MCPs **9a** and **9b** were obtained by the simple demethylation of **8a**, but only dihydroxymethoxy[2.2.1]MCP **10** was obtained from the demethylation of **8b**, the remaining methoxy group presumably being sterically shielded from the BBr$_3$ attack. Treatment of the mixture of **9a** and **9b** with NaH/ethyl bromoacetate afforded only MCP **11**. On the other hand, a 2:1 mixture of both MCPs **12a** and **12b** was obtained from similar treatment of *2,2-alternate* **10**. The single-crystal X-ray structures of both **8a** and **8b** further confirmed their structures. Tashiro's group had previously reported the synthesis of trimethyl [2.2.2]MCPs [29]. These molecules, however, contained only intra-annular methyl groups and thus cannot be considered to be calix[3]arene-analogous MCPs, which, of course, typically contain hydroxyl or *O*-alkylated functional groups in the corresponding intra-annular positions.

The efficient syntheses of larger macrocyclic MCPs in which the aromatic groups are linked by varying larger numbers of –CH$_2$– groups can also be effectively accomplished using two other non-sulfur extrusion processes, namely the low-valent zinc McMurry coupling reaction, or the TosMIC (*p*-tolylsulfonylmethyl isocyanide) [30,31] methodology that was used to great advantage by Vogtle's group, for the synthesis of [3_n]MCPs [32]. The synthesis, conformations and properties of several types of the larger homocalix[3]arene-analogue MCPs [33–38], using TosMIC has been reported by our group. For example, the synthesis of the symmetrical all-trimethylene-bridged hexahomocalix[3]arene analogue MCP, namely tri-*tert*-butyl-trimethoxy[3.3.3]MCP **13a**, was achieved via the TosMIC-NaH mediated cyclotrimerization reaction of 2,6-bis(bromomethyl)-4-*tert*-butylanisole **14**, followed by acid hydrolysis, to form trione **15** along with dione **16** in 22% and 10% yields, respectively (Scheme 3) [33], although better yields of **15** (68%) could be obtained using the 1 + 1 coupling reaction of bis-TosMIC adduct **17** with **14**, instead, as shown in Scheme 3. Wolff–Kishner reduction of **15** produced the desired tri-*tert*-butyl-trimethoxy[3.3.3]MCP **13a** in 86% yield. The demethylation of **13a** with BBr$_3$ in dichloromethane gave the tri-*tert*-butyl-trihydroxy[3.3.3]MCP **13b** in 89% yield. The *O*-Alkylation of **13b** could be achieved in high yields with several alkyl halides (RX: R = Et, Pr, and *n*-Bu), in the presence of Cs$_2$CO$_3$ under reflux conditions in acetone, to predominantly yield the *partial-cone* conformers of derivatives **13c–g**. The *partial-cone* conformers of **13c–e** were formed exclusively, but, in the cases of **13f** and **13g**, the *partial-cone:cone* ratio fell from 95:5 to 67:33. On the other hand, when NaH was used instead as the base to form **13f** and **13g** also in high yields (>95%), the *cone* conformers were now exclusively formed. The authors rationalized their results by a mechanism in which ring-inversion of one of the rings could occur at different rates, leading to the formation of the *partial-cone* isomers with the ethyl, propyl and butyl derivatives. However, with the corresponding larger ethoxycarbonylmethoxy and *N,N*-diethylaminocarbonylmethoxy derivatives, a "metal-templating" effect with Cs^{2+} ion and more strongly with the Na$^+$ ion, prevents complete ring inversion. Similar metal-templating effects had been noted previously in the case of *O*-alkylation reactions with calix[4]arenes. "Doubly-bridged" derivatives of **13b** [34] in which two of the hydroxy groups were capped using 3,5-bis(bromomethyl)toluene, or 1-*tert*-butyl-3,5-bis(bromomethyl)-benzene, Cs$_2$CO$_3$ and acetone under reflux conditions in acetone, was deduced to also be in a *partial-cone* conformation.

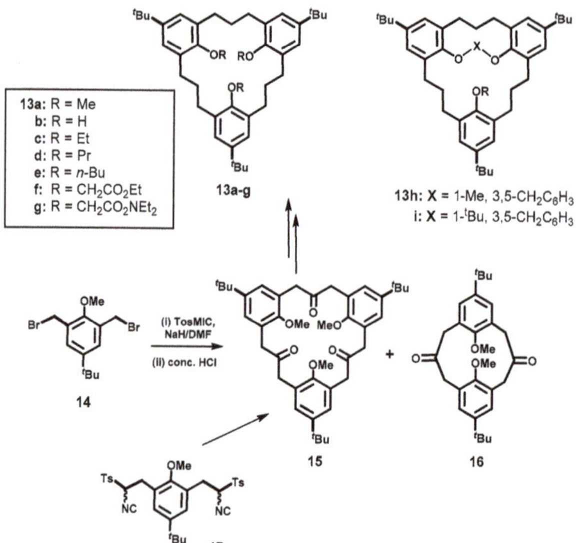

Scheme 3. Hexahomocalix[3]arene analogue MCPs **13a–g** and capped **13h–i**.

In a similar manner as with the synthesis of **13a**, the tetrahomocalix[3]arene-analogue MCP, tri-*tert*-butyl-trimethoxy-[3.3.1]MCP **20a** was synthesized by a NaH/DMF-mediated cyclocondensation of **17** and **18** followed by Wolff–Kishner reduction of the diketone intermediate **19** (Figure 5). The inherent chirality of the resulting C_1-symmetrical **20e** was confirmed by its room temperature

^{1}H-NMR spectrum with added Pirkle's chiral shift reagent, which caused a splitting of the OMe groups and AB patterns corresponding to the methylene protons indicating a *2-partial-cone* conformation.

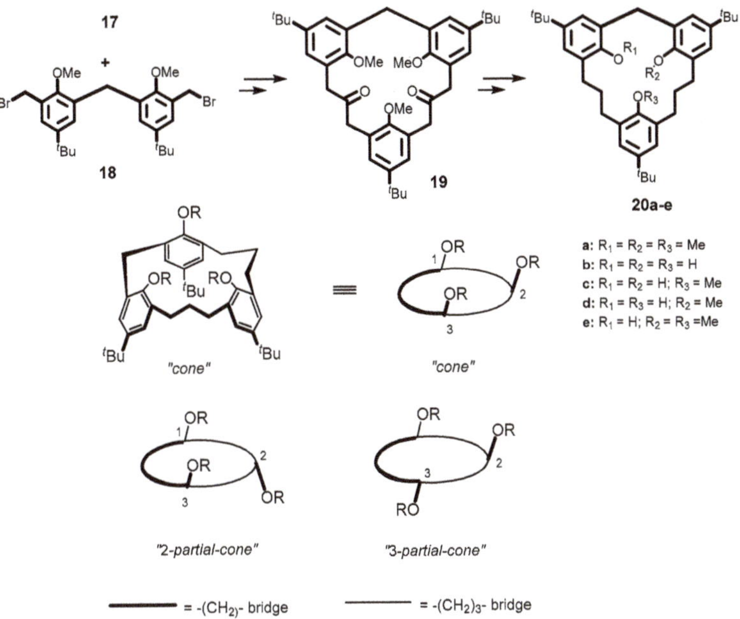

Figure 5. Synthesis and conformers of tetrahomocalix[3]arene analogue MCPs **20a–e**.

The demethylation of **20a** using tetramethylsilyl iodide (TMSI) in acetonitrile produced the corresponding trihydroxy[3.3.1]MCP, 20b, as expected, but when BBr$_3$ in dichloromethane was used, the partial demethylation of **20a** resulted, forming only the dihydroxy[3.3.1]MCP, **20c**. Upon further treatment with TMSI/MeCN, however, **20c** afforded the trihydroxy[3.3.1]MCP **20b** [38]. Its ^{1}H-NMR spectrum (in CDCl$_3$) exhibits the signals for the hydroxyl groups at ~ δ 3.33 and 6.25 ppm, which is evidence for the formation of intramolecular hydrogen bonding and a *cone* conformation [38]. In contrast, both the ^{1}H-NMR spectrum and a single crystal analysis confirmed the *2-partial-cone* conformation of macrocycle **20c**. Its ^{1}H-NMR spectrum shows two sets of doublets at δ 3.61 and 4.36 ppm (*J* = 13.3 Hz) for the ArCH$_2$Ar methylene protons and a single peak at δ 1.72 ppm for the methoxy protons consistent with a *2-partial-cone* conformation. To date, no further studies on either the computational or supramolecular properties of these compounds have been reported.

4. Calixbenzofuran-Analogous MCPs: Calix[3]benzofuran Analogues

The demethylation of [3.3.1]MCP-dione **19** in acetonitrile with in-situ-generated TMSI produced a mixture of benzofuran ring-containing products, the symmetrical **22a** and unsymmetrical **23a** and a new spirobisdihydro-furan **24** in 24, 45 and 5% yields, respectively (Figure 6) [39]. It is presumed that the trihydroxy-diketo intermediate **21** is first formed, and then undergoes intramolecular TMSI-mediated cyclizations to form the observed products [39–41]. Sawada and co-workers had previously reported that the treatment of tetramethoxy [2.1.2.1]MCPs with TMSI formed hemisphere-shaped calixarene analogues containing a dihydrobenzofuran ring [42,43]. The CDCl$_3$ ^{1}H-NMR spectrum of the symmetrical **22a** shows the hydroxyl signal at δ 6.54 ppm, indicating intramolecular hydrogen bonding between the hydroxyl and the oxygen of one of the benzofuran rings. The single-crystal X-ray structure of **22a** confirmed the intramolecular hydrogen bond as predicted from the ^{1}H-NMR spectroscopic data, with a distance of 2.182 Å between the hydroxyl proton and one of the benzofuran oxygens. The unsymmetrical regioisomer **23a** is fixed in an asymmetrical hemispherically shaped *cone* conformation and the spirobisdihydrofuran **24** is presumed to have been formed by a nucleophilic attack on one of the formed benzofurans in **22a**.

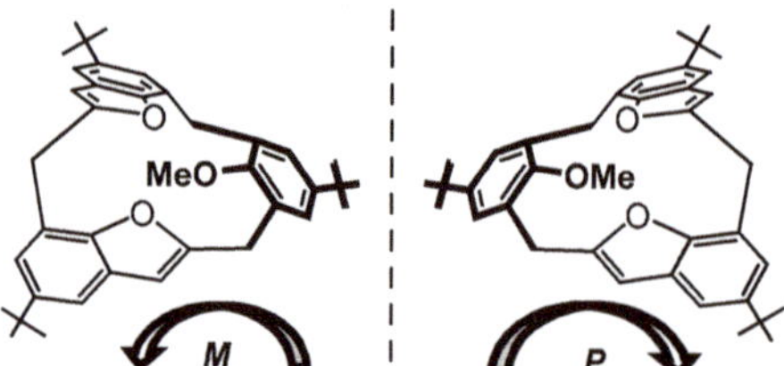

Figure 6. Calix[3]*di*benzofuran analogue MCPs **22–24**.

The *O*-methylation of **23a** with MeI/K$_2$CO$_3$ in acetone resulted in the folding of the arene ring into the macrocycle to form the non-symmetrical *partial-cone* methoxy[1.1.1]MCP **23b**, which is evident from the ^{1}H signals observed for the bridge methylene hydrogen atoms. The methoxy group of **23b** is shifted to the high field as a singlet at δ 1.97 ppm due to its shielding by the benzofuran rings; these ^{1}H signals correspond to an unsymmetrical *partial-cone* structure as depicted from density functional theory (DFT)-optimized energy structures. Its ^{1}H-NMR spectra with added Pirkle's reagent revealed the racemic mixture of *P*- and *M*-enantiomers (Figure 7). Single-crystal X-ray analysis of **23b** revealed that the macrocyclic skeleton adopts a highly asymmetric hemisphere-shaped *cone*-type conformation in which the methoxy group is pointed upwards and is exo to the two benzofuran rings, predicted from the ^{1}H-NMR spectra. These molecules, therefore, by adopting curvature in their planar structures that have no symmetry planes in their three-dimensional representations, fit Szumna's expanded definition of inherent chirality [44,45]. Others, including Böhmer [46,47] and Mandolini [48], had previously proposed and studied inherent chirality in calixarenes.

Figure 7. Schematic diagram of *M*-**23b** (**left**) and *P*-**23b** (**right**). Figure adapted from [39].

The DFT calculations conducted on these molecules using *Gaussian-09* [49] showed that the energy-minimized structures of the hemisphere-shaped *cone* conformers were in agreement with the observed single-crystal X-ray structures. The DFT gas phase calculations using B3LYP/6-31G(d) showed that of the regioisomers **22a** and **23a**, the latter of which was energetically more favored by

3.791 kJ mole^{-1}. The DFT calculations of the corresponding conformations of methoxy derivatives **22b** and **23b** showed that the latter was also similarly more favored by 2.358 kJ mole^{-1}, with the methoxy groups being favored to be in *exo* rather than *endo* orientations to the benzofuran rings, as shown in Figure 8.

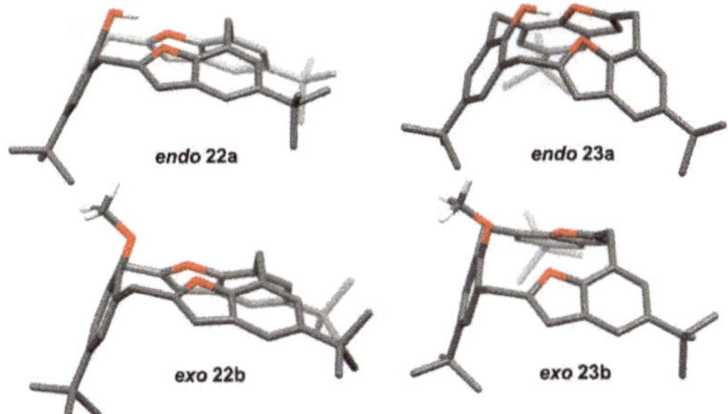

Figure 8. Geometry-optimized structures of *endo* **22a** and *endo* **23a**; *exo* **22b** and *exo* **23b**. All hydrogens except methoxy and phenolic hydrogens are omitted for clarity. Figure adapted and re-calculated from [39].

A series of derivatives **25b–e** of calix[3]benzofuran **25a** were synthesized using typical electrophilic aromatic substitution reactions to investigate the influence of the substituents on the conformations of the calix[3]benzofurans. The ^{1}H-NMR spectra of **25a–e** reveal that they adopt diverse conformations in solution and in some cases undergo very fast conformational changes relative to the NMR time scale. For example, **25a** freely interconverts between *cone* and *saddle* conformers (Figure 9) in solution, but the tribromocalix[3]benzofuran **25b**, adopts a rigid *cone* type hemisphere-shaped symmetrical structure in the solid state. The triformyl derivative **25c** showed fast *cone-saddle* interconversion in solution, but, when the three formyl groups were reduced to form the corresponding trihydroxymethyl derivative **25d**, the molecule adopted a fixed-*cone* conformation as with **25b**. The acylation of **25a** to form **25e** once again led to the formation of a slowly-interconverting mixture of *cone-saddle* conformers (Table 1) [40].

Figure 9. Calix[3]benzofuran analogue MCPs, **25a–e** and *cone* and *saddle* conformers of **25a** [40].

Table 1. Influence of substituents on the conformations of benzofurans **25a–e** [40].

Compound	T_c	$\Delta G^{\neq}$ kJ mol^{-1}	*Cone:saddle*($-30\,^\circ$C)
25a	40 [a]	61.9	80:20
25b	45 [a]	62.8	100:0
25c	28 [a]	58.6	40:60
25d	50 [a]	68.2	100:0
25e	75 [b]	69.0	20:80

[a]: CDCl$_3$; [b]: CDBr$_3$ (300 MHz).

Solvent and gas-phase DFT determinations with calix[3]benzofuran MCPs **25a–e** [40] reveal that both the *saddle* and corresponding *cone* conformers have lower ground-state energies in the solvent system than in the gas phase (Figure 10).

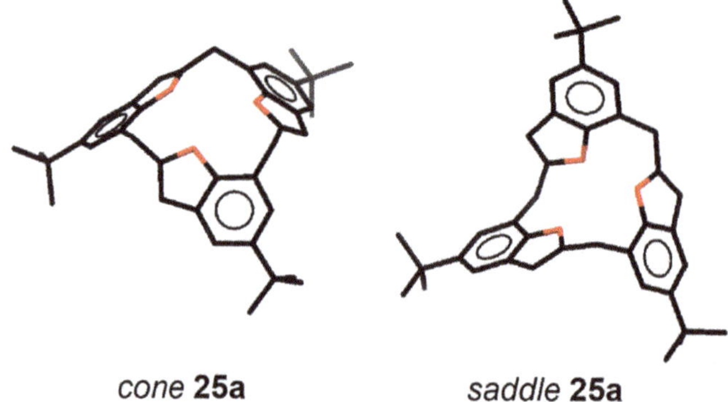

Figure 10. Geometry-optimized structures of the *cone* and *saddle* conformers of **25a**. Similar structures were obtained for the *cone* and *saddle* conformers of **25b–e** and are not shown.

The calculations also show that, among the calix[3]benzofurans, the *saddle* conformers are energetically more favored than their corresponding *cone* conformers in both the solvent and gas-phase. The energy differences are in the following order: **23a** > **25e** > **25d** > **25c** > **25a** > **23b** > **22b**. Thus, by introducing the different groups at the furan moieties, their *saddle* conformers become energetically more favored, roughly according to the increasing size of the groups (i.e., COMe > CH$_2$OH > CHO), except for **25b**. In the case of **25b**, there may be two factors influencing the stability of the *cone* conformers: bromine is electronegative in nature and also has greater electron density due to multiple lone-pairs of electrons. A single-crystal structure of **25b**, however, revealed it to be in a *cone* conformation in the solid state, but it should be noted that it co-crystallized with a well-defined chloroform molecule and with disordered solvent methanol molecules.

The cycloaddition reaction of **17** and **26**, followed by acidic work-up afforded [3.3.3]MCP **27a** (Figure 11) which, in turn, when treated with TMSI, generated in situ from chlorotrimethylsilane/sodium iodide in CH$_3$CN, generated, instead of the expected trihydroxy **27b**, two calix[3]benzofuran[3.1.1]MCP-analogues **28a** and **29** in 52% and 7% yields, respectively [41]. The regioisomer of **28a**, namely the unsymmetrical **29**, however, could not be isolated, although it was detected in the ^{1}H NMR spectra of the crude reaction mixture. The ^{1}H-NMR spectrum of **28a** exhibits a single peak at δ = 4.09 ppm for the ArCH_2Ar methylene bridge protons and the trimethylene protons appeared as a broad multiplet at δ 2.01 and 2.98 ppm. The position of the hydroxyl group at δ 6.34 ppm is consistent with the existence of intramolecular hydrogen bonding between the hydroxyl group and the oxygen of one of the benzofuran rings. The formation of **30**, which contains the spirobisdihydrofuran moiety, is analogous to the spirobishydrofuran **24** described previously [41].

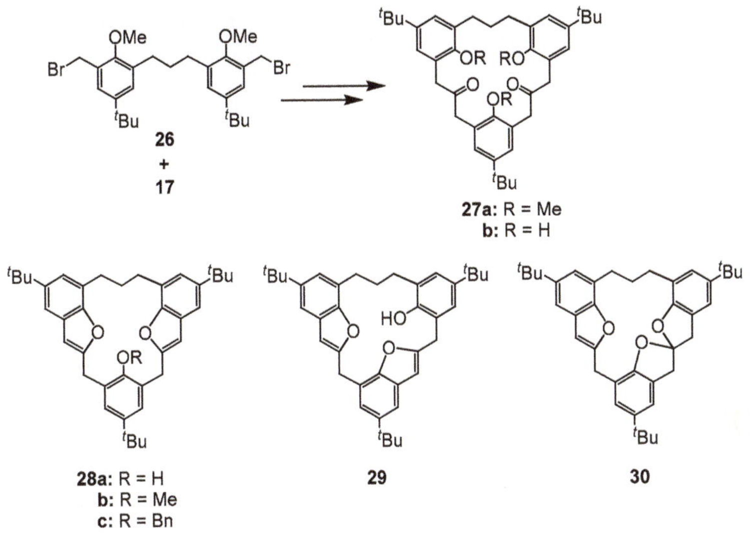

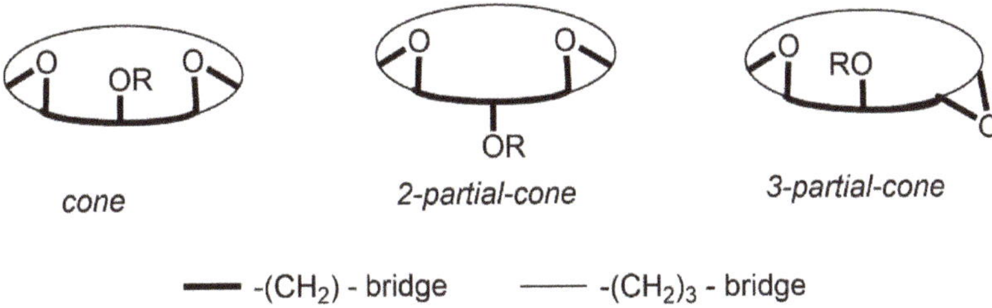

Figure 11. Structures of calix[3]benzofuran analogue MCPs **27–30**.

When **28a** was treated with MeI in the presence of K_2CO_3 in acetone methoxy[3.1.1]MCP **28b** was produced as its *cone*-conformer. This is evident from the ^{1}H signals observed for the bridge methylene hydrogen atoms that are split and appear as a pair of doublets at δ = 3.79 and 4.07 ppm. Previously, we had noted that the *O*-methylation of **23a** under the same conditions resulted instead in the inversion of one of the benzofuran rings, affording the non-symmetrical *partial-cone* **23b** (Figure 6). Thus, the size of the linking methylene chains can have significant effects on the resulting conformers, which are formed [41]. The *O*-benzylation of **28a**, produced the *cone*-benzyloxy[3.1.1]MCP **28c** with no isomerization being observed. This is evident from the ^{1}H signals for the bridge methylene hydrogen atoms that are split and appear as a pair of doublets at δ = 3.77 and 4.45 ppm [41]. In principle, these molecules could adopt three possible conformations, which are schematically represented in Figure 12.

Figure 12. Schematic representations of the three basic possible conformers of **28** and **29**.

Gas-phase DFT computational analysis of the three basic types of conformers of compounds **27a–30** was undertaken using the geometry-optimized structures of each of these conformers [41]. The calculated optimized-energy differences (kJ mol^{-1}) for **27a–30** showed that, of the various conformational isomers of **27a–30**, the *cone*-shaped structures are the most favored energetically, in the following order: *cone* > *2-partial-cone* > *1,3-alternate*. For example, the *cone* conformer of **28a** is −13.4 and −23.9 kJ mol^{-1} more stable than its corresponding *2-partial-cone* and *1,3-alternate* conformers, respectively. The findings were also consistent with the experimental evidence that the phenolic hydrogen forms weak intramolecular hydrogen bonding with an oxygen atom of one of the benzofuran rings. The distances between the hydroxyl hydrogen and the benzofuran oxygens are, respectively, 2.014 Å and 3.374 Å, 2.698 Å and 4.253 Å, and 2.010 Å and 3.493 Å for the *cone*, *2-partial-cone* and *1,3-alternate* conformers of **28a**. For the corresponding conformers of **29**, the respective distances are 1.836 Å vs. 3.277 Å, 2.605 Å vs. 3.065 Å, and 3.660 Å vs. 3.836 Å.

5. Homooxacalixarene-Analogue MCPs Containing Three Aromatic Rings: "Oxacalix[3]arene" Analogues

Hexahomotrioxacalix[3]arene **31** (Figure 13; R = H), commonly also called "oxacalix[3]arene", was first reported in 1962 by Hultzsch who isolated it in less than 1% yield from the reaction of formaldehyde with *p-tert*-butylphenol [50], and the evolution of the chemistry of this and related analogues have been well reviewed recently [17,18]. Hexahomotrioxacalix[3]arene **31** can be considered to be an example of a [3.3.3]MCP in which the phenolic units are linked by $-CH_2OCH_2-$ bridges. In 1983, Dhawan and Gutsche described detailed reproducible syntheses of **31** [51], which can be conducted in relatively large-scale. Hampton and co-workers also have described an alternative smaller-scale acid-catalyzed procedure which has been used for the synthesis of several different "wide-rim" *para*-substituted analogues of **31** [52]. Shinkai's group and others have extensively studied **31** as a platform to generate several versatile hosts [53–56] since, in comparison with the basic calix[4]arene structure, **31** offers some potential advantages: (1) its intra-annular cavity consists of a 18-membered ring, whereas that of calix[4]arene is a 16-membered ring; (2) the rate of ring inversion for derivatives of **31** should be much faster than that for calix[4]arenes because of the flexibility of the dimethyleneoxa linkages; (3) conformational isomerism is much more simplified, as there are only two types of formal conformations possible, namely *cone* and *partial-cone* (Figure 13), in contrast to the four possible formal conformations in calix[4]arenes; (4) the ethereal ring oxygens may act cooperatively with the phenolic oxygens upon binding with metal ions; and (5) its *cone* and *partial-cone* conformers can have C_{3v} and/or C_s symmetry, which is particularly useful for the design of receptors for biologically relevant $RNH_3{}^+$ ammonium ions.

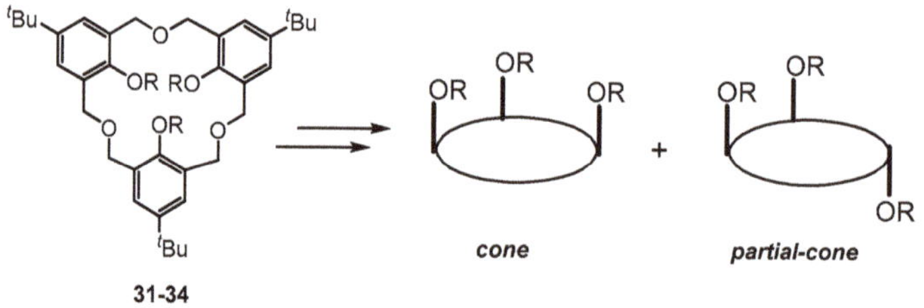

Figure 13. Structures of hexahomotrioxacalix[3]arene **31** (R = H) and schematic representations of the *cone* and *partial-cone* conformers of its tri-*O*-functionalized derivatives **32** (R = alkyl); **33** (R = ethoxycarbonylmethoxy) and **34** (R = *N,N*-diethyl *O*-methylacetamido).

In 1993, Shinkai and coworkers reported that treatment of **31** with four alkyl halides in DMF in the presence of NaH, K_2CO_3, Cs_2CO_3, *t*-BuOK or K afforded the corresponding *O*-methoxy, *O*-ethoxy, *O-n*-propyloxy and *O-n*-butyloxy derivatives that can adopt *cone* and/or *partial-cone* conformations (Figure 13) in varying yields ratios. They showed that a metal templating and solvent effect could lead to preferential thermodynamic and/or a kinetic preference for the *partial-cone* conformers of the corresponding tri-*O*-alkylated derivatives. In particular, tri-*O-n*-butylated **32** (R – *n*-Bu) was formed exclusively with the use of NaH in DMF but also in a 99:1% ratio over the *cone* conformer when Cs_2CO_3 in DMF was used [53]. Shinkai [54–56] also reported the synthesis of the *tris-O*-ethoxycarbonylmethoxy derivative **33** (R = CH_2CO_2Et) by the reaction of excess ethyl bromoacetate with **31** in acetone under reflux conditions with K_2CO_3 or Cs_2CO_3, and that an alkali-metal template effect led exclusively to the formation of the *partial-cone*-**33** conformer. With NaH or *t*-BuOK in THF, however, a mixture of *cone* and *partial-cone* products was formed, with the *cone*-**33** conformer being formed in only a

20–22% yield. Shinkai was also able to show both a metal templating and solvent effect for the efficient formation of the tris-*O*-amido derivatives **34** (R = CH$_2$CONEt$_2$) from the reactions of **31** with diethylchloroacetamide [55]. With NaH in THF, *cone*-**34** was formed exclusively, but, with K$_2$CO$_3$ or Cs$_2$CO$_3$ in acetone, *partial-cone*-**34** was exclusively formed. With NaH in DMF, a lower overall yield (29 and 26%, respectively) of the *cone* and *partial-cone*-**34** was obtained. However, with *t*-BuOK in THF a mixture of *cone* and *partial-cone*-**34** products was formed in **60** and 38% overall yield.

Recently, **35** the C_3-symmetric *N*-pyridyl *O*-methylacetamido functionalized derivative of **31** (Figure 14) was reported to be able to selectively and cooperatively bind Ag$^+$ and *n*-butylammonium ions and be controlled by the metal ion [57]. The geometries of the molecular structures were optimized in the gas phase using the PBE0 functional theory with the LANL2DZ basis set. The calculated binding or interaction energies (IE) for *cone*-**35** ⊃ ∩*n*-BuNH$_3^+$, *cone*-**35** ⊃ ∩*t*-BuNH$_3^+$, *cone*-**35** ⊃ ∩Ag$^+$, and *n*BuNH$_3^+$ ⊂ [*cone*-**35** ⊃ ∩Ag$^+$] are −298.8 kJ mol^{-1}, −268.3 kJ mol^{-1}, −457.1 kJ mol^{-1}, and −525.8 kJ· mol^{-1}, respectively, and were in agreement with the trend for the complexation data determined by the ^{1}H-NMR spectroscopic titration experiments with the corresponding perchlorate salts [57]. A conceptualization of the complexation of *n*BuNH$_3^+$ by the receptor *cone*-**35** is shown in Figure 15.

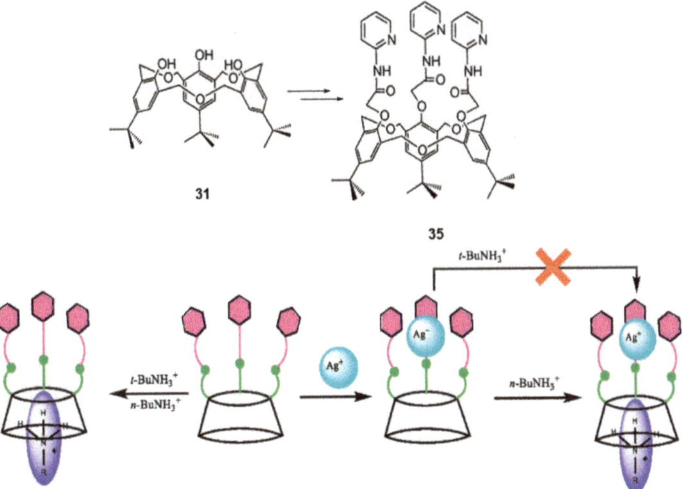

Figure 14. A schematic representation showing the cooperative binding mode of **35** (shown in a cone conformation) with and without Ag$^+$ complexation and selective binding with *n*-BuNH$_3^+$ and *t*-BuNH$_3^+$ cations. The green circles represent the CONH− and the hexagons represent the pyridyl groups. (Figure take from [57]).

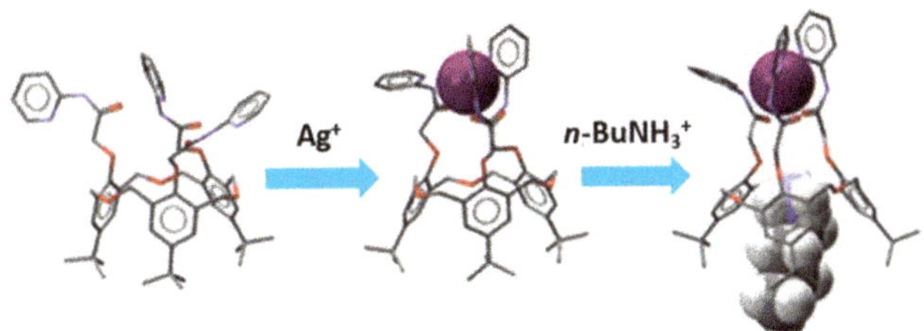

Figure 15. Geometry-optimized (PBE0/LANL2DZ) structures of *cone*-**35** and its complex with Ag+ and *n*BuNH$_3^+$. Left: The free *cone*-**35**, Middle: 1:1 *cone*-**35** ⊃ ∩Ag$^+$ complex, and right: *n*BuNH$_3^+$ ⊂ [*cone*-35 ⊃ ∩Ag$^+$] complex. Figure taken from reference [57].

Chirality is also possible with unsymmetrically substituted hexahomooxacalix[3]arenes. Araki et al. had previously reported the *"pseudo C_2-symmetric inherent chirality"* of the di-*O*-benzylated derivative **36a** and the di-*O*-benzylated-mono-*O*-methoxy derivative (Figure 16) **36b** of oxacalix[3]arene **31**, by *O*-alkylation at the lower rim of **31** and that these molecules were useful for the recognition of α-amino acid derivatives [58].

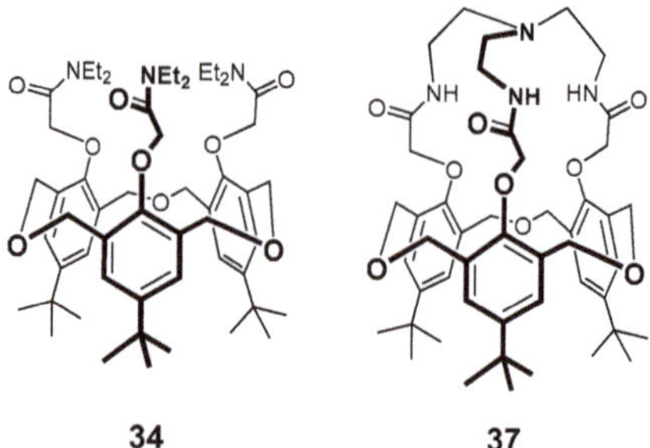

Figure 16. Synthesis of chiral oxacalixarene MCP analogues **36a–b** showing the chiral atropisomers. Figure adapted from reference [58].

In a very recent contribution, Marcos and co-workers reported their latest results with cation and anion ditopic receptors based upon trisubstituted derivatives of **31** with lower-rim *O-tert*-butylurea, phenylthiourea or phenylurea functionalities, which are connected to the hexahomotrioxacalix[3]arene scaffold via butyl spacers. [59] Both *cone* and *partial-cone* conformers were obtained with the phenylurea but only *partial-cone* products, as determined by solution NMR studies and their binding properties towards several relevant anions with different geometries were assessed by proton NMR titrations. The *cone* conformation of the *tris*phenylurea derivative was also confirmed by single-crystal X-ray crystallography and also proved to be the best an anion receptor, with the highest affinity being shown toward the trigonal planar acetate and benzoate anions (log K_{assoc} = 4.12 and 4.00, respectively) amongst the spherical, linear and tetrahedral anions tested. It also proved to be an effective ditopic receptor for several biogenic amine hydrochlorides monoamine neurotransmitters and trace amine hydrochlorides, in different solvents.

Jabin's group [60] compared the binding of a series of ammonium ions with two previously reported tris-*O*-amido derivatives, **34** [54] and the *tris*-mediated cryptand **37** previously reported by Yamato and coworkers [61] in 10% yields, and later in an improved 50% yield by Jabin and coworkers [62]. Both of these receptors are locked in *cone*-conformations (Figure 17) and ^{1}H-NMR studies with both formed endo-complexes with seventeen different primary ammonium ion guests, including biomolecules, even in protic media.

Figure 17. Structures of hexahomotrioxacalix[3]arene-analogue-MCPs **34** and **37**.

The ammonium ions of each of these guests inserted deeply into the polyaromatic cavity of the host receptor molecules. The NH_3^+ moieties were situated very close to the 18-atom lower-rim macrocyclic plane, forming H-bonds with the three oxygen atoms of the macrocycle. The overall topological complementarity between primary ammonium ions and the two cavity-based receptors resulted in the previously unreported complexation specificity for the primary ammonium ions over the secondary, tertiary and quaternary ones that were tested in this study. The authors were able to take advantage of this and demonstrated selective liquid–liquid extraction of primary ammonium salts from water solutions and for the selective recognition of lysine-containing peptides, with obvious potential for peptide sensing. Interestingly, also, **34** showed better binding than **37** presumably since the capping reduced the flexibility for the polyaromatic cavity to accommodate the guests.

To date, there is only one report of a 2-naphthol ring-analogue of **31**, namely oxacalix[3]naphthalene **38**, which has its naphthol units linked by -CH_2OCH_2- bridges and can be considered to be an example of a [3.3.3]metanaphthalenophane analogue of the more common [3.3.3]homooxacalixarenes described above. This compound that was reported in 2001 by our group, was formed in 25% yield along with its unsymmetrical regioisomer in lower yield, via a fragment-based approach in which the linear precursor underwent cyclo-condensation under acidic conditions [63] similar to those described Hampton [51]. Only a limited study of its complexation properties with metal ions was conducted; however, a ^1H NMR study of its complexation properties with C_{60} fullerene was reported; additionally, it formed a solid state 2:1 supramolecular complex in its *cone* conformation with C_{60} (Figure 18) [64] in contrast to the 1:1 complexes observed for C_{60} with **31** and several derivatives thereof reported earlier by Tsubaki et al. [65].

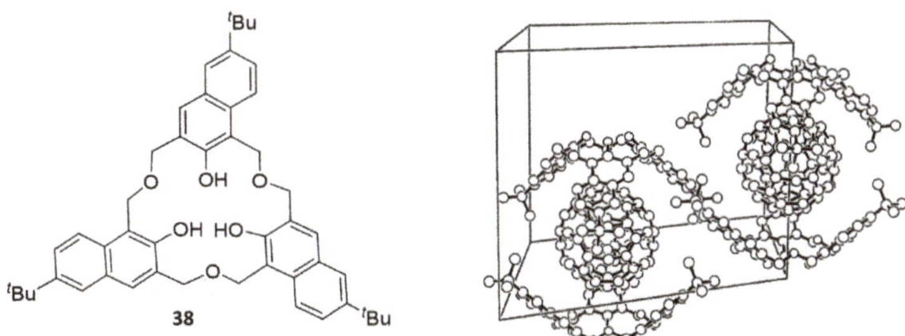

Figure 18. Hexahomotrioxacalix[3]naphthalene **38** and the single-crystal X-ray structure of its 2:1 complex with fullerene-C_{60}. Figure adapted from Reference [64].

6. Conclusions

In this review article, we have focused mainly upon the synthesis, structure and conformational properties of calixarene-analogous metacyclophanes, which contain three aromatic rings. These molecules can be considered to be analogues of the relatively less-known "calix[3]arenes". This group of molecules also encompasses homocalix[3]arenes and homooxacalix[3]arenes and can be considered as homocalix[3]arene- and homooxacalix[3]arene-analogous metacyclophanes. To date, the latter, including their supramolecular properties, have been the most extensively studied. This review highlights the origins, synthesis, structural aspects including their inherent chirality, and some of the different modifications that are possible with these molecules as a group that can lead to diverse application possibilities. We have mostly considered and cited the more recent literature that has been published and, where available, the more recently published reviews. The supramolecular properties of the new synthetic homocalix[3]arene-analogous metacyclophanes, which the Yamato group has mainly been focused on, however, have been untapped and will be subjected to further investigation.

Funding: This research received no external funding.

Acknowledgments: M.M.I. and T.Y. thank the OTEC at Saga University for financial support. This work was performed under the Cooperative Research Program of "Network Joint Research Center for Materials and Devices (Institute for Materials Chemistry and Engineering, Kyushu University)". S.R. would like to thank A. Alodhayb and the Deanship of Scientific Research, King Saud University, Saudi Arabia, ongoing support.

Conflicts of Interest: The authors have no conflicts of interest to declare.

References

1. Gleiter, R.; Hopf, H. *Modern Cyclophane Chemistry*; Wiley-VCH: Weinheim, Germany, 2004; ISBN 978-3-527-30713-5.
2. Ghasemabadi, P.G.; Yao, T.; Bodwell, G.J. Cyclophanes containing large polycyclic aromatic hydrocarbons. *Chem. Soc. Rev.* **2015**, *44*, 6494–6518. [CrossRef]
3. Pellegrin, M.M. Contribution à l'étude de la réaction de Fittig. *Recl. Trav. Chim. Pays-Bas* **1899**, *18*, 457–465. [CrossRef]
4. Cram, D.J.; Steinberg, H. Macro rings. I. Preparation and spectra of the paracyclophanes. *J. Am. Chem. Soc.* **1951**, *73*, 5691–5704. [CrossRef]
5. Zincke, A.; Ziegler, E. Zur kenntnis des härtungsprozesses von phenol-formaldehyd-harzen, X. mitteilung. *Ber. Dtsch. Chem. Ges.* **1944**, *77*, 264–272. [CrossRef]
6. Gutsche, C.D.; Iqbal, M. *p-tert*-Butylcalix[4]arene. *Org. Synth.* **1990**, *68*, 234. [CrossRef]
7. Stewart, D.R.; Gutsche, C.D. The one-step synthesis of *p-tert*-butylcalix[5]arene. *Org. Prep. Proc. Intl.* **1993**, *25*, 137–139. [CrossRef]
8. Gutsche, C.D.; Dhawan, B.; Leonis, M.; Stewart, D.R. *p-tert*-Butylcalix[6]arene. *Org. Synth.* **1990**, *68*, 238. [CrossRef]
9. Munch, J.H.; Gutsche, C.D. *p-tert*-Butylcalix[8]arene. *Org. Synth.* **1990**, *68*, 243. [CrossRef]
10. Haase, C.H.W. Path to industrial production of calix[8 and 4]arenes. *J. Org. Chem.* **2020**, *85*, 603–611. [CrossRef]
11. Neri, P.; Sessler, J.; Wang, M.-X. *Calixarenes and Beyond*; Springer International Publishing: Cham, Switzerland, 2016; ISBN 978-3-319-31867-7.
12. Agrawal, Y.K.; Kunji, S.; Menon, S.K. Analytical applications of calixarenes. *Rev. Anal. Chem.* **1998**, *17*, 69–139. [CrossRef]
13. Van Loon, J.D.; Verboom, W.; Reinhoudt, D.N. Selective functionalization and conformational properties of calix[4]arenes, a review. *Org. Prep. Proced. Int.* **1992**, *24*, 437–462. [CrossRef]
14. Gaeta, C.; Talotta, C.; Neri, P. The calixarene *p*-bromodienone route: From a chemical curiosity to an useful synthetic tool. *J. Incl. Phenom. Macrocycl. Chem.* **2014**, *79*, 23–46. [CrossRef]
15. Tlusty, M.; Eigner, V.; Babor, M.; Kohouta, M.; Lhotak, P. Synthesis of upper rim-double-bridged calix[4]arenes bearing seven membered rings and related compounds. *RSC Adv.* **2019**, *9*, 22017–22030. [CrossRef]
16. Nakamura, Y.; Fujii, T.; Inokuma, S.; Nishimura, J. Homocalixarenes. In *Calixarenes 2001*; Asfari, Z., Böhmer, V., Harrowfield, J., Vicens, J., Eds.; Kluwer Academic Publishers: Dordrecht, The Netherlands, 2001; pp. 219–234. [CrossRef]
17. Cottet, K.; Marcos, P.M.; Cragg, P.J. Fifty years of oxacalix[3]arenes: A review. *Beilstein J. Org. Chem.* **2012**, *8*, 201–226. [CrossRef] [PubMed]
18. Masci, B. Homooxa- and homoaza-calixarenes. In *Calixarenes 2001*; Asfari, Z., Böhmer, V., Harrowfield, J., Vicens, J., Saadioui, M., Eds.; Springer: Dordrecht, The Netherlands, 2001; pp. 235–249. [CrossRef]
19. Wang, D.-X.; Wang, M.-X. Azacalixaromatics. In *Calixarenes and Beyond*; Neri, P., Sessler, J., Wang, M.-X., Eds.; Springer International Publishing: Cham, Switzerland, 2016; pp. 363–397. [CrossRef]
20. Kaewtong, C.; Pulpoka, P. Azacalix[3]arenes: Chemistry and recent developments in functionalization for specific anion and cation recognition. *J. Incl. Phenom. Macrocycl. Chem.* **2009**, *65*, 129–136. [CrossRef]
21. Chen, C.-F.; Han, Y. Triptycene-derived macrocyclic arenes: From calixarenes to helicarenes. *Acc. Chem. Res.* **2018**, *51*, 2093–2106. [CrossRef]
22. Kohnke, F.H. Calixpyrroles: From anion ligands to potential anticancer drugs. *Eur. J. Org. Chem.* **2020**, *2020*, 4261–4272. [CrossRef]

23. Moshfegh, A.A.; Beladi, E.; Radnia, L.; Hosseini, A.S.; Tofigh, S.; Hakimelahi, G.H. The synthesis of 5,11,17-trihalotetracyclo [13.3.1.1^{3,7}.1^{9,13}]henicosa-1 (19),3,5,7,9,11,13, 15,17-nonaene-19,20,21-triols and 5,11,17-trihalo-19,20,21-trihydroxytetracyclo [13.3.1.1^{3,7}. 1^{9,13}]henicosa-1(19),3,5,7(20),9,11,13(21), 15,17-nonaene-8,14-dione. Cyclo-derivatives of phloroglucide analogues. *Helv. Chim. Acta* **1982**, *65*, 1264–1270. [CrossRef]

24. Yamato, T.L.; Doamekpor, K.; Tsuzuki, H.; Tashiro, M. A novel calixarene-analogous macrocyclic metacyclophane 'molecular pendulum'. *Chem. Lett.* **1995**, *24*, 89–90. [CrossRef]

25. Yamato, T.L.; Doamekpor, K.; Tsuzuki, H. Synthesis and conformational studies of novel calixarene-analogous macrocyclic [3.1.1]metacyclophanes. *Liebigs Ann. Recl.* **1997**, *1997*, 1537–1544. [CrossRef]

26. Tsuge, A.; Sawada, T.; Mataka, S.; Nishiyama, N.; Sakashita, H.; Tashiro, M. Preparation and structural properties of hydroxy- and alkoxy-[2.2.1]metacyclophanes. *J. Chem. Soc. Perkin Trans. 1* **1992**, 1489–1494. [CrossRef]

27. Laufenberg, S.; Feuerbacher, N.; Pischel, I.; Börsch, O.; Nieger, M.; Vögtle, F. "New biphenylenophanes and biphenylophanes–1,8-dimethylbiphenylene by continuous vacuum pyrolysis. *Liebigs Ann. Chem.* **1997**, 1901–1906. [CrossRef]

28. Vögtle, F. *Cyclophane Chemistry: Synthetics, Structures and Reactions*; John Wiley & Sons: Chichester, UK, 1993; ISBN 10: 0471931993.

29. Tashiro, M.; Watanabe, T.; Tsuge, A.; Sawada, T.; Mataka, S. Metacyclophanes and related compounds. 24. Preparation and reaction of trimethyl[2.2.2]- and tetramethyl[2.2.2.2]metacyclophane. *J. Org. Chem.* **1989**, *54*, 2632–2638. [CrossRef]

30. Possel, O.; van Leusen, A.M. Tosylmethyl isocyanide employed in a novel synthesis of ketones. A new masked formaldehyde reagent. *Tetrahedron Lett.* **1977**, *18*, 4229–4231. [CrossRef]

31. van Leusen, A.M.; Boerna, G.J.; Helholdt, R.B.; Siderius, H.; Strating, J. Chemistry of sulfonylmethylisocyanides. Simple synthetic approaches to a new versatile chemical building block. *Tetrahedron Lett.* **1972**, *13*, 2367–2368. [CrossRef]

32. Breitenbach, J.; Vögtle, F. Macrocyclizations with TosMIC-Yielding [3$_n$]metacyclophanes. *Synthesis* **1992**, 41–43. [CrossRef]

33. Yamato, T.; Doamekpor, L.K.; Koizumi, K.; Kishi, K.; Haraguchi, M.; Tashiro, M. Synthesis and conformational studies of calixarene-analogous trihydroxy[3.3.3]metacyclophanes and their O-alkylated derivatives. *Liebigs Ann.* **1995**, *7*, 1259–1267. [CrossRef]

34. Yamato, T.; Zhang, F. The synthesis and probable conformation of *m*-xylenyl capped homocalix[3]arenes derived from a trihydroxy[3.3.3]metacyclophane. *J. Chem. Res.* **1999**, *23*, 34–35. [CrossRef]

35. Yamato, T.; Kohno, K.; Tsuchihashi, K. Synthesis, structures and ion selectivity of homocalix[3]arene thioketals derived from homocalix[3]arene ketones. *J. Incl. Phenom. Macrocycl. Chem.* **2002**, *43*, 137–144. [CrossRef]

36. Yamato, T.; Haraguchi, M.; Nishikawa, J.; Ide, S. Synthesis, conformational studies and inclusion properties of O-benzylated calixarene analogues of trihydroxy[3.3.3]metacyclophanes. *J. Chem. Soc. Perkin Trans. 1* **1998**, 609–614. [CrossRef]

37. Yamato, T. Synthesis, conformations and inclusion properties of homocalix[3]arenes. *J. Incl. Phenom. Macrocycl. Chem.* **1998**, *32*, 195–207. [CrossRef]

38. Islam, M.M.; Tomiyasu, H.; Thuery, P.; Matsumoto, T.; Tanaka, J.; Elsegood, M.R.J.; Redshaw, C.; Yamato, T. Synthesis and conformational studies of calixarene analogue chiral [3.3.1]metacyclophane. *J. Mol. Struct.* **2015**, *1098*, 47–54. [CrossRef]

39. Islam, M.M.; Tomiyasu, H.; Matsumoto, T.; Tanaka, J.; Rahman, S.; Georghiou, P.E.; Redshaw, C.; Yamato, T. Synthesis and conformational studies of chiral macrocyclic [1.1.1]metacyclophanes containing benzofuran rings. *Org. Biomol. Chem.* **2015**, *13*, 9055–9064. [CrossRef] [PubMed]

40. Islam, M.M.; Akther, T.; Ikejiri, Y.; Matsumoto, T.; Tanaka, J.; Rahman, S.; Georghiou, P.E.; Hughes, D.L.; Redshaw, C.; Yamato, T. Synthesis, structural properties, electrophilic substitution reactions and DFT computational studies of calix[3]benzofurans. *RSC Adv.* **2016**, *6*, 50808–50817. [CrossRef]

41. Islam, M.M.; Wang, C.-Z.; Feng, X.; Rahman, S.; Georghiou, P.E.; Alodhayb, A.; Yamato, T. Synthesis, structures and DFT computational studies of [3.1.1]metacyclophanes containing benzofuran rings. *ChemistrySelect* **2018**, *3*, 13542–13547. [CrossRef]

42. Sawada, T.; Hongo, T.; Matsuo, N.; Konishi, M.; Kawaguchi, T.; Ihara, H. Hemisphere-shaped calixarenes and their analogs: Synthesis, structure, and chiral recognition ability. *Tetrahedron* **2011**, *67*, 4716–4722. [CrossRef]

43. Sawada, T.; Nishiyama, Y.; Tabuchi, W.; Ishikawa, M.; Tsutsumi, E.; Kuwahara, Y.; Shosenji, H. Novel calixarene hemisphere synthesized via pinacol rearrangement of [2.1.2.1]metacyclophane. *Org. Lett.* **2006**, *8*, 1995–1997. [CrossRef]

44. Szumna, A. Inherently chiral concave molecules-from synthesis to applications. *Chem. Soc. Rev.* **2010**, *39*, 4274–4285. [CrossRef]

45. Wierzbicki, M.; Jędrzejewska, H.; Szumna, A. Chiral calixarenes and resorcinarenes. In *Calixarenes and Beyond*; Neri, P., Sessler, J., Wang, M.-X., Eds.; Springer International Publishing: Cham, Switzerland, 2016; pp. 13–42. ISBN 978-3-319-31867-7. [CrossRef]

46. Wolff, A.; Böhmer, V.; Fogt, W.; Ugozzoli, F.; Andreetti, G.D. Dissymmetric calix[4]arenes with C2 and C4 symmetry. *J. Org. Chem.* **1990**, *55*, 5665–5667. [CrossRef]

47. Pickard, S.T.; Pirkle, W.H.; Tabatabai, M.; Vogt, W.; Böhmer, V. Dissymmetric calix[4]arenes: Optical resolution of some conformationally fixed derivatives. *Chirality* **1993**, *5*, 310–314. [CrossRef]

48. Mandolini, L.; Ungaro, R. *Calixarenes in Action*; Imperial College Press: London, UK, 2000. [CrossRef]

49. Frisch, M.J.; Trucks, G.W.; Schlegel, H.B.; Scuseria, G.E.; Robb, M.A.; Cheeseman, J.R.; Scalmani, G.; Barone, V.; Petersson, G.A.; Nakatsuji, H.; et al. *Gaussian 09, Revision D.01*; Gaussian, Inc.: Wallingford, CT, USA, 2013.

50. Hultzsch, K. Ring condensates in alkyl phenol resins. *Kunststoffe* **1962**, *52*, 19–24.

51. Dhawan, B.; Gutsche, C.D. Calixarenes. 10. Oxacalixarenes. *J. Org. Chem.* **1983**, *48*, 1536–1539. [CrossRef]

52. Hampton, P.D.; Bencze, Z.; Tong, W.; Daitch, C.E. A new synthesis of oxacalix[3]arene macrocycles and alkali-metal-binding studies. *J. Org. Chem.* **1994**, *59*, 4838–4843. [CrossRef]

53. Araki, K.; Inada, K.; Otsuka, H.; Shinkai, S. Conformational isomerism in and binding properties to alkali-metals and an ammonium salt of *O*-alkylated homooxacalix[3]arenes. *Tetrahedron* **1993**, *49*, 9465–9478. [CrossRef]

54. Araki, K.; Hashimoto, N.; Otsuka, H.; Shinkai, S. Synthesis and ion selectivity of conformers derived from hexahomotrioxacalix[3]arene. *J. Org. Chem.* **1993**, *58*, 5958–5963. [CrossRef]

55. Matsumoto, H.; Nishio, S.; Takeshita, M.; Shinkai, S. Syntheses and ion selectivities of tri-amide derivatives of hexahomotrioxacalix[3]arene. Remarkably large metal template effect on the ratio of *cone* vs. *partial-cone* conformers. *Tetrahedron* **1995**, *51*, 4647–4654. [CrossRef]

56. Takeshita, M.; Inokuchi, F.; Shinkai, S. C$_3$-Symmetrically-capped homotrioxacalix[3]arene. A preorganized host molecule for inclusion of primary ammonium ions. *Tetrahedron Lett.* **1995**, *36*, 3341–3344. [CrossRef]

57. Jiang, X.-K.; Ikejiri, Y.; Wu, C.; Rahman, S.; Georghiou, P.E.; Zeng, X.; Elsegood, M.R.J.; Redshaw, C.; Teat, S.J.; Yamato, T. A hexahomotrioxacalix[3]arene-based ditopic receptor for alkylammonium ions controlled by Ag$^+$ ions. *Molecules* **2018**, *23*, 467. [CrossRef]

58. Araki, K.; Inada, K.; Shinkai, S. Chiral recognition of α-amino acid derivatives with a homooxacalix[3]arene: Construction of a *pseudo*-C$_2$-symmetrical compound from a C$_3$-symmetrical macrocycle. *Angew. Chem. Int. Ed. Engl.* **1996**, *35*, 72–74. [CrossRef]

59. Teixeira, F.A.; Ascenso, J.R.; Cragg, P.J.; Hickey, N.; Geremia, S.; Marcos, P.M. Recognition of anions, monoamine neurotransmitter and trace amine hydrochlorides by ureido-hexahomotrioxacalix[3]arene ditopic receptors. *Eur. J. Org. Chem.* **2020**, 1930–1940. [CrossRef]

60. Lambert, S.; Bartik, K.; Jabin, I. Specific binding of primary ammonium ions and lysine-containing peptides in protic solvents by hexahomotrioxacalix[3]arenes. *J. Org. Chem.* **2020**, *85*, 10062–10071. [CrossRef]

61. Jiang, X.-K.; Deng, M.; Mu, L.; Zeng, X.; Zhang, J.X.; Yamato, T. Synthesis and crystal structure of a novel hexahomotrioxacalix[3]cryptand. *Asian J. Chem.* **2013**, *25*, 515–517. [CrossRef]

62. Zahim, S.; Ajami, D.; Laurent, P.; Valkenier, H.; Reinaud, O.; Luhmer, M.; Jabin, I. Synthesis and binding properties of a tren-capped hexahomotrioxacalix[3]arene. *ChemPhysChem* **2019**, *21*, 83–89. [CrossRef] [PubMed]

63. Ashram, M.; Mizyed, S.; Georghiou, P.E. Synthesis of hexahomotrioxacalix[3]naphthalenes and a study of their alkali-metal cation binding properties. *J. Org. Chem.* **2001**, *66*, 1473–1479. [CrossRef] [PubMed]

64. Mizyed, S.; Ashram, M.; Miller, D.O.; Georghiou, P.E. Supramolecular complexation of [60]fullerene with hexahomotrioxacalix[3]naphthalenes: A new class of naphthalene-based calixarenes. *J. Chem. Soc. Perkin Trans. 2* **2001**, 1916–1919. [CrossRef]
65. Tsubaki, K.; Tanaka, K.; Kinoshita, T.; Fuji, K. Complexation of C_{60} with hexahomooxacalix[3]arenes and supramolecular structures of complexes in the solid state. *Chem. Commun.* **1998**, 895–896. [CrossRef]

MDPI

St. Alban-Anlage 66

4052 Basel

Switzerland

Tel. +41 61 683 77 34

Fax +41 61 302 89 18

www.mdpi.com

Molecules Editorial Office

E-mail: molecules@mdpi.com

www.mdpi.com/journal/molecules